Geoheritage, Geoparks and Geotourism

Conservation and Management Series

Series editor

Franz W. Eder, Paris, France

For further volumes:
http://www.springer.com/series/11639

Patricia Erfurt-Cooper
Editor

Volcanic Tourist Destinations

Springer

Editor
Patricia Erfurt-Cooper
James Cook University
Cairns
Australia

ISBN 978-3-642-16190-2 ISBN 978-3-642-16191-9 (eBook)
DOI 10.1007/978-3-642-16191-9
Springer Heidelberg New York Dordrecht London

Library of Congress Control Number: 2014939052

Printed on acid-free paper

Springer is part of Springer Science+Business Media (www.springer.com)

This book is dedicated to my family, who have endured my preoccupation with this work on many occasions

Foreword

For centuries people have been attracted by volcanic landscapes. Whether active or dormant, the unique dynamics of volcanic environments have encouraged travellers to visit to the remotest locations. Artists have been inspired to paint volcanoes such as the perfect shape of Mt. Fuji or Mt. Vesuvius erupting. Volcanoes have provided archaeologists with history and scientists with landscapes buried and preserved under layers of volcanic ash with examples including Pompeii and Herculaneum (Vesuvius), the Caribbean island of Montserrat (Soufrière Hills), the landscape that was buried by Mt. St. Helens and many others. Volcanoes have featured as dramatic backdrops in movies and as major topics in the literature, scientific as well as fiction, and have been regarded as an unforgettable visitor experience. Some volcanoes are easy to access tourist attractions, whereas others can only be found in more remote locations (e.g. Alaska, Antarctica or Kamchatka). While some volcanoes are visited for their special individual features, others are included in tour packages as highlights of the trip agenda.

More than 30 years ago (1980), Mount St. Helens in the USA (see Chap. 15) was making the news worldwide with a cataclysmic eruption which consequently claimed 57 lives. Apart from raising awareness about volcanic activity, this eruption opened a window of opportunity for volcanologists to study eruptive events and their precursors as never before possible. Some of the more destructive volcanoes require ongoing monitoring and there is constant pressure for reliable data on all mountains which have been identified as active. Despite all efforts, unfortunately, lives are still lost and economies are affected when volcanic eruptions take place and cause a natural disaster. The more recent events that affected not just the local population, but also disrupted air travel worldwide, were created by the volcanoes Eyjafjallajökull in Iceland and Puyehue in Chile. These and other eruptive events (Table 1) have contributed to further advancing research into volcanic processes and their impact on the environment, as well as leading to the development and upgrade of essential risk management strategies.

Heightened volcanic activity frequently attracts an increased number of visitors (e.g. Kilauea, Hawaii in early 2011; see Chap. 21) compared to 'normal' activity or quiescence. Chaitén volcano for example (Chile; see Chap. 26) was uncharted and not recognised as a potentially active volcano until its unexpected reawakening in early May 2008. The last eruption of this volcano is thought to have happened several thousand years ago and the recent events did not just take the locals but also the scientific community completely by surprise. On the economic side Chaitén's eruption stimulated the local tourism industry with a sudden increase of visitors to the area, an additional reason of concern for the local authorities.

When it comes to monitoring tectonic activity, funding is one of the main issues. This leaves dormant volcanoes with the potential to reactivate largely unsupervised and possibly even unrecorded. Although more volcano observatories are gradually established, some remote areas or volcanoes in developing countries are not yet monitored sufficiently to detect early warning signs indicating eruptions due to logistical, financial and political restrictions (Tilling 1989). Figure 1 gives an overview of the major distribution of active volcanoes worldwide.

Table 1 Examples of recent volcanic eruptions (countries in alphabetical order)

Recent major eruptive events	
2009 Redoubt, Alaska (ash)	2011 Bromo, Indonesia (ash, pyroclasts)
2006 Augustine, Alaska (lava, ash)	2012 Etna, Italy (lava, ash, strombolian)
2011 Mt. Erebus, Antarctica (lava lake)	2012 Stromboli, Italy (ash eruption)
2012 Soufriere Hills, Montserrat Caribbean (ash, lava dome, pyroclastic flows)	2001 Mt. Usu, Japan (lava domes, ash)
2010 Chaitén, Chile (ash, pyroclastic flows, lava)	1996 Mt. Unzen, Japan (lava dome, lahars, pyroclastic flows)
2012 Puyehue Volcano, Chile (ash eruption)	2012 Mt. Sakurajima, Japan (ongoing ash eruptions)
2010 Galeras, Colombia (ash, gas)	2012 Karymsky, Kamchatka, Russia (lava, ash)
2012 Nevado del Ruiz, Colombia (ash, lahars)	2007 Ruapehu, New Zealand (lahars, current unrest)
2007 Mt. Karthala, Comoros Islands (lava flows)	2012 White Island, New Zealand (crater lake, ash eruptions)
2012 Rincón de la Vieja, Costa Rica (increased activity)	2012 Tongariro, New Zealand (ash, pyroclasts)
2011 Nyiragongo, DR Congo (lava lake)	2011Ulawun, Papua New Guinea (gas, steam)
2012 Nyiamulagira, DR Congo (lava flows, gas)	2010 Mayon, Philippines (ash, gas, lava)
2012 Reventador, Ecuador (ash, pyroclastic flows)	1993 Pinatubo, Philippines (ash, pyroclastic flows)
2012 Tungurahua, Ecuador (lava, gas, ash)	2010 Piton de la Fournaise, Reunion (lava flows)
2012 Santa María/Santiaguito, Guatemala (ash, lava)	2012 El Hierro, Spain (submarine eruption)
2011 Eyjafjallajökull, Iceland (fissure, ash eruption)	2011 Kilauea, Hawaii, USA (lava lake)
2011 Grímsvötn, Iceland (subglacial)	2008 Mt. St. Helens, Washington State USA (ash, pyroclastic flows)
2000 Hekla, Iceland (tephra, lava)	2011 Ambrym, Vanuatu (lava lake)
2000 Mt. Batur, Indonesia (lava flows)	2008 Lopevi, Vanuatu (ash, pyroclasts)
2010 Merapi, Indonesia (lava dome, pyroclastic flows)	2012 Yasur, Vanuatu (strombolian, ash)
2012 Mt. Lokon, Sulawesi, Indonesia (ash eruption)	
2012 Krakatau, Indonesia (lava dome growth)	

Source Compiled by Author from various sources

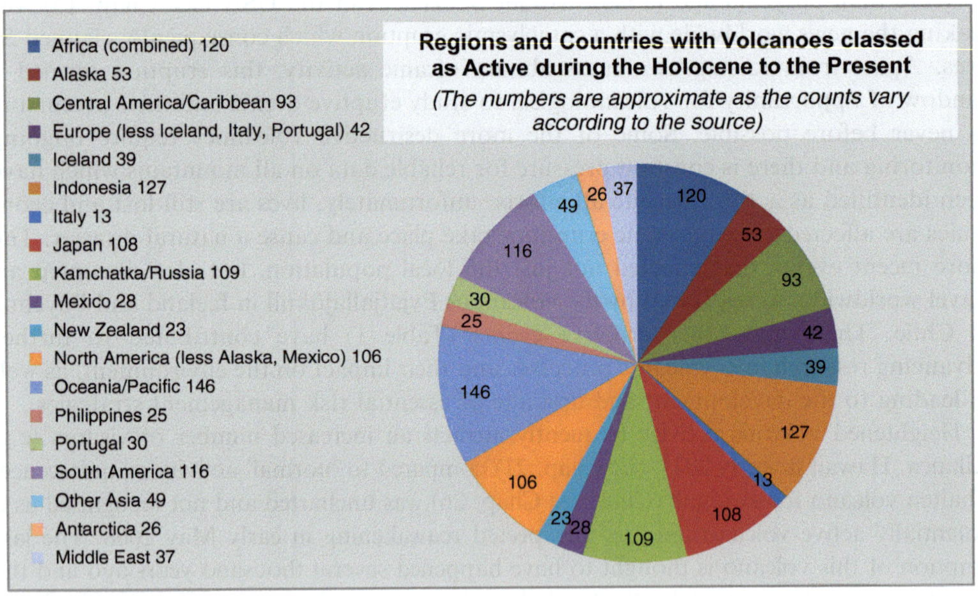

Fig. 1 Worldwide distribution of active volcanoes. Compiled by Patricia Erfurt-Cooper

Reference

Tilling, R. I. (1989). Volcanic hazards and their mitigation: Progress and problems. Reviews of Geophysics, 27, 237–269.

Preface

Volcanic landscapes, both active and dormant, and their related features provide priceless natural resources, which are used worldwide as tourist attractions that are popular with people from a wide range of backgrounds.

The preparation for this book has taken rather longer than anticipated but combining the collective information was a complex task. The delay, however, turned into a blessing, as in the meantime several more volcanic areas acquired protected site status, either as inclusion on the UNESCO World Heritage list or as members of the growing geopark associations such as the European Geopark Network, the Japanese Geopark Network and the UNESCO Global Geoparks Network.

Throughout this book, earth science and social science go hand in hand and some chapters may be more 'technical' than others. However, the aim is to provide a broad insight into volcano tourism and all factors involved. Some chapters focus predominantly on the geoscience aspects of the area described, while others emphasise the overall geodiversity of the natural and cultural heritage and its integration into various tourism sectors from geotourism, health tourism based on volcanic hot springs, adventure tourism and ecotourism. Therefore this is not an earth science text book as such, although the chapters are contributed by highly qualified scientists from various disciplines including volcanologists, seismologists, geologists, geomorphologists, geographers, tourism researchers and experts specialising in protected site management.

The main focus of this book is on the rich geodiversity in volcanic areas, which includes their geological heritage, with features and attractions of particular landforms as an important part of the whole geoexperience individual regions have to offer.

This book aims to contribute to the emerging geotourism literature as well as to the existing literature about protected sites. The individual chapters will enhance the knowledge of volcanic and geothermal environments and raise awareness about hazards and potential risks in these particular natural settings.

The chapters are separated into individual parts and provide a broad perspective on active and dormant volcanic destinations, although this is by no means exhaustive. Detailed descriptions of volcanic heritage and how it is showcased in geoparks are discussed in some chapters, while others include a focus on local biodiversity or cultural heritage as additional attractions and as an essential part of the overall learning experience. Widespread evidence suggests that volcanoes, dormant as well as active, have been and still are an important focal point within cultural and religious settings and in local histories. In many regions or countries (e.g. Japan, Indonesia), events and festivals related to volcanoes are held on a regular basis and attract large numbers of visitors.

Additional natural resources are frequently seen alongside the volcanoes with associated volcanic hot springs successfully used by resorts and spas for health and wellness purposes. These are another major draw cards for visitors, particularly in countries such as Japan, Iceland and New Zealand, where active volcanoes are common as a scenic backdrop to the spa experience.

Despite the time it took to edit this book it was an enjoyable and rewarding, although challenging undertaking, encouraged by the need for new and advanced information about an unusual tourism sector and the opportunity to fill a gap in the current tourism literature. I am sure that every reader will find something of interest and enjoy the different geosites and their aspects of conservation, sustainability and education based on their volcanic heritage as they are portrayed in this book.

June 2014 Patricia Erfurt-Cooper

Acknowledgments

With grateful thanks I wish to acknowledge all contributors to this book for sharing their knowledge, research findings and field experience to increase the awareness about the multitude of volcanic destinations. Without their efforts, time and commitment this book would not have been possible and their cooperation and patience throughout the duration of this project is greatly appreciated.

I also like to acknowledge the editorial support and encouragement from the staff at Springer, particularly Chris Bendall, Janet Sterrit and Annett Buettner, during the time of preparing this book.

With [illegible] their knowledge, research [illegible] and field experience [illegible] the [illegible] of volcanic [illegible]. Without their efforts, time and commitment this book would not have been [illegible] and their [illegible] of this project [illegible].

[illegible] to acknowledge [illegible] editorial [illegible] and [illegible] from [illegible] Springer [illegible]

Contents

Part III The Pacific Ring of Fire

Part IV Volcanic Destinations Worldwide

Contributors

Gülpınar Akbulut Cumhuriyet University Sivas, Sivas, Turkey

Alireza Amrikazemi Department of Geoheritage, Geological Survey of Iran, Tehran, Iran

Emili Bassols Technicians of the Management Team of the Natural Park, Garrotxa, Spain

Deanne K. Bird Risk Frontiers, Macquarie University, North Ryde, Australia

Ted A. Brattstrom Korea International School (KIS), Seoul, Korea

Nathalie Cayla Universe and Environment, EDYTEM Laboratory, Environnements, DYnamiques et Territoires de la Montagne, Chambéry, France

Malcolm J. M. Cooper Ritsumeikan Asia Pacific University, Beppu, Japan

Ross Dowling Edith Cowan University, Joondalup, WA, Australia

Carolyn Driedger USGS Cascades Volcano Observatory, Vancouver, WA, USA

Patricia Erfurt-Cooper James Cook University, Cairns, QLD, Australia

Richard Field School of Geography, University of Nottingham, Nottingham, UK

Peter Frenzen US Forest Service, Mount St. Helens National Volcanic Monument, Amboy, WA, USA

Kalliopi Gaki-Papanastassiou Faculty of Geology and Geoenvironment, Department of Geography and Climatology, National University of Athens, Athens, Greece

Henry Gaudru SVE (Société Volcanologique Européenne), Volcanic Risk Mitigation UNISDR (United Nations International Strategy for Disaster Reduction), Geneva, Switzerland

Guðrún Gísladóttir Department of Geography and Tourism, Institute of Life and Environmental Sciences, Earth Science Institute, University of Iceland, Reykjavík, Iceland

Gilson Burigo Guimarães Geography Department, Ponta Grossa State University, Ponta Grossa, Brazil

Szabolcs Harangi MTA-ELTE Volcanology Research Group, Budapest, Hungary; Department of Petrology and Geochemistry, Eötvös Loránd University, Budapest, Hungary

Harry Keys Department of Conservation and Fire and Ice Consultants, Turangi, New Zealand

Lisa M. King Curtin University Malaysia, Miri, Sarawak, Malaysia

Jiun-Chuan Lin Department of Geography, National Taiwan University, Taipei, Taiwan

Rosaly M. C. Lopes Jet Propulsion Laboratory, California Institute of Technology, Pasadena, CA, USA

Julie Margueritte Edith Cowan University, Joondalup, WA, Australia

Roberto Rendeiro Martín-Cejas Facultad de Economía, Empresa y Turismo, Departamento de Análisis Económico Aplicado, Universidad de Las Palmas de Gran Canaria, Las Palmas, Spain

Guillermo Melendez-Hevia Departamento de Geología (Paleontología), Universidad de Zaragoza, Zaragoza, Spain

Jasmine Cardozo Moreira Tourism Department, Ponta Grossa State University, Ponta Grossa, Brazil

Chris Newhall Earth Observatory of Singapore and formerly with US Geological Survey and University of Washington, Santo Domingo, Albay, Philippines

David Newsome School of Environmental Science, Murdoch University, Perth, WA, Australia

João Carlos Nunes Department of Geosciences, University of the Azores, Azores Geopark Scientific Coordinator, Ponta Delgada, Portugal

Dimitrios Papanastassiou Institute of Geodynamics, National Observatory of Athens, Thissio, Athens, Greece

Josep M. Prats Technicians of the Management Team of the Natural Park, Garrotxa, Spain

Bruce Prideaux James Cook University, Cairns, QLD, Australia

Jani Radebaugh Brigham Young University, Provo, UT, USA

Angus M. Robinson Geotourism Sub Committee, Geological Society of Australia, Sydney, NSW, Australia

Francesco Stoppa Earth Sciences Department, Gabriele d'Annunzio University, Chieti, Italy

Karen Williams Department of Conservation and Fire and Ice Consultants, Taupo, New Zealand

Part I

Introduction

Fig. 1 Volcano Tourists exploring the 'young' volcano Eldfell on the island of Heimaey, Iceland. *Photo* Patricia Erfurt-Cooper

1 Introduction

Patricia Erfurt-Cooper

1.1 Volcano Tourism

1.1.1 Geotourism and Volcanic Environments

Volcano tourism involves the exploration and study of active volcanic and geothermal landforms. Volcano tourism also includes visits to dormant and extinct volcanic regions where remnants of volcanic activity attract visitors with an interest in geological heritage (Erfurt-Cooper 2010).

In April 2010 the whole world was abruptly made aware of the power wielded by one volcano in Iceland—*Eyjafjallajökull*. The effects were far reaching. Volcanic ash clouds produced during the explosive eruptions of this volcano stayed in the atmosphere for several weeks and forced the closure of many European airports, which subsequently brought international air travel to its knees. There was no precedent for management action as this had never happened before to this extent. Travellers, unable to continue to their destinations by air were stranded around the world. Some had to wait weeks for the backlog of flights to clear before they could continue their journey, unless they were able to employ alternative travel options at considerable extra cost.

However, the news of the eruption of Eyjafjallajökull caused great interest, not just among scientists, but also attracted adventure seeking tourists, who arrived in Iceland to witness the initial eruption stage and its fiery activity. As the eruption entered its second phase, ash clouds erupted into the atmosphere, blanketing large parts of European airspace. When the eruption finally stopped, visitors kept travelling to Iceland to see the source of the chaos that almost shut down international travel. Suddenly volcano tourism was mentioned throughout the media, putting Iceland firmly on the map, and tour operators hit the ground running to profit from the potentially short-lived excitement. Global headlines indicated economic benefits for Iceland caused by the volcanic eruption with CNN reporting *'Erupting volcano boosts tourism in Iceland'* (Theodorou 2010) and National Geographic (2010) claimed that *'Eruption sparks tourist boom'*. A year later the Sydney Morning Herald newspaper described Iceland as a *'Hot Destination'* and the influx of adventure tourists with headlines such as *'After the ash comes volcano tourist cash'* (SMH 2011).

In May 2011 *Grímsvötn*, another active volcano in Iceland's south, erupted violently and again volcano tourism received a priceless boost, which no normal advertising campaign could have achieved. Grímsvötn is located under the Vatnajökull Glacier located in Europe's largest national park, the Vatnajökull National Park (established in 2008) and is hailed as Iceland's newest geological wonder along with Eyjafjallajökull —the star volcano of 2010 (IceNews 2011).

Tourist attractions and activities in Iceland are predominantly based on volcanic and geothermal environments set in dynamic landscapes which are celebrated as popular travel destination.

Not far from the action of Eyjafjallajökull and Grímsvötn a third volcano, Hekla, is expected to erupt any time based on reported signs of unrest. Since 1970, when Hekla's repose intervals changed from several decades to more frequent activity, eruptions have taken place on a more regular basis (approximately 10 years apart) with the most recent event in 2000 (Dekadevolcano.net 2011). Due to this pattern there is a common perception that another eruption may be 'overdue'. Any future activity of Icelandic volcanoes is certain to attract increasing numbers of visitors, who are prepared to take a calculated risk. Visitor guidelines and recommendations are essential and should be in place to assist with safe access to all active volcanic environments worldwide.

P. Erfurt-Cooper (✉)
James Cook University, Cairns, QLD, Australia
e-mail: patricia.erfurt@my.jcu.edu.au

P. Erfurt-Cooper (ed.), *Volcanic Tourist Destinations*, Geoheritage, Geoparks and Geotourism,
DOI: 10.1007/978-3-642-16191-9_1,

Admittedly Iceland is a special showcase for volcano tourism due to the highly publicised activity and the relatively easy access to eruption sites. However, volcano tourism is not restricted to active environments; many protected sites worldwide such as National Parks, World Heritage areas and Geoparks, feature dormant or extinct remnants of volcanic activity, although expectations of heightened activity or possible eruptions seem to attract larger visitor numbers (Sigurdsson and Lopes-Gautier 2000) and do not appear to deter people with an interest in volcanism (Struck 2010). Apart from Iceland many destinations with active volcanoes (e.g. Sakurajima, Japan; Etna, Italy; Ruapehu, New Zealand; Arenal, Costa Rica) cater specifically for the demand of volcano tourists to be as close to the action as possible with eruption viewing from observation points (including hotel balconies and terraces) at night as an unforgettable experience. It is quite common that Hotels, Lodges, Bed & Breakfasts and other tourist facilities are located within close vicinity of active mountains to offer access to these unique environments, both physical and visual. Tourist activities like skiing or hiking on the flanks of active volcanoes are often combined with other local attractions.

In June 2011 the southern hemisphere was affected by the eruption of a volcano in Chile—*Puyehue*. When the ash cloud from this eruption drifted with the trade winds eastwards around the globe and reached New Zealand and the southern parts of Australia, closures of airspace lasting several days were the result. On 21 June 2011 the Australian media reported an AUD10 million loss of income per day for the tourism industry due to the impact of the volcanic ash clouds from Puyehue. This particular type of volcanic hazard can occur without much warning and potentially affects all countries within reach of the drifting ash clouds.

In Italy Mount Etna (Sicily) attracts approximately 1 million visitors annually, many of whom stay in accommodation around the volcano (Struck 2010). In New Zealand the 'Chateau Tongariro', a hotel next to the active volcano *Ruapehu*, caters for winter sport visitors as well as summer guests. And Hawaii is one of the greatest draw cards for volcano connoisseurs, offering a range of both active and dormant environments. These destinations are examples of volcanic areas where tourism is based on the existence of a particular geological heritage and is generally combined with cultural attractions of the area as well as with regional cuisine (e.g. food and wine tourism). Volcano tourism can be the primary revenue winner for a region with an active volcano and developed tourism infrastructure or it can be a secondary attraction as volcanic landforms are commonly used as a marketing feature to offer additional uniqueness to a destination (Table 1.1).

As an activity, volcano tourism can be broadly categorised into three groups: (a) tours and day trips which often include the visit to volcanic environments as a highlight of the trip agenda; (b) field trips and excursions of longer duration specially designed and offered worldwide on the internet for special interest groups; and (c) explorative travel or expeditions to more extreme destinations for participants in search of unusual experiences—adventure guaranteed. Activities in volcanic regions can include a wide variety of action packed excitement, from hot air ballooning above bizarre landscapes of eroded tuff layers as in Cappadocia, Turkey, through hiking across the Tongariro National Park in New Zealand, to climbing Mt Fuji at night to see the sun rise from the summit. For dedicated volcano tourists these experiences remain unforgettable travel memories.

1.2 The Aim and Structure of This Book

This comprehensive book addresses the increasing interest in up-to-date literature about volcanic destinations (active and dormant) and their role in tourism worldwide. It

Table 1.1 Active volcanic environments are tourist attractions worldwide

Examples of active volcanoes regularly visited by tourists		
Chile	*Iceland*	*Peru*
Chaitén, El Tatio Geysers, Villarica, Lascar, Osorno, Copahue (Border Chile/Argentina)	Askja, Eldfell, Eyjafjallajökull, Hekla, Katla, Krafla	El Misti
Colombia	*Indonesia*	*Portugal*
Galeras	Mt Batur (Bali), Mt Agung (Bali), Mt Rinjani (Lombok), Mt Merapi (Java)	Azores Volcanoes
Costa Rica	*Italy*	*Russia*
Arenal, Poás	Campi Flegrei, Mt Etna, Ischia, Stromboli, Vesuvius	Kamchatka Volcanoes
DR Congo	*Japan*	*Spain*
Nyiragongo	Mt Aso, Mt Fuji, Mt Sakurajima, Mt Unzen, Mt Usu	Canary Islands
Ecuador	*Mexico*	*Tanzania*
Galápagos Islands, Cotopaxi, Tungurahua	Colima, Popocatépetl, Iztaccíhuatl	Ol Donyo Lengai
Ethiopia	*New Zealand*	*USA*
Erta Ale	Ruapehu, Tongariro, White Island	Hawaii Volcanoes, Mt Rainier, Mt St Helens, Yellowstone
Greece		*Vanuatu*
Nisyros, Santorini, Milos		Ambrym, Mt Yasur

includes a range of significant volcanic areas and draws attention to the potential of volcano tourism by establishing the global extent of this particular tourism sector. The book aims to present a balanced view of volcano-based tourism worldwide and to discuss important issues such as the different volcanic hazards, the potential for disasters and accidents, as well as safety recommendations for visitors. Individual chapters and case studies are contributed by a group of international authors with expertise in geology and volcanology, risk management, environmental science and other relevant disciplines associated with volcanoes as well as in geotourism based on geological resources.

Even though the scientific study of volcanoes (volcanology) is not a new discipline, scientists in recent decades had several opportunities to experience some extraordinary learning curves. The 1980 eruption of Mt St Helens in the State of Washington (USA) heralded a new era of volcano research, which was followed in 1991 by the cataclysmic eruption of Mt Pinatubo in the Philippines. This book however is not competing with the earth science literature, but takes a look at volcanic regions and their role as tourist destinations. To cover all of these would require several books and therefore this volume is building on the first book published about volcano tourism by Erfurt-Cooper and Cooper (2010) with up-to-date assessments of the current situation related to this interesting tourism sector. Apart from providing information, especially regarding the eruptive events from 2010 and 2011, this book also includes the latest research in geotourism with a special focus on volcanic environments.

The decision where to start was difficult, but after considering the impact of the more recent events the geographical region of Europe took preference, with the remainder of the book divided into parts including the Pacific Ring of Fire and other volcanic regions spread over several continents including Asia, Australia, the Americas and Africa. Each contributing author was faced with the challenge of presenting current information about volcanic areas and their role in the local tourism industry. To produce an indispensable reference book for readers interested in this field of research, discussions include the demand for volcano tourism, geotourism and adventure tourism, some historical facts related to volcanoes, as well as case studies of interesting socio-cultural settings. Protected sites such as National Parks, World Heritage areas and Geoparks are the major locations where volcano tourism takes place and are listed and described. Moreover, additional aspects of volcano tourism such as risk perception, risk management and public safety in volcanic environments are discussed in Chaps. 26 and 27 and contribute to the existing risk management literature about nature based tourism with special consideration of volcanic environments.

1.3 Further Reading

In addition to the material contained in this book, the following titles are recommended for readers interested in the geological and cultural heritage of volcanic environments. They provide essential information for people intending to visit volcanoes and also present valuable references for the scientific minded audience:

- Edelmann, J. (2000). *Vulkane Besteigen und Erkunden. Reise know how*. Bielefeld: Verlag Peter Rump GmbH.
- Erfurt-Cooper, P. & Cooper, M. (2010). *Volcano and geothermal tourism: Sustainable geo-resources for leisure and recreation*. London: Earthscan.
- Fisher, R. V., Heiken, G. & Hulen, J. B. (1997). *Volcanoes: Crucibles of change*. New Jersey: Princeton University Press.
- Krafft, M. (1991). *Volcanoes—Fire from the Earth*. London: Thames and Hudson.
- Lockwood, J. P. & Hazlett, R. W. (2010) *Volcanoes—Global perspectives*. Oxford: Wiley-Blackwell.
- Lopes, R. (2005). *The volcano adventure guide*. Cambridge: Cambridge University Press.
- Lopes, R. (2010). *Volcanoes: A beginner's guide*. Oxford: Oneworld Publications.
- Rosi, M., Papale, P., Lupi, L. & Stoppato, M. (2003). *Volcanoes: A firefly guide*. New York: Firefly Books.
- Sigurdsson, H., Houghton, B., McNutt, S., Rhymer, H. & Stix, J. (Eds.) (2014). *Encyclopedia of volcanoes* (2nd edn.). San Diego: Academic Press-Elsevier.

1.4 Background: Volcanoes in History and Culture

The fascination of people with volcanoes goes back to ancient civilisations, to people who were in awe of the unpredictable natural forces generated by these geological phenomena which they could not explain. Many cultures have linked volcanoes and related natural features such as hot springs, geysers and hissing steam vents directly to the mythological 'underworld', where humankind was supposed to suffer in the eternal fires of hell. In fact, to envisage the grand entrance to everlasting damnation there is hardly a better place for the imagination to work overtime than the crater of an active volcano. The devastated landscapes shaped by lava flows, layers of ash and steaming fissures are less than inviting and present a hostile environment for both flora and fauna (Fig. 1.1). On a cold and rainy day a hike through the mist shrouded Eldgjá fissure in southern Iceland feels like being in a different world (Fig. 1.2), akin to a scene out of the Nordic mythology. In contrast a warm summer day on a tropical volcano summit in the Philippines

Fig. 1.1 The area of the Krafla Fires in the north east of Iceland. The fissure eruptions lasted from 1975 to 1984. *Photo* Patricia Erfurt-Cooper

Fig. 1.2 Eerie scenery in the cold and mist shrouded Eldgjá fissure, Southern Iceland. *Photo* Patricia Erfurt-Cooper

or Indonesia covered in ominous clouds can also evoke a disturbing atmosphere not just for superstitious minds. It is no wonder that local history and culture are often linked closely to past eruptive activity with examples like Pompeii and Herculaneum coming to mind. Both cities were buried by pyroclastic flows from Mt Vesuvius in 79 AD and are today famous tourist destinations with Pompeii attracting 2.3 million visitors in 2010 (Bailey 2012).

Volcanoes are still dominating cultural and religious traditions in many areas. In volcanic regions such as Indonesia, New Zealand, Iceland, Hawaii and Japan local legends and mythologies are closely connected to volcanoes and their unique geothermal features. Sacrifices in the form of food, flowers, live animals and money are a common means of trying to influence the 'mood' of active volcanoes; customs which are still practiced in many areas. In the past even human sacrifice was used in an attempt to appease the 'angry mountain', although no doubt the success rate was rather unsatisfactory (Alvarado and Soto 2008; Erfurt-Cooper 2010a, b; Henderson 2007; Viramonte and Incer-Barquero 2008). Based on these traditions some communities that reside close to an active volcano, still carry out appeasement ceremonies under the guidance of their religious leaders. In some areas this is now also considered a tourist attraction and contributes to the local economy. In Indonesia for example special appeasement ceremonies conducted by ritual experts are held to *'placate the spirits believed to dwell in the mountain'* to withhold an imminent eruption (Chester and Duncan 2010; Marshall 2008). *Gunung Merapi* (meaning Fire Mountain), one of the most active volcanoes in Central Java, is considered by local residents as *'the heart of the universe'*. On the other hand, Merapi has erupted at least 13 times over the last thousand years, resulting in many human casualties, and offerings to placate the mountain are therefore by no means a thing of the past. Formal sacrifices to Mount Merapi are made annually on the 29th day of the seventh month of the Islamic calendar. This is based on the Javanese belief that a miniature world inside the crater of Mount Merapi resembles the everyday life of the Javanese people (Chester and Duncan 2010). To prevent further loss of lives the Indonesian government has tried unsuccessfully to resettle residents from endangered villages such as Turgo on the slopes of the mountain (Dove 2010). The villagers resisted to the point, that the government eventually agreed to allow the communities to continue to live in Turgo, but insisted that the village would be officially removed from government maps (Chester and Duncan 2010).

The case of Mount Merapi is not that uncommon. Villagers in volcanic areas often prefer to stay on their land, with one of the reasons the fertile volcanic soils. Being independent in their natural traditional environment is also preferable to living somewhere else. Despite the imminent danger of volcanic activity, ancestral homes, temples and family kinship are further strong determinants to stay compared with uncertain resettlement conditions. While the risk is certainly a calculated and accepted one based on experience, appeasement ceremonies are playing an important role for the peace of mind of the local populations.

Other examples of appeasement rituals come from Hawaii, where the 'goddess of fire', Madame Pele, is honoured with sacrificial ceremonies, including dancing and singing and special gifts like floral offerings to prevent disastrous eruptions. According to Knipe (1982) *'nowhere else in the world does volcanic mythology exist as it does in Hawaii'* with Pele as the central figure. All kinds of offerings have been made to Pele including chickens, pigs, money, chewing gum, sugarcane, leaf tobacco, taro roots, and sweet potatoes (Nimmo 1986). Despite initial scepticism, some tourists now believe that Pele is the cause of bad luck for people who have removed pieces of volcanic rock from Hawaii. Allegedly after returning home they started to suffer from misfortunes, which confirmed their belief that negative consequences can only be avoided by returning the rock to Pele in Hawaii (Hammond 1995).

Throughout the world volcanoes feature strongly in local legends and mythology. During archaeological excavations in Mexico near the volcano *Popocatépetl*, small effigy volcanoes were discovered and are thought to have been used commonly to appease the volcano which figured prominently in village rituals (Plunket and Uruñuela 2000). Even though the terms are mostly dictated by volcanic activity, people have learnt to co-exist with nature by focussing on advantages such as fertile soils and hot springs, and more increasingly in recent times as a 'priceless' attraction for tourists.

1.5 The Attraction of Volcanoes

1.5.1 Why are Volcanoes so Popular?

The popularity of volcanic environments has grown as a result of their featuring in movies such as *Dante's Peak* (1997)*; Volcano* (1997)*; Lord of the Rings Trilogy* (2001–2003)*; Volcanoes of the Deep Sea* (2003) *and Krakatoa* (2006). In the literature volcanoes are the main topic with books including *Krakatau* by Winchester (2003); *Montserrat* by Davison (2003) and *A journey to the center of the earth* by Verne (1871). TV documentaries for example resulted in Yellowstone now commonly known as a 'Supervolcano'. But more importantly the interest in volcano tourism grows with every eruption that is covered in the international media and on the Internet. With access to remote destinations becoming easier and more affordable, essential infrastructure to cater for larger numbers of visitors is improving exponentially in many countries. Even unstable political conditions do not stop volcano tourism, but merely cause interruptions for a while. As a result of all these factors and trends, visitor numbers at individual destinations are an indication that volcano tourism is by no means a small niche market. And if the volcanic environment is not the main destination determinant, as an additional drawcard it fits in well with other local attractions.

In fact, volcanoes do not have to be active to be a tourist attraction as long as they are reasonably easy to access. To reach more remote locations to visit volcanoes with special individual features (e.g. lava lakes) requires more determination and planning. But often volcanoes are part of a tour package featuring local attractions and are included in the trip agenda. On the other hand research has shown, that heightened volcanic activity attracts more visitors (e.g. Chaitén volcano—Chile, 2008 (Chap. 26); Kilauea—Hawaii, 2011 (Chap. 21)), compared to times of 'normal' activity or quiescence; however, this activity can be potentially dangerous (Gaudru and Erfurt-Cooper 2010). It is therefore essential for tour operators to provide sufficient information about any possible risks and hazards that can be encountered, even in quiescent times.

To better understand the role of volcanic environments as tourist destinations a conceptual framework (Fig. 1.3) was developed, which outlines not only the elements of volcano tourism, but also shows the integration of these dynamic environments in various tourism sectors (e.g. Ecotourism, Geotourism, Adventure Tourism). The model incorporates several tourism groups, each of which takes into account the distinctive characteristics of volcanic environments as a tourism resource. The individual sectors are geotourism, adventure tourism, ecotourism and dark tourism, and are based on specific visitor expectations and demand for unique settings. Although these four tourism sectors can be distinguished from each other, the potential for overlap from one sector to another frequently does exist; generally with a complementary result. The model also establishes correlations between the various integrated elements of volcano tourism, which are associated with environmental settings and natural resources, recreational attractions, cultural traditions and customs, as well as access and infrastructure. Together these represent the particular attractiveness of the dynamics of volcanic settings as a tourism resource. For example volcano tourists not only partake in sightseeing tours, but they also appreciate the opportunity to visit hot spring spas and resorts as well as other geothermal features of interest if these are present and accessible. In Iceland visitors can undertake daytrips to volcanic attractions including geyser fields and geothermal power stations, but they also have the opportunity to visit the geothermal Blue Lagoon for a relaxing soak after hiking over lava fields or viewing spectacular waterfalls. In New Zealand visitors of volcanoes and geothermal nature parks can take tours to nearby hot spring spas, and in Japan tourists commonly visit every worthwhile attraction at their destination as well as sampling the local hot springs (Erfurt-Cooper 2011).

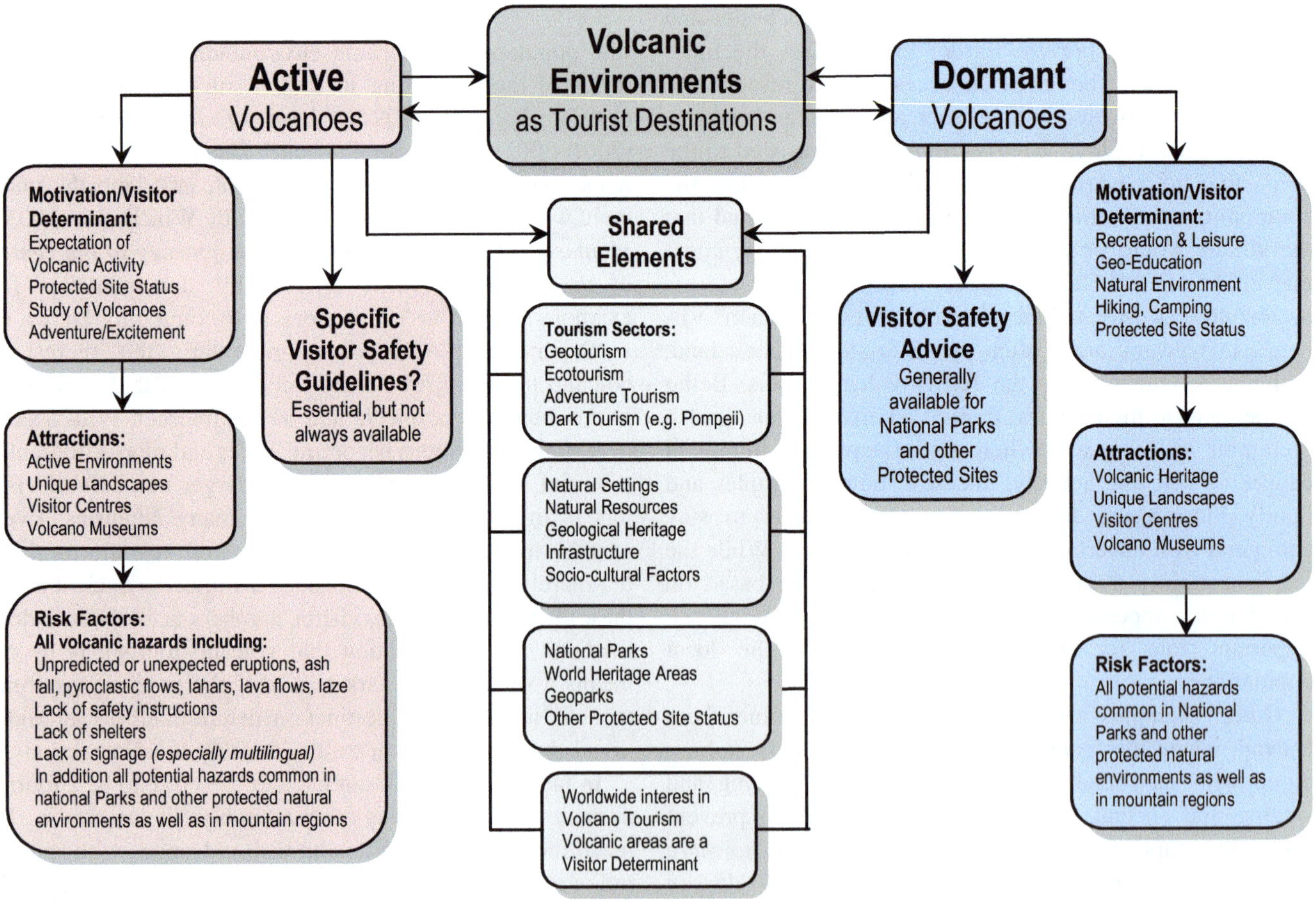

Fig. 1.3 A conceptual framework demonstrating the role of volcanic environments as tourist destinations. By Erfurt-Cooper (2012)

Table 1.2 Examples of volcano tours advertised on the internet

Volcano tours	
Volcano Tours, Grindavík	*Arenal Volcano Tours—Costa Rica Tours*
www.volcanotours.is	www.arenal.net/arenal-volcano-tour.htm
Unity Airlines—Vanuatu Volcano Discovery and Charter Services	*Volcano Tours*
www.unity-airlines.com/index.html	www.icelandtravel.is/our-tours/tours/daytours-activities/volcano-tours/
Volcano adventure travel	*Hawaii volcano hiking*
http://www.volcanolive.com/travel.html	www.volcanotours.com/
O.X. Outdoor Excursions—Guatemala	*Hawaii Volcano Tours*
www.oxexpeditions.com/	www.lavatours.com/
White Island Tours—New Zealand	*Lava Ocean Adventures*
www.whiteisland.co.nz/	www.seelava.com
VolcanoDiscovery—Study trips, photo expeditions	*Etna Excursions and Volcano Tours*
www.volcanodiscovery.com/volcanotours.html	www.volcanotrek.com/
Undara Experience—North Queensland	*Masaya Volcano National Park—Nicaragua*
http://undara.com.au/	www.virtualnicaragua.com/masaya-volcano-national-park.php
Volcano tours and photography—Indonesia	*Vulkan Reisen*
www.firevolcano.com/	www.v-e-i.de/
Viva Ethiopia Tour—Erta Ale	*Vulkane aktiv erleben*
www.vivaethiopiatour.com/	www.vulkankultour.de/
Vulkan Expeditionen International	*The Lakes and Volcanoes Adventure Tour—Chile*
www.vulkane.net/	http://www.southamerica.travel/

Source Compiled by Author from various sources

Table 1.3 Elements of destination marketing

Volcanic and geothermal environments		
Attractions	Activities	Motivations
Eruptive events	Eruption viewing	Adventure
Active lava flows	Hiking, trekking, skiing, canoeing	Curiosity
Crater lakes	Sightseeing	Field research
Lava lakes	Learning	Scientific interest
Lava tunnels	Hot spring bathing	Study, educational
Fumaroles, steam vents	Volcano museums	Part of trip agenda
Geysers, hot springs	Interpretive centres	Part of outdoor activity
Boling mud pools	Guided tours	Photography
Hot streams	Camping	
Unique rock formations		

Source Compiled by Author from various sources

1.5.2 Supply and Demand

The supply is impressive with thousands of active and dormant volcanoes worldwide. And as we have seen, the demand for volcano tourism is fuelled by the publicity from eruptions like Eyjafjallajökull in 2010, which was broadcast worldwide. A year later the eruption of the Puyehue volcano in Chile raised similar awareness about volcanic activity and its far reaching effects, this time in the southern hemisphere. While both eruptions caused substantial problems for the travel industry, they also highlighted the potential of unexpected volcanic activity anywhere in the world. Catering for this growing demand are volcano tours run by specialised tour operators to the most popular volcanic regions on Earth. Their offers include guidance by geologists and/or volcanologists while their marketing is Internet based to reach a worldwide audience (Table 1.2).

The local economic benefits from volcano tourism also play a central role in many areas. Visitor motivation is therefore important and encouraged with successful marketing of local attractions and activities (Table 1.3). Volcanoes, together with unique geothermal features are powerful drawcards for visitors and many regions (e.g. New Zealand, Iceland) have invested significantly in volcano and hot spring tourism as a viable market in combination with other local attractions. Package deals including snowboarding, ice climbing and skiing on active volcanoes are not unusual and are combined in many areas with hot spring spa resorts. Accommodation such as hotels and resorts are built as close as 'safely' possible to active volcanoes to benefit from the

Table 1.4 Educational facilities are in demand at volcanic sites

Volcano museums and interpretive science centres	
Penshurst Volcanoes Discovery Centre, Australia	Aso Volcano Museum, Japan
Vulcania—Volcanic Theme Park, France	Mt Unzen Disaster Memorial Hall, Shimabara, Japan (plus several memorial sites)
Eifel-Volcano Museum Daun, Maar-Museum Manderscheid and Vulkanhaus Strohn, Vulkaneifel, Germany	Sakurajima Volcano Museum, Japan
Nisyros Volcanological Museum, Greece	Usu Volcano Science Museum, Toyako, Japan
Ság [Volcano] Museum, Hungary	Jeju National Museum, Korea
Eldfjallasafn Volcano Museum Stykkishólmur, Iceland	Te Papa Museum Wellington, New Zealand
Eldheimar ('World of Fire') Museum Vestmannaeyjar, Iceland	Auckland Museum Volcano House, New Zealand
Eyjafjallajökull Visitor Centre Þorvaldseyri, Iceland	Maison du Volcan, Reunion
Batur Volcano Museum Kintamani, Indonesia	Museu del Volcans, Garrotxa, Spain
Merapi Volcano Museum, Indonesia	Hawaii Volcanoes NP Jaggar Museum, Hawaii, USA
Volcano Museum Etna, Italy	Mt St Helens National Volcanic Monument (several visitor centres and observatories), Washington State, USA
Museo Archeologico Eoliano, Lipari, Italy	Yellowstone NP Trailside Museums, Wyoming, USA
Museo Archeologico Virtuale Ercolano, Italy	
Museo dell'Osservatorio Vesuviano, Italy	
Nicolosi Vulcanological Museum, Sicily, Italy	
San Venanzo Volcano Museum, Italy	

Source Compiled by Author from various sources

Fig. 1.4 Volcano science museum in Toyako near Mt Usu, Japan. *Photo* Patricia Erfurt-Cooper

visual attraction, especially if eruptions can be viewed from a (hopefully) secure distance (e.g. Hekla, Iceland; Etna, Sicily; Arenal, Costa Rica; Sakurajima, Japan).

1.5.3 Visitor Centres and Volcano Museums

Volcanic and geothermal exhibitions at museums and visitor centres (Table 1.4) contribute significantly to educating the public about these unique natural environments. Most regions with active volcanism offer visitors extensive information through documentaries, geologic exhibits, guided tours and entertaining activities for all age groups. Also of interest are local cultural and historical connections with volcanic and geothermal events with retail areas usually offering books about the local geology along with general souvenirs and creations by local artists. Popular in Cappadocia in central Turkey are small pumice carvings of the Fairy Chimneys while at Mt Aso in Japan sulphur blocks and volcanic rock samples are sold on the viewing platform next to the crater.

Major volcanic events often result in a new exhibition or visitor centre being built near an eruption site to cater for the increasing demand for visitor information. Japan for instance has many volcano science museums (Fig. 1.4), which are frequently connected to the local volcano

Table 1.5 A selection of volcano national parks from countries around the world

Protected volcanic sites—National parks	
Lanín National Park, Argentina	Krakatau Volcano National Park, Indonesia
Crater Lakes National Park, Australia	Mt Rinjani National Park, Lombok, Indonesia
Mt Kaputar National Park, Australia	Mt Etna Provincial Park, Sicily, Italy
Undara Volcanic National Park, Australia	Vesuvius National Park, Italy
Warrumbungles National Park, Australia	Aso-Kuju National Park, Kyushu, Japan
Mt Cameroon National Park, Cameroon	Fuji-Hakone-Izu National Park, Honshu, Japan
Lauca National Park, Chile	Kirishima-Yaku National Park, Kyushu, Japan
Villarica National Park, Chile	Shikotsu-Toya National Park, Hokkaido, Japan
Haikou Shishan National Park, China	Unzen-Amakusa National Park, Kyushu, Japan
Volcano National Geological Park, China	Kilimanjaro National Park, Kenya
Zhangzhou Coastal Volcano Park, China	Tongariro National Park, New Zealand
Los Nevados National Park, Colombia	Masaya Volcano National Park, Nicaragua
Nevado del Huila National Park, Colombia	Mayon Volcano Natural Park, Philippines
Puracé National Park, Colombia	Volcanoes National Park, Rwanda
Arenal Volcano National Park, Costa Rica	Teide Volcanoes National Park, Canary Islands, Spain
Poas Volcano National Park, Costa Rica	Yangmingshan National Park, Taiwan
Rincon de la Vieja Volcano National Park, Costa Rica	Göreme National Park, Turkey
Galápagos Islands National Park, Ecuador	Crater Lakes National Park, Oregon, USA
Cotopaxi National Park, Ecuador	Hawaii Volcanoes National Park, USA
Virunga National Park, Democratic Republic of the Congo	Katmai National Park, Alaska, USA
Volcano Regional National Park, Auvergne, France	Lassen Volcanic National Park, California, USA
Eyjafjallajökull, Vatnajökull National Park, Iceland	Lava Beds National Park, California, USA
Þingvellir National Park, Iceland	Mt Rainier National Park, Washington, USA
Snæfellsjökull National Park, Iceland	Mt St Helens National Volcanic Monument, USA
Bromo-Tengger-Semeru National Park, Indonesia	Sunset Crater Volcano National Monument, Arizona, USA
Gunung Merapi National Park, Indonesia	Yellowstone National Park, USA

Source Compiled by Author from various sources

observatory, where visitors are provided with up-to-date data (e.g. web cams, seismographs), as well as exhibits relevant to the local geology. Film footage of past eruptive events and simulated pyroclastic flows complete with special effects as well as bird's eye views of eruption columns are a fascinating way of educating the public about the explosive powers of a volcano.

At the base of Eyjafjallajökull at Þorvaldseyri a new visitor centre and museum was established, which opened on the eruption's first anniversary in April 2011 (Edelstein 2011; Eyjafjallajökull Visitor Centre 2012; IceNews 2011; Visitor Centre Þorvaldseyri 2011). In May 2012 the Eldfjallasafn Volcano Museum in Stykkishólmur opened with an exhibition featuring international art and objects related to volcanic eruptions and their impact (Eldfjallasafn Volcano Museum 2012). Also in May 2012 plans were announced for a new volcano museum (Eldheimar, The Pompei of the North) on the island of Heimaey, were in 1973 a volcanic eruption forced the evacuation of all residents. Currently the opening is scheduled for early 2015 (Bjarnason, 2014). From the increasing demand for information the educational value of volcano museums and information centres cannot be disputed as a safe learning experience. They are a perfect option for visitors who are interested in the geological background, but prefer to watch the action from a safe distance via webcam and video footage, especially if the volcano is too active for a visit (Erfurt-Cooper 2011). Volcano museums also serve as an opportunity to ask questions and to put into perspective what has been observed 'in the field'.

1.6 Protected Areas: Volcanic National Parks, World Heritage Sites and Geoparks

Looking at the global distribution of volcanoes we discover that in many cases they are located within protected areas. With more than 1,500 volcanoes classed as active worldwide and many more dormant, large numbers of these volcanic landforms are integrated as unique attractions in protected sites such as National Parks (Table 1.5), World Heritage Areas (Table 1.6) and Geoparks (Table 1.7). Probably the most famous volcanic and geothermal national park is Yellowstone in the State of Wyoming, USA, where visitor numbers have reached around 3.5 million people annually in recent years.

Since the first estimated figures from 1872 of 300 visitors to Yellowstone there has been a steady increase over 140 years (The Total Yellowstone Page 2012). Although

Yellowstone visitor numbers			
2007	3,151,343	2010	3,640,205
2008	3,066,580	2011	3,394,322
2009	3,295,192	2012	3,447,729

Table 1.6 World Heritage areas with active and dormant volcanic landforms (in alphabetical order of countries and regions)

Active and dormant volcanic World Heritage areas		
Volcanic World Heritage sites	Country	Year listed
1. Ngorongoro Conservation Area	Tanzania, Africa	1979, Ext. 2010
2. Taï National Park	Ivory Coast, Africa	1982
3. Gondwana Rainforests	Australia	1868
4. Heard and McDonald Islands	Australian Territory	1997
5. Lord Howe Island Group	Australia	1982
6. Macquarie Island	Australia	1997
7. Wet Tropics of Queensland	Australia	1988
8. Fernando de Noronha Archipelago	Brazil	2001
9. Area de Conservación Guanacaste (*Rincon de la Vieja*)	Costa Rica	1997
10. Cocos Island National Park	Costa Rica	1997
11. Morne Trois Pitons National Park	Dominica, Lesser Antilles	1997
12. Rapa Nui National Park	Easter Islands, Chile	1995
13. Mt Huangshan	China	1990
14. Virunga National Park	Democratic Republic of the Congo	1979
15. Galápagos Islands	Ecuador	1978
16. Sangay National Park	Ecuador	1983
17. Joya de Cerén	El Salvador	1993
18. Pitons, Cirques and Remparts of Reunion Island	Reunion, France	2010
19. Þingvellir National Park	Iceland	2004
20. Surtsey Island	Iceland	2008
21. Bali Cultured Landscape	Bali, Indonesia	2012
22. Borobudur Temple (*Merapi*)	Java, Indonesia	1991
23. Komodo National Park	Indonesia	1991
24. Tropical Rainforest Heritage	Sumatra, Indonesia	2004
25. Ujung Kulon National Park	Indonesia	1991
26. Giants Causeway	Ireland	1986
27. Aeolian Islands	Italy	2000
28. Mount Etna	Sicily, Italy	2013
29. Pompeii and Herculaneum	Italy	1997
30. Mount Fuji	Honshu, Japan	2013
31. Ogasawara Islands	Japan	2011
32. Shiretoko National Park	Hokkaido, Japan	2005
33. Yakushima	Japan	1993
34. Mount Kenya National Park	Kenya	1997; Ext. 2013
35. Jeju Volcanic Island and Lava Tubes	Korea	2007
36. El Pinacate &d Gran Desierto de Altar Biosphere Reserve	Mexico	2013
37. Tongariro National Park	New Zealand	1993
38. Rock Islands Southern Lagoon	Republic of Palau (Micronesia)	2012
39. Pico Island	Azores, Portugal	2004
40. Volcanoes of Kamchatka	Russia	1996; Ext. 2001
41. Brimstone Hill Fortress National Park	Saint Kitts and Nevis, Lesser Antilles	1999
42. Pitons Management Area	Saint Lucia, Lesser Antilles	2004
43. Old and New Towns of Edinburgh (*Arthurs Seat*)	Scotland	1995
44. Garajonay National Park	La Gomera, Canary Islands, Spain	1986

(continued)

Table 1.6 (continued)

Active and dormant volcanic World Heritage areas		
Volcanic World Heritage sites	Country	Year listed
45. Teide National Park	Tenerife, Canary Islands, Spain	2007
46. Kilimanjaro National Park	Tanzania	1987
47. Göreme NP and the Rock Sites of Cappadocia	Turkey	1985
48. Pamukkale - Hierapolis	Turkey	1988
49. Hawaii Volcanoes National Park	Hawaii, USA	1987
50. Yellowstone National Park	Wyoming, USA	1978
51. Yosemite National Park	California, USA	1984

Source Compiled by Author from various sources

Table 1.7 Examples of volcanic Geoparks from Europe and Asia. The Geopark concept is rather new and so far no 'global' geoparks have been recognised in American and African countries, although regional and national geoparks may be under consideration or have been established as a first step towards global recognition

Protected volcanic sites—Geoparks (national and global)	
Kanawinka Geopark, Australia	Toya Caldera/Usu Volcano Global Geopark, Japan
Hong Kong Global Geopark, China	Unzen Volcanic Area Global Geopark, Japan
Leiqiong Volcano Geopark, Guangdong, China	Jeju Island, Korea
Wudalianchi Volcanoes Geopark, Heilongjiang, China	Magma Geopark, Norway
Vulkaneifel European Geopark, Germany	Azores Geopark, Portugal
Katla Geopark, Iceland	Penghu Geopark, Taiwan
Itoigawa Global Geopark, Japan	Geopark Shetland, Scotland, UK
Mount Aso Geopark, Japan	

Source Compiled by Author

Table 1.8 Examples of annual visitor numbers at volcanic and geothermal destinations. Precise visitor statistics are difficult to obtain as many parks do not check how many people enter the area

Destinations	Visitor numbers (p/a)
Fuji-Hakone-Izu National Park, Japan (Honshu)	103+ million[a]
Aso-Kuju Volcanics, Japan (Kyushu)	20+ million[a]
Shikotsu-Toya, Japan (Hokkaido)	16.3 million[a]
Kirishima-Yaku, Japan (Kyushu)	14.4 million[a]
Bandai—Asahi National Park, Japan (Honshu)	14.4 million[a]
Teide National Park, Spain	2.73 million (2011)[b]
Yellowstone National Park, USA (Wyoming)	3,18 million (2013)[c]
Hawai'i Volcanoes National Park, USA	1.58 million (2013)[d]
Vesuvius National Park, Italy	1 million[e]
Pompeii, Italy	2.3 million (2010)[f]
Tongariro National Park, New Zealand	1 million[g]
Giant's Causeway, Ireland	600,000 (2011)[h]
Mt Etna, Italy	1 million (2010)[i]
Haukadalur with original Geyser, Iceland	407,230 (2011)[j]
Arenal Volcano National Park, Costa Rica	273,000 (2005)[k]
Sunset Crater National Park, USA (Arizona)	184,864 (2013)[d]

[a] International EMECS Center 2012; [b] TM Tenerife Magazine 2012; [c] Yellowstone up close and personal 2014; [d] NPS Public Use Statistics Office 2014a, b; [e] Karkut 2010; [f] Bailey 2012; [g] www.nationalpark.co.nz; [h] Moriarty 2012; [i] Struck 2010; [j] Icelandic Tourist Board 2012; [k] Aguirre and Ahern 2007

Table 1.9 Examples of cities in close proximity of volcanoes classed as active

Country	City	Volcano(es)	Population
Costa Rica	San Jose	Arenal	365,799 (metro 1.7 million)
Colombia	Pasto	Galeras	399,723
DR Congo	Goma	Nyiragongo	377,112
Guatemala	Guatemala City	Pacaya	1.1 million (metro 4.1 million)
Ecuador	Quito	Cotopaxi, Guagua Pichincha, Pululagua	2.24 million (metro 3.34 million)
Indonesia	Jogjakarta	Merapi	388,088
Italy	Catania	Etna	291,000
Italy	Naples	Vesuvius	960,000 (metro 4 million)
Japan	Nagasaki	Unzen	446,007
Japan	Shimabara	Unzen	48,461
Japan	Kumamoto	Mt Aso and Unzen	731,286
Japan	Kagoshima	Sakurajima	605,855
Japan	Tokyo	Mt Fuji	8.7 million (metro 35.6 million)
Mexico	Colima	Volcán de Colima	662,500
Mexico	Mexico City	Popocatépetl	8.85 million
New Zealand	Auckland	Rangitoto, Auckland volcanic fields	1.37 million
New Zealand	Rotorua	Mt Edgecumbe, Tarawera	68,900
Nicaragua	Managua	Masaya	927,087 (metro 2.4 million)
Philippines	Manila, Luzon	Pinatubo and Taal	1.6 million
Philippines	Legazpi City, Luzon	Mt Mayon	182,000
Spain	Santa Cruz, Tenerife	Mt Teide	222,271
USA	Anchorage, Alaska	Redoubt	291,826
USA	Portland, Oregon	Mt Hood	593,820 (metro 2.26 million)
USA	Seattle	Mt Rainier	620,778 (metro 3.5 million)
USA	Vancouver, Washington	Mt Baker	603,562 (metro 2.3 million)

Source Compiled by Author from various sources

visiting Yellowstone is not officially promoted as volcano tourism, the geothermal activity based on volcanism is what the majority of the people come to see. Yellowstone has an unfortunate reputation as a super-volcano with publications and documentations reflecting increased media attention aimed at a potential future disaster site. Confirming these concerns the United States Geological Survey (USGS) states in their web-based information that *'Yellowstone's volcanic and hydrothermal history suggests the potential for various kinds of eruptions in the future'* (USGS Volcano Hazards Program 2012). Growing media interest in Yellowstone's earthquake activity in early 2014 has contributed to further public concern. To compare popular destinations, examples of annual visitor numbers of volcanic and geothermal areas were collected (Table 1.8), although it is nearly impossible to obtain accurate figures as many protected areas do not keep count of their visitor numbers.

Many cities, large and small, are located near active volcanoes and their populations are accustomed to the permanent risk to their personal living environment. Table 1.9 shows examples of cities and their population figures; both for the actual city as well as the greater metropolitan area at risk from eruptive activity.

1.7 Volcanic Processes

Volcanology is a complex field of earth science that investigates volcanoes and monitors their internal processes. Specific fields of study include plate tectonics, seismology, structural geology and geomorphology as well as geology disciplines such as igneous petrology, geophysics, geochemistry and mineralogy. For more in-depth information it is recommended to consult the earth science literature with a special focus on volcanology. A short overview of the principal volcano types is included here and describes briefly shield volcanoes, strato volcanoes, cinder cones and calderas. Additional information in the form of a glossary is attached as Appendix 2.

Fig. 1.5 The Shield Volcano Rangitoto (meaning bloody sky) in New Zealand is part of the Auckland volcanic field and can be reached by ferry from Auckland. The last eruption was possibly 600–700 years ago, although no accurate data is available. *Photo* Patricia Erfurt-Cooper

Fig. 1.6 Mt Fuji (stratovolcano) is located on the Japanese island of Honshu close to Tokyo and Yokohama and is a major tourist destination. *Photo* Patricia Erfurt-Cooper

Shield volcanoes are built from high temperature lava flows (800–1200 °C) which can spread out multi directional over vast areas. The Hawaiian volcanoes are an example of shield volcanoes which are located over a hot spot and still 'under construction'. Other shield volcanoes include Etna in Italy, Jeju Island in Korea, Erta Ale in Ethiopia and Rangitoto in New Zealand (Fig. 1.5).

Composite or strato volcanoes are built from layers of lava, tephra and pyroclastic debris and have steep sides compared to the rather broad shape of shield volcanoes. Strato volcanoes are known for explosive eruptions caused by built up pressure as magma moves through conduits slowly upwards. In their active state they generate pyroclastic flows and lahars and can be potentially dangerous to nearby residents and visitors. Strato volcanoes are located all over the world and some well-known mountains include Vesuvius (Italy), Damavand (Iran), Popocatépetl (Mexico), Mt St Helens (USA), Copahue (Chile), Tungurahua (Ecuador), Krakatau and Merapi (Indonesia), Fuji (Fig. 1.6) and Sakurajima (Japan), Pinatubo and Mayon (Philippines), Mt Ararat (Turkey), Hekla (Iceland) and White Island (New Zealand). These are only a few examples from a long list of

Fig. 1.7 Strombolian spatter covers the ground near cinder cones which were created by lava fountains from parasitic vents on the flanks of Mt Etna. *Photo* Patricia Erfurt-Cooper

Fig. 1.8 Scoria cones created from spatter during the Krafla fires in Iceland. The view is slightly obstructed by mist. *Photo* Patricia Erfurt-Cooper

strato volcanoes currently classed as active and which are popular tourist destinations.

Cinder cones are steep, cone shaped hills of fragmented lava particles (cinder, scoria) which have accumulated around a single vent (Fig. 1.7). They are a common form of volcano and can range in size from small spatter cones to hundreds of metres in elevation. Cinder cones can grow rapidly in groups on the flanks of parent volcanoes (Fig. 1.8) or they grow into one single mount like Paricutín in Mexico or Eldfell on the Westman Islands in Iceland (Fig. 1.9).

A *caldera* (Spanish for cauldron) is a collapsed magma chamber resulting from an explosive eruption during which the magma chamber was emptied. The overlying chamber roof is left without support and gives way by collapsing into the void below. Calderas can be any size and some of the largest calderas include the Taupo Caldera (New Zealand), Toba Caldera

Fig. 1.9 The cinder cone or composite volcano Eldfell (200 m) on the island of Heimaey, Iceland. *Photo* Patricia Erfurt-Cooper

Fig. 1.10 Inside view of a small lava tube in the Krísuvík area, Southern Iceland. *Photo* Patricia Erfurt-Cooper

(Indonesia), Yellowstone (USA), Caldera de las Cañadas (Tenerife, Spain), Santorini (Greece), Aso Caldera and Aira Caldera (Japan) and the Crater Lake in Oregon (USA).

Lava tunnels, lava tubes or pyro-ducts occur in areas where lava of low viscosity flows from a vent or fissure during an eruption. While the top layers of the channelled lava are cooling and solidifying the lava underneath continues to flow inside the conduit pulled by gravity and draining downhill. This can continue for long distances if the lava keeps flowing from the eruption vent, covered by a layer of hardened crust. The diameters of lava tunnels can vary greatly from very small (Fig. 1.10) to big enough for a large bus to drive though (Fig. 1.11). Roof or ceiling collapses provide sometimes access to passages of lava tunnels which are frequently inhabited by certain types of fauna such as bats who do not rely on daylight (Fig. 1.12). Lava

Fig. 1.11 The size of this Undara Lava cave gives an impression of the enormous amount of lava that was flowing through the former conduits. Roof collapses in many areas have left sections of the lava tubes accessible for visitors to explore. *Photo* Patricia Erfurt-Cooper

caves and tunnels are not as rare as some people may think; they occur worldwide and are utilised by the local population for various reasons including shelter, storage and occasionally as secret hiding places. Lava caves exist on the flanks of Mount Etna, in several states of North America, in Iceland, New Zealand and Australia. Examples of their distribution show that they can be found in many regions with past and present volcanic activity (Table 1.10).

Fig. 1.12 Bats are disturbed by the torch of a tour guide to give visitors a chance to take photos. *Photo* Patricia Erfurt-Cooper

Table 1.10 Examples of lava caves and tunnels

The longest lava tubes and caves worldwide			
Country		Approximate length	
USA, Hawaii	Kazamura Cave	65.5 km	40.69 mi
Spain, Canary Islands Tenerife	Cueva del Viento and Cueva del Sobrado System	17.03 km	10.58 mi
Kenya	Leviathan Lava Tube	9.15 km	5.68 mi
South Korea, Jeju Island	Manjang Gul Lava Cave	8.92 km	5.54 mi
Chile, Easter Islands	Roiho Lava Cave System	6.5 km	4.03 mi
Spain, Canary Islands, El Hierro Island	Cueva de Don Justo	6.31 km	3.92 mi
Mexico	Ferrocarril-Mina Inferior Lava Cave	6.19 km	3.84 mi
	Iglesia-Mina Superior Lava Cave	5.14 km	3.19 mi
Portugal, Azores	Gruta das Torres Pico	5.4 km	3.35mi
Rwanda	Ubuvumo Bibihonga	4.5 km	2.79 mi
Iceland	Surtshellir-Stepanshellir Cave System	5 km	3.1 mi
Kenya	Mount Suswa Lava Tube Cave System	3.5 km	2.17 mi
Western Samoa, Upolu Island	Ana Pe'ape'a at Salamumu	3.48 km	2.16 mi
Ecuador, Galápagos Islands, Santa Cruz	Cueva del Cascajo	3 km	1.86 mi
	Cueva de Gallardo	2.25 km	1.39 mi
Japan	Mitsuike-Ana Lava Cave	2.2 km	1.36 mi
Saudi Arabia	Umm Jirsan Lava Tube System	1.4 km	0.86 mi
Australia	Undara—Bayliss Cave	1.35 km	0.83 mi
Argentina	Cueva de Dona Otilla	838 m	916 yd
France, Reunion Island	Caverne du Brule De Citron Galet	680 m	743 yd

Source Compiled by Author from various sources

All of the volcanic areas discussed here are popular tourist destinations which offer numerous points of interest; not just volcanic, but also cultural and historical aspects. Visitors may not always see themselves as volcano tourists, but there is no doubt that volcanic environments play an important role in the local tourism industries all over the world.

Volcanic processes also include a wide range of *geothermal phenomena*. Geysers, hot springs, steam vents and boiling lakes are caused by heat emitting from an underlying magma chamber, where the temperature of subterranean water basins can be increased beyond boiling point. Geothermal activity based on volcanism is observed by millions of visitors every year in Yellowstone National Park with the largest concentration of geothermal features in the world (Rhinehart 1980). Around 100 different hot spring groups create over 10,000 individual thermal features (Bryan 1986) which are controlled by the regional fault systems and the Yellowstone Caldera (Eaton et al. 1975; White et al. 1988).

Earthquakes are closely related to volcanism and a precursor of imminent eruptions as seismic activity increases during the pre-eruption stage. Another important process caused by volcanic activity is sea floor spreading, a process that is confined to the divergent plate boundaries on the seafloor. A few locations where these boundaries are exposed above sea level can be observed in Iceland (Fig. 1.13). Here the Mid-Atlantic Ridge (MAR) has risen out of the sea and presents an opportunity to study this unusual process (Carey 2005). Several areas bear evidence where the Eurasian and North American continents are drifting apart and due to their uniqueness are promoted as impressive must-see destinations for visitors (Figs. 1.14 and 1.15), especially for volcano tourists with an interest in geology.

Fig. 1.13 Seafloor spreading visible and in real time, although very slow (about 2.5 cm or 1 in per year). The Mid-Atlantic Ridge emerges from the sea at Reykjanes, South Iceland with the Eurasian plate on the left side facing east. *Photo* Patricia Erfurt-Cooper

Fig. 1.14 At Þingvellir in South-western Iceland the rift between the continents has widened and visitors walk on the former seafloor of the Mid-Atlantic Ridge. The Þingvellir National Park is one of Iceland's most popular tourist sites and was World Heritage listed in 2004. *Photo* Patricia Erfurt-Cooper

1.7.1 Eruption Types

The majority of volcano tourists would consider viewing an eruptive event from a safe distance as the highlight of their trip. Night views of fiery *Strombolian* eruptions as well as ash columns and lava flows are the most popular volcanic attractions. Scientists and researchers have grouped the different eruption types into several categories with Rosi et al. (2003) suggesting a division into seven basic eruption types:

- Hawaiian
- Strombolian

Fig. 1.15 This bridge across the rift caused by the Mid-Atlantic Ridge has to be extended regularly to match the seafloor spreading. A sign on the bridge marked MIÐLINA indicates where the Eurasian and North American continental plates meet. *Photo* Patricia Erfurt-Cooper

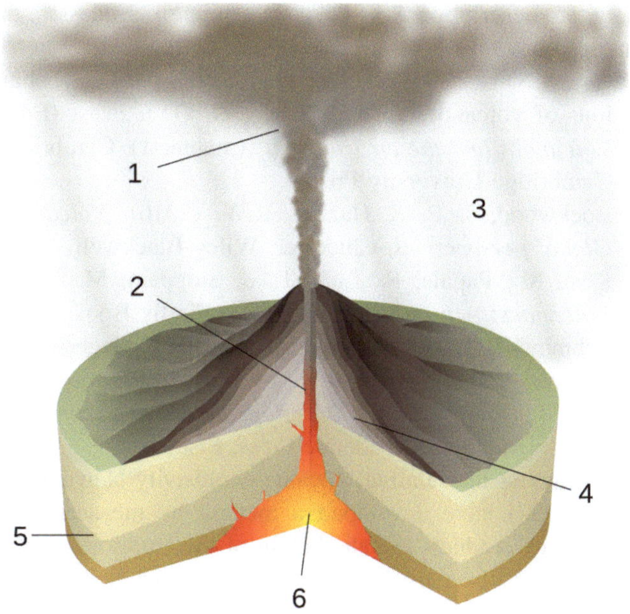

Fig. 1.16 Plinian eruption: *1* Ash plume, *2* Magma conduit, *3* Volcanic ash rain, *4* Layers of lava and ash, *5* Stratum, *6* Magma chamber. *Source* © Sémhur/CC-BY-SA-3.0

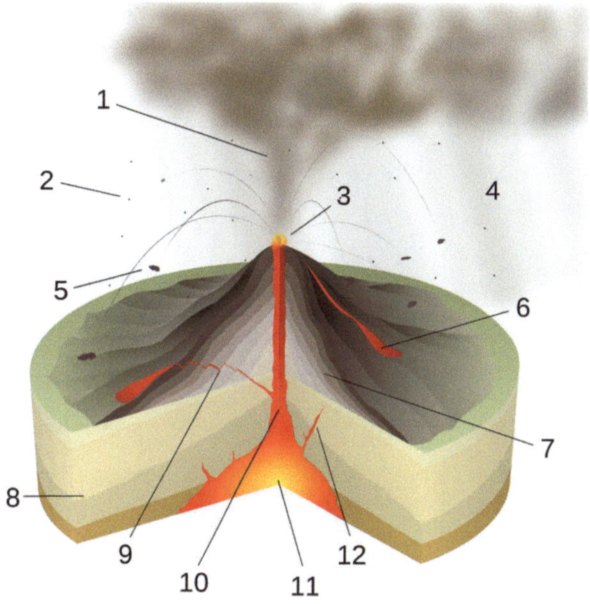

Fig. 1.17 Vulcanian eruption: *1* Ash plume, *2* Lapilli, *3* Lava fountain, *4* Volcanic ash rain, *5* Volcanic bomb , *6* Lava flow, *7* Layers of lava and ash, *8* Stratum, *9* Sill, *10* Magma conduit, *11* Magma chamber, *12* Dike. *Source* © Sémhur/CC-BY-SA-3.0

- Sub-plinian
- Plinian
- Vulcanian
- Surtseyan
- Phreato-plinian.

For detailed descriptions of all eruption types their guide book *Volcanoes* is highly recommended. In this chapter four major eruption types are briefly described to provide an overview of possible situations involving active volcanoes. The most dangerous and violent eruptions are called *Plinian*, which are highly explosive and can generate pyroclastic density currents (PDC), large eruption columns of ash and pyroclastic debris (air fall deposits) and in the worst case scenario can lead to a caldera collapse

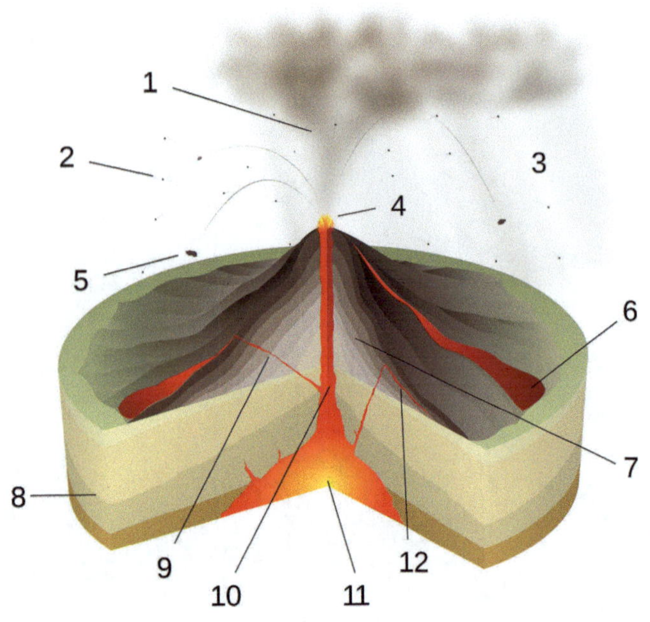

Fig. 1.18 Strombolian eruption: *1* Ash plume, *2* Lapilli, *3* Volcanic ash rain, *4* Lava fountain, *5* Volcanic bomb, *6* Lava flow, *7* Layers of lava and ash, *8* Stratum, *9* Dike, *10* Magma conduit, *11* Magma chamber, *12* Sill. *Source* © Sémhur/CC-BY-SA-3.0

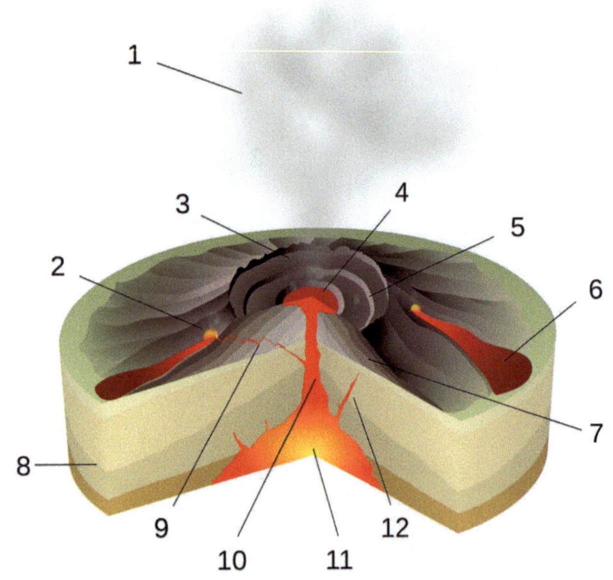

Fig. 1.19 Hawai'ian eruption: *1* Ash plume, *2* Lava fountain, *3* Crater, *4* Lava lake, *5* Fumaroles, *6* Lava flow, *7* Layers of lava and ash, *8* Stratum, *9* Sill, *10* Magma conduit, *11* Magma chamber, *12* Dike. *Source* © Sémhur/CC-BY-SA-3.0

(Fig. 1.16). Volcanoes known for this type of eruption include Vesuvius, Krakatau, Pinatubo, Mt St Helens, Santorini, Tambora and more recently Eyjafjallajökull.

Vulcanian eruptions are also explosive and can cause pyroclastic flows and base surges from collapsing ash columns (Fig. 1.17). Volcanoes in this category can form lava domes by emitting viscous lava which leads to pressure build-up in the vent. Representative examples of vulcanian eruptions are the volcanoes Bandai and Sakurajima (Japan), Soufrière Hills (Montserrat), Mt Pelée (Martinique), Galeras (Colombia), Ngauruhoe (New Zealand), Vulcano (Italy) and Arenal (Costa Rica).

Strombolian eruptions are explosions at intervals emitting fiery pyroclastic material such as pulverised rock (ash) and cinders (Fig. 1.18). A typical example for this type of eruption is the Italian volcano Stromboli, a famous destination for guaranteed eruption viewing. Strombolian activity is also associated with Paricutín (Mexico), Eldfell (Iceland) and Mt Yasur (Vanuatu).

Hawai'ian eruptions are effusive with low viscosity lava either flowing 'quietly' from a fissure or erupting as lava fountains from a vent, which is characteristic for fissure eruptions common in Hawaii (Fig. 1.19). They are a great visual attraction and the best example of Hawaiian volcanic activity is Kilauea in Hawaii.

Additional subcategories have been proposed by a number of scientists to measure and compare different eruptions. For further information on this subject and more detailed descriptions of volcanic eruption processes the interested reader is referred to the following texts:

- Carey, S. N. (2005). Understanding the physical behaviour of volcanoes. In J. Marti & G. G. J. Ernst (Eds.), *Volcanoes and the environment* (Chapter 1). Cambridge: Cambridge University Press.
- Lockwood, J. P. & Hazlett, R.W. (2010). *Volcanoes: Global perspectives*. Chicester: Wiley-Blackwell.
- Rosi, M., Papale, P., Lupi, L. & Stoppato, M. (2003). *Volcanoes: A firefly guide*. Buffalo: Firefly Books Ltd.
- Schmincke, H. U. (2006) *Vulkanism*. Berlin: Springer Verlag.
- Sigurdsson, H., Houghton, B., McNutt, S., Rhymer, H., & Stix, J. (Eds.) (2014). *Encyclopedia of volcanoes* (2nd Edn). San Diego: Academic Press, Elsevier.

To assess the relative volcanic explosivity Newhall and Self (1982) have developed a method by measuring the volume of products and the height of the eruption cloud as well as by observation of the intensity and force of the eruption. An open-ended scale from 1 to 8 is used to determine the value of the volcanic explosivity index (VEI) with the largest historical eruptions placed at a magnitude of 8 (e.g. Yellowstone, Toba). By comparison the eruptions of Vesuvius in 79 AD and Mt St Helens in 1980 'only' registered at magnitude 5. The recent eruption of Eyjafjallajökull reached magnitude 4 according to the VEI, the same estimate that was assigned to the eruption of Mt Pelée (Martinique) in 1902. Hawaiian and Strombolian eruptions generally register at the lowest magnitude of 1 or even 0 which includes lava lakes.

Table 1.11 The most common volcanic hazards and their potential dangers

Acid rain is created when rain passes through clouds containing volcanic gas or acid particle emissions. Acid rain can irritate eyes and skin and affects vegetation and water quality (Hansell et al. 2006)
Air fall deposits containing volcanic ash or tephra are created during phreatic eruptions when blocks, bombs and lapilli are ejected from an eruption column or plume and can be deposited over great distances. The size of the pyroclasts can range from tephra (<2 mm) and lapilli (>2 mm) to blocks and bombs (>64 mm), presenting a significant hazard depending on the proximity to the volcano
Jökulhlaup is the Icelandic name for a glacier burst which can cause severe flooding by melt water when a volcano erupts underneath a glacier or an ice cap
Lahars are gravity controlled mudflows loaded with volcanic debris and follow either valleys or river beds downhill. Lahars are extremely dangerous due to their momentum and velocity carrying large boulders and have the power to destroy foundations of bridges and buildings. Lahars can be triggered long after volcanic ash eruptions through heavy rainfall, snow melt or rim failure of crater lakes (Lockwood and Hazlett 2010)
Lava flows are the most common volcanic hazard from effusive or Hawaiian eruptions (Tilling 2005). Lava flows consist of red-hot molten rock and usually follow the topography. If the lava flows are slow they are not necessarily life threatening and there may be time to evacuate, although anything in their flow path such as property and vegetation will be destroyed through burial and burning. Severe thermal burns from lava flows are not uncommon in areas where tourists have the opportunity to hike across lava fields. Fast flowing lava with low viscosity is extremely dangerous
Laze (*Lava Haze*) consists of Hydrochloric Acid (HCl) gas clouds or mist which form when lava flows enter the ocean and come in contact with seawater. This mist affects the respiratory system, irritates eyes and skin and can be fatal (Heggie et al. 2010)
Pyroclastic flows or pyroclastic density currents are depositions from glowing clouds (Nuée Ardente) containing a high temperature mixture of gas, ash and fragmented magma (Lockwood and Hazlett 2010). Pyroclastic flows can travel at several hundred km per hour and can reach over 800 °C (Schmincke 2006). They can jump over ridges and even travel uphill causing death and injury, the destruction of buildings, impact damage and bury the landscape in layers of what is commonly known as volcanic ash or more correctly tephra
Pyroclastic surges are caused by the collapse of an eruption column and result in a lateral or ground hugging blast. They are considered as one of the major volcanic hazards and have become a main focus of volcanological research (Schmincke 2006)
Tsunami, also known as 'sea wave' or 'harbour wave', are caused by the displacement of water in the ocean either by an earthquake or a volcanic eruption (Tilling 2005). A tsunami is *not* a tidal wave but can inundate vast stretches of low lying coastal areas
Volcanic ash or tephra is a collective term for the most common and most hazardous fall process containing airborne fragments of pyroclasts—size < 2 mm in diameter and which can affect the respiratory and cardiovascular system of humans and animals as well as being an irritant to eyes and skin (Hansell et al. 2006; Tilling 2005). Tephra deposits can destroy vegetation, agricultural crops, block roads and drains as well as watercourses and can cause damage to equipment. In large eruptions, the danger to aircraft is considerable and ash particles can impact on climate and weather (Chester et al. 2001). Of all eruptive hazards tephra falls can affect the most people because of the wide areas that can be covered (Blong 1996). Injuries and death through impact of larger rock fragments are possible.
Volcanic bombs and blocks are rocks and lava lumps which are ejected during volcanic eruptions and can cause serious injuries including death from impact and burns (Hansell et al. 2006)
Volcanic gas emissions represent a serious hazard to humans as well as to animals and vegetation. The effects of SO_2, HCl, CO_2 and H_2S are most dangerous close to a volcano or near hydrothermal steam vents (fumaroles). Hydrogen sulphide (H_2S) also occurs near hot springs and can cause fatalities in a very short time and without warning

Source Compiled by Author from various sources

1.8 Volcanic Hazards

All active volcanoes can be potentially dangerous. People who live in their proximity should be aware of the most common hazards related to different types of eruption (Table 1.11). Unlike temporary visitors, who have the option to leave a volcanic area at any time, local residents do not have that choice because their homes and work places may be located close to volcanic activity. The Italian city of Naples comes to mind (Fig. 1.20) with a population of more than 4 million people in the overall metropolitan area surrounding Mt Vesuvius. Since the last eruption in 1944 suburban developments have encroached upon the flanks of Vesuvius and residents are taking advantage of the fertile volcanic soils and the views over the Bay of Naples. Perhaps there is not enough serious consideration of the potential for further eruptions? Similar to the villagers on the slopes of Mt Merapi in Indonesia, the people living in the shadow of Mt Vesuvius are dominated by economic and social reasoning, which appears to be the major basis for their acceptance of the potential risk.

The city of Kagoshima on the southern island of Kyushu (Japan) is another example. Separated from the city by a mere 15 min ferry ride the active volcano Sakurajima is a close neighbour to Kagoshima's population of over 600,000 people. The residents of the actual volcanic island itself are used to eruptions with ash falls on an almost daily basis and many have their own fallout shelters in their gardens (Fig. 1.21). For the people living on Sakurajima Island it is a matter of choice and risk acceptance for a variety of reasons which, similar to Indonesia or the Philippines, include ancestral homes and temples, family bonds and

Fig. 1.20 View across the Bay of Naples with Vesuvius in the background. Clearly visible are the residential areas on the slopes of the volcano. *Photo* Patricia Erfurt-Cooper

profitable agriculture courtesy to the rich volcanic soil. While Sakurajima is constantly erupting with smaller explosions accompanied by ash clouds, the last 'big' eruption took place in 1914 (VEI 4) and covered the surrounding local areas in deep layers of tephra while at the same time extensive lava flows connected the island to the mainland of the *Osumi* Peninsula (Fig. 1.22) on the eastern side of *Kinko* Bay. Because of the potential danger of Sakurajima the volcano was included in the list of 'Decade Volcanoes', a group of 16 active volcanoes, which were chosen for close monitoring under the United Nations International Decade for Natural Disaster Reduction (IDNDR) study program.

Local residents have a definite advantage over temporary visitors in case of an emergency when it comes to risk management strategies. Living in the area generally comes with a higher level of awareness regarding volcanic hazards and the subsequent risks involved. In most volcanic areas residents have the opportunity through their municipal governments to access up-to-date information about the state of the volcano in their 'backyard'. However, an additional risk factor for local residents and one not to be underestimated is complacency. Depending on aspects including the socio-cultural background of the population, their level of education and awareness, access to information about emergencies, as well as interest and mentality, it is apparent that not all volcanic regions offer the same standard of safety for their residents.

While risk management strategies are developed to suit individual volcanic regions, these strategies generally do not take into account a) the frequently very large numbers of temporary visitors and b) their lack of local knowledge (Erfurt-Cooper 2010a, b). Consequently there are two different groups of people who could be potentially endangered by volcanic activity. While the visiting tourists, who accept the risk voluntarily, have a much shorter time of exposure to potential threats, people who live near

Fig. 1.21 Private concrete shelters (*inset*) for protection against ash fall and pyroclastic debris are common all over Sakurajima Island. *Photo* Patricia Erfurt-Cooper

Fig. 1.22 Lava flows on Sakurajima Island from the Taisho eruption in 1914 with the city of Kagoshima in the background. *Photo* Patricia Erfurt-Cooper

Table 1.12 Potential risks for tourists in volcanic environments

General risks in natural areas	Additional risks in volcanic areas
Hiking accidents	Unexpected eruptions
Sudden weather changes—Fog, snow, rain	Gas emissions and toxic fumes
Snow avalanches and rock falls	Earthquakes
Getting lost	Lahars, mudslides, landslides
Hypothermia and/or altitude sickness	Thermal burns from lava flows
Bushfires, wildlife	Thermal burns from extreme hot springs

The dangers for volcano tourists can include one or more of the problems listed above in the left column indicating risks common in mountain areas. These can be compounded by the additional dangers listed in the right column, which are frequently encountered in active volcanic or geothermal environments

volcanoes are exposed on a permanent basis. For example the inhalation of airborne ash particles can seriously affect the respiratory system leading to chronic medical problems.

Volcano tourism is never advertised as potentially dangerous. Tour operators appeal to the adventurous nature of visitors and promote active volcanoes as 'once in a lifetime experiences' with irresistible features, rather than focusing on possible threats. Warnings are more subtle in the form of disclaimers to be signed by tour members in case of accidents. A comparative overview in Table 1.12 sums up the potential risks for tourists in volcanic and non-volcanic areas, which are often associated with a mountainous environment. Some adventurous tourists may also overestimate their ability to cope with hostile environments and/or are unprepared for unexpected situations, which may require the use of rescue services. Especially at protected sites such as national parks visitors may have a false sense of security which can lead to tragic accidents. Several hundred people have died in Yellowstone National Park alone, mainly due to negligence and human mistakes with the cause of death in many cases resulting from thermal burns after falling into the hot springs that are the most dangerous natural features in Yellowstone (Arrandale 2012; Whittlesey 1995).

1.9 Risk Management and Volcanoes

On average between 50 and 60 volcanoes erupt worldwide every year and affect the people who live in their vicinity (SVE 2012). Loss of life could possibly be minimised or even prevented with more rigorous safety regulations; not just for residents of volcanic areas, but especially for temporary visitors, who are either not aware of any danger or ignore the risks associated with visiting active volcanoes. Because risk perception and risk assessment is a personal matter for volcano tourists, millions of people who are visiting active volcanic regions every year make their own decisions about health and safety while travelling. Tourists have been caught up in dangerous situations because they either ignored warning signs and advice or because they did not follow the recommended escape routes. Emergency strategies cannot guarantee the rescue of all visitors because of their often uncertain numbers. This is an area that would benefit from more research into safety issues related to active volcanic environments to improve the safety standards for volcano tourists.

Looking at the bigger picture of international air travel, a risk management strategy of great consequence includes the decision whether to close airspace and ground aircraft in the case of volcanic ash plumes affecting air travel safety. Encounters between aircraft and volcanic ash are dangerous and costly, as examples from the past have shown. Air traffic can be significantly disrupted as ash clouds drift with the wind over great distances from their source (Gardner and Guffanti 2007; USGS Fact Sheet 030-97 2004). Opinions are often divided about the impact and the potential damage volcanic ash can cause for jet engines. However, to avoid any risk to the lives of passengers and crew as well as expensive repairs to the aircraft it is preferable to err on the side of caution. Economic losses due to flight cancellations are unavoidable and occur when airlines decide to suspend flying during potentially unsafe conditions (Neal et al. 2010), as was shown during the Icelandic eruptions of 2010 when some 30,000 flights a day had to be cancelled or diverted.

Following the Augustine volcano eruption of 2005–2006 in Alaska, the public interest in volcano hazard information greatly increased. This encouraged the Alaska Volcano Observatory (AVO) to focus on publishing specific eruption related information to be made available to the concerned public by providing real time information for communities close to the volcano (Adleman et al. 2010). In this context, several risk management plans are available on the Internet, although most of them are more than ten years old and do not include policies relating to temporary visitors. One mitigation plan related to Mt Rainier National Park (USA) does mention tourism by stating that *'park service personnel and tourists in the impacted area will have a very low chance of survival'* and that *'any citizens or park personnel in the vicinity of a pyroclastic flow will experience the devastating impact and heat created by the flowing hot ash*

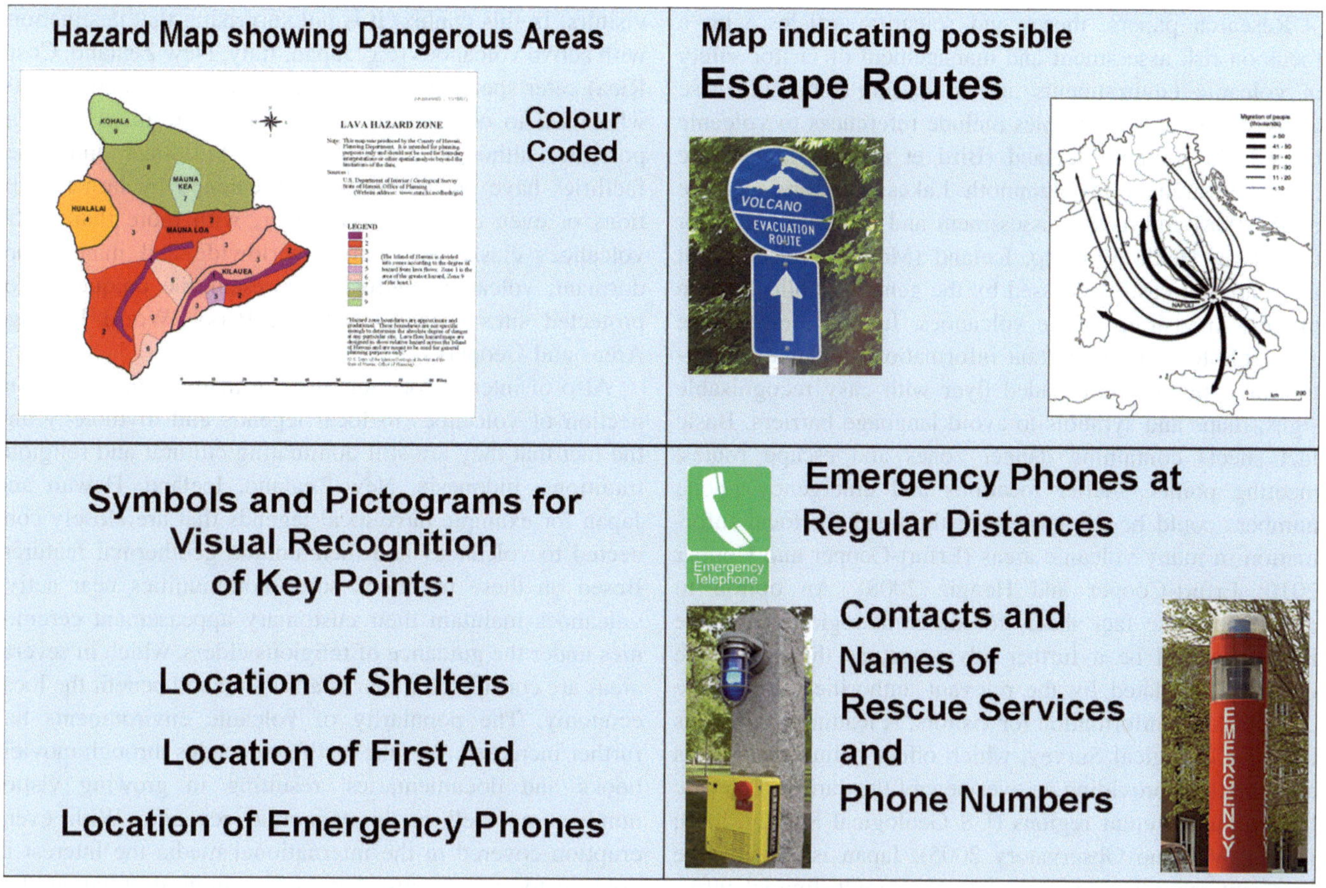

Fig. 1.23 Sample fact sheet which can be used as a template for individual volcanic regions. By Patricia Erfurt-Cooper

and rock with the usual result of possible death, or if surviving, then major burns and or partial asphyxiation' (Meyers 2004). Impact scenarios such as these give an impression what to expect in case of emergency. But it still does not provide advice for the volcano tourist how to handle unexpected situations.

In New Zealand the Waikato Regional Council developed a Volcanic Risk Mitigation Plan to minimise the adverse effects associated with natural hazards and to meet natural disaster reduction requirements. The document included the provision of mitigation plans involving long term prevention, preparedness and community awareness (Munro and Parkin 1999). While reference is made to disruptions of the ski seasons (1995–1996) due to ash eruptions, Munro and Parkin also acknowledge positive side effects such as growing numbers of tourists interested in observing volcanic activity.

In 1994 the IAVCEI (International Association of Volcanology and Chemistry of the Earth's Interior) presented a report titled *'Safety Recommendations for Volcanologists and the Public'*. This report was first published in the Bulletin of Volcanology and looked at the planning and logistics, operations and equipment and provided recommendations for scientists and residents (Aramaki et al. 1994). Aimed at scientists and the general public temporary visitors are not considered despite their great numbers at many volcanic destinations. In response to the absence of international guidelines it appears that individual regions have developed their own risk management strategies which include visiting tourists. Most volcano national parks have information centres which provide visitors with at least some form of safety instructions or with updated information which can be accessed online before travelling. However, there is no international standard or code of conduct which would be beneficial for many areas where safety concerns should be a priority. To make volcano tourism as safe as possible fact sheets (Fig. 1.23) with useful guidelines, hazards maps, emergency information and how to prevent accidents and injuries should be made available to visitors of active volcanic environments (Erfurt-Cooper and Cooper 2010). It is crucial to improve the understanding of safety standards in volcanic and geothermal environments by creating guidelines for best practice management to cover ALL regions. A list of revised safety recommendations developed by the author is attached at the end of this book as Appendix 1.

Research papers, theses and scientific articles with a focus on risk assessment and management of visitor safety in volcanic environments are addressing some of these areas. Noteworthy examples include references to volcanic hazards in southern Iceland (Bird et al. 2010), volcanic hazard awareness in Mammoth Lakes, California (Kaye et al. 2008) and hazard assessment and risk mitigation for tourists at Hekla Volcano, Iceland (Montalvo 2010). But these are not easily accessed by the general public with an interest in visiting active volcanoes. It may therefore be advisable to make important information available to visitors as a plain double sided flyer with easy recognisable signs, maps and symbols to avoid language barriers. Basic fact sheets containing danger zones and escape routes, meeting points, shelter locations and emergency phone numbers could be the answer to the need for local information in many volcanic areas (Erfurt-Cooper and Cooper 2010; Erfurt-Cooper and Heggie 2008). An option to download these fact sheets for different regions from the internet would be a further advantage, as they could be constantly updated by the relevant authorities to provide current safety information for visitors. A leading example is the US Geological Survey, which offers online fact sheets and booklets, providing an overview of the various volcanic hazards in different regions (US Geological Survey 2000; Hawaii Volcano Observatory 2005). Japan is also on the forefront regarding visitor safety with multi-lingual information available online for most volcanic destinations.

1.10 Conclusion

Active and dormant volcanoes and their dynamic environments are widely recognised as unique visitor attractions which have fascinated people for centuries. Even remote places such as the Russian Kamchatka, the Galápagos Islands (Ecuador), Antarctica and Alaska's volcanic island chain (Aleutians) are attractive destinations for determined volcano tourists. Recent eruptive events including Eyjafjallajökull in Iceland (2010), Chaitén (2008) and Puyehue (2011) in Chile contribute to further advancement in the research of volcanic processes and their impact on the environment and have lead to improvements in risk management, especially as heightened volcanic activity commonly attracts increased numbers of visitors. The 2010 eruption of Eyjafjallajökull in Iceland for example was not only exciting for scientists, but also for thousands of adventure seeking tourists. Therefore in areas, where eruptions are in progress, it is essential for tour operators and local authorities to provide sufficient information about any potential risks and hazards that may be encountered by visitors. In this context it is not surprising that destinations with active volcanoes (e.g. Japan, Italy, New Zealand, Costa Rica) cater specifically for the demand of volcano tourists, who wish to observe an active volcano from as close as possible, although accommodation buildings and other facilities have been frequently threatened by larger eruptions or even destroyed. However, with more than 1,500 volcanoes classed as active worldwide and many more dormant, volcanic landforms represent the uniqueness of protected sites such as National Parks, World Heritage Areas and Geoparks.

Also of interest for tourism is in many regions the connection of volcanoes to local legends and mythology and the fact that they are still dominating cultural and religious traditions. Indonesia, New Zealand, Iceland, Hawaii and Japan for example have local legends that are closely connected to volcanoes and their unique geothermal features. Based on these traditions some communities near active volcanoes maintain their customary appeasement ceremonies under the guidance of religious elders, which in several areas are considered a tourist attraction and benefit the local economy. The popularity of volcanic environments has further increased over the last few decades through movies, books and documentaries, resulting in growing visitor numbers as well as in economic revenue. With every eruption covered in the international media the interest in volcano tourism grows. Access to remote destinations has become easier and more affordable with vital infrastructure improvements in many countries to cope with increasing tourist numbers. Major volcanic events often result in new volcano museums or visitor centres being built near an eruption site as a safe option for visitors to learn about the geological background. Here tourists can watch a volcano from a safe distance or via webcams; especially if the volcano is too active for a visit.

The number of volcanoes erupting worldwide (50–60 every year) affects millions of people living in their vicinity and makes risk management a challenging task for the authorities involved. While local residents are living with an active volcano on a permanent basis they may have a slight advantage over temporary visitors in case of an emergency. When it comes to risk management involving an evacuation their better knowledge of the area will be valuable. Because strategies are developed to suit individual volcanic regions, they rarely take into account the multitude of visitors and their lack of local knowledge. This can lead to tourists getting caught up in dangerous situations because they cannot locate the recommended escape routes.

Another problem affecting volcano tourism in some volcanic regions is the lack of experts and sufficient funding to monitor potential activity accurately enough to provide up-to-date advice for residents and visitors. The example of Chaitén (Chile) has shown that dormant volcanoes with the potential to re-activate are largely unsupervised until there are clear signs of seismic activity. And although more volcano observatories are established as funding becomes available, volcanoes in remote areas or in some developing countries are not yet monitored continuously. To assist with disaster prevention planning warning signs require early detection. Investment in scientific knowledge and education about volcanic hazards and risk avoidance is therefore extremely important, not just for local residents, but for all temporary or seasonal visitors. In this context, the following chapters introduce various aspects of volcano tourism through showcasing locations around the world, with a diversity of approaches of integrating volcanic landscapes as natural resources for sustainable tourism. The book's aim is to present an interesting and informative reference work and to enhance public awareness about the 'beauty of the volcanic beast', combined with the clear message that special care should be taken while getting close to active volcanoes.

References

Adleman, J. N., Cameron, C. E., Snedigar, S. F., Neal, C. A. & Wallace, K. L. (2010). Public outreach and communications of the Alaska Volcano Observatory during the 2005–2006 eruption of Augustine Volcano. In J. A. Power, M. L. Coombs & J. Freymueller (Eds.), *The 2006 eruption of Augustine Volcano, Alaska* (Chapter 27, pp. 621–644). U.S. Geological Survey Professional Paper 1769. http://pubs.usgs.gov/pp/1769/chapters/p1769_chapter27.pdf. Accessed on 2 June 2012.

Aguirre, J. A., & Ahearn, M. (2007). Tourism, volcanic eruptions, and information: lessons for crisis management in National Parks, Costa Rica, 2006. *Revista de Turismo y Patrimonio Cultural, 5*(2), 175–191.

Alvarado, G. E., & Soto, G. J. (2008). Volcanoes in the pre-Columbian life, legend, and archaeology of Costa Rica. *Journal of Volcanology and Geothermal Research, 176*, 356–362.

Aramaki, S., Barberi, F., Casadevall, T., & McNutt, S. (1994). Safety recommendations for volcanologists and the public. *Bulletin of Volcanology, 56*, 151–154.

Arrandale, T. (2012). *Caution: Geothermal wonders*. www.yellowstonepark.com/2007/01/cautionary-tale/. Accessed on 20 Oct 2012.

Bailey, M. (2012). Italy allows UNESCO into Pompeii. The Roman city risks joining the world heritage in danger list. *The Art Newspaper*. www.theartnewspaper.com/articles/Italy-allows-Unesco-into-Pompeii/25422. Accessed on 12 Nov 2012.

Bird, D. K., Gísladóttir, G., & Dominey-Howes, D. (2010). Tourism, volcanic hazards and education in Southern Iceland: Implications for hazard, risk and emergency response education and training. *Journal of Volcanology and Geothermal Research, 189*(1/2), 33–48.

Bjarnason, F. (2014). *Eldheimar munu kosta 890 miljónir*. www.visir.is/eldheimar-munukosta-890-milljonir/article/2014701029944. Accessed 4 April 2014.

Blong, R. J. (1996). Volcanic hazards risk assessment. In R. Scarpa & R.I. Tilling (Eds.), *Monitoring and Mitigation of Volcano Hazards* (pp. 675–700). Berlin: Springer-Verlag.

Bryan, T. S. (1986). *The geysers of Yellowstone*. Boulder: Colorado Associated University Press.

Carey, S. N. (2005). Understanding the physical behaviour of volcanoes. In J. Marti & G. G. J. Ernst (Eds.), *Volcanoes and the environment* (Chapter 1, pp. 1–54). Cambridge: Cambridge University Press.

Chester, D. K., Degg, M., Duncan, A. M. & Guest, J. E. (2001). The increasing exposure of cities to the effects of volcanic eruptions: A global survey. *Environmental Hazards, 2*, 89–103.

Chester, D. K., & Duncan, A. M. (2010). Responding to disasters within the Christian tradition, with reference to volcanic eruptions and earthquakes. *Religion, 40*, 85–95.

Davison, P. (2003). *Volcano in paradise: Death and survival on the Caribbean island of Montserrat*. London: Methuen Publishing.

Dekadevolcano.net (2011) *Hekla Volcano, Iceland*. www.decadevolcano.net/volcanoes/iceland/hekla.htm. Accessed on 29 Dec 2012.

Dove, M. R. (2010). The panoptic gaze in a non-western setting: Self-surveillance on Merapi Volcano, Central Java. *Religion, 40*, 121–127.

Edelstein, J. H. (2011) Volcano country. *Condé Nast Traveller*. www.cntraveller.com/news/2011/april/iceland-volcano-museum. Accessed on 21 May 2012.

Eaton, G. P., Christiansen, R. L., Iyer, H. M., Pitt, A. M., Mabey, D. R., Blank, H. R, Jr, et al. (1975). Magma beneath Yellowstone National Park. *Science, 188*, 787–796.

Eldfjallasafn Volcano Museum (2012). *Eldfjallasafn Volcano Museum*. Stykkishólmur, Iceland. www.eldfjallasafn.is/e/index.html. Accessed on 10 Oct 2012.

Erfurt-Cooper, P. (2010a). Introduction to volcano and geothermal tourism. In P. Erfurt-Cooper & M. Cooper (Eds.), *Volcano and geothermal tourism: Sustainable geo-resources for leisure and recreation*. London: Earthscan.

Erfurt-Cooper, P. (2010). Active geothermal and volcanic environments as tourist destinations. In R. Dowling & D. Newsome (Eds.), *Global geotourism perspectives* (Chapter 3, pp. 33–48). Oxford: Goodfellow Publishers.

Erfurt-Cooper, P., & Cooper, M. (2010). *Volcano and geothermal tourism: Sustainable geo-resources for leisure and recreation*. London: Earthscan.

Erfurt-Cooper, P. & Heggie, T. (2008). *Risk perception, prevention, management and mitigation in volcanic environments: Workshop held at The Inaugural Global Geotourism Conference, Perth, Australia*, 17–20 Aug 2008.

Erfurt-Cooper, P. (2011). Geotourism in volcanic and geothermal environments: Playing with fire? *Geoheritage Journal, 3*(3), 187–193.

Erfurt-Cooper, P. (2012). An assessment of the role of natural hot and mineral springs in health, wellness and recreational tourism. Thesis submitted for the degree of Doctor of Philosophy, James Cook University, Cairns, Australia.

Eyjafjallajökull Visitor Centre (2012) *Eyjafjallajökull erupts—The eruption that put the world's flights on hold*. http://www.icelanderupts.is/. Accessed on 12 May 2012.

Gardner, C. A. & Guffanti, M. C. (2007) *U.S. Geological Survey's alert notification system for volcanic activity*. U.S. Geological Survey. http://pubs.usgs.gov/fs/2006/3139/. Accessed on 3 May 2012.

Gaudru, H. & Erfurt-Cooper, P. (2010). *Tourism and volcanic risk during eruption—A challenge between security and local economy: Presented to the Cities on Volcanoes Conference CoV7 in Tenerife, Spain*, 31 May–4 June 2010.

Hammond, J. D. (1995). The tourist folklore of Pele: Encounters with the other. In B. Walker (Ed.), *Out of the ordinary* (Chapter 9, pp. 159–179). All USU Press Publications, Book 51. http://digitalcommons.usu.edu/usupress_pubs/51. Accessed on 10 June 2012.

Hansell, A. L., Horwell, C. J. & Oppenheimer, C. (2006). The health hazards of volcanoes and geothermal areas. *Occupational Environmental Medicine, 63*, 149–156.

Hawaii Volcano Observatory (2005). *Volcanic hazards: Types of volcanic hazards on the Island of Hawaii*. US Geological Survey. www.hvo.wr.usgs.gov/hazards/types/main.html. Accessed on 12 Nov 2012.

Heggie, T. W., Heggie, T. M. & Heggie, T. J. (2010). Death by volcanic laze. In P. Erfurt-Cooper & M. Cooper (Eds.), *Volcano and Geothermal Tourism: Sustainable Geo-Resources for Leisure and Recreation* (pp 118–119). London: Earthscan.

Henderson, M. (2007). *Incas fattened up their children before sacrifice on the volcano*. www.thetimes.co.uk/tto/news/world/americas/article1996643.ece. Accessed on 28 Dec 2012.

Icelandic Tourist Board (2012). *Tourism in Iceland in Figures, April 2012*. http://www.ferdamalastofa.is/static/files/ferdamalastofa/talnaefni/tourism_in_iceland_in_figures_may_-2012.pdf. Accessed 4 Apr 2014.

IceNews. (2011). *Iceland already cleaning up after volcano: sees good summer ahead*. www.icenews.is/2011/05/26/iceland-already-cleaning-up-after-volcano-sees-good-summerahead/. Accessed 4 Apr 2014.

International EMECS Center (2012) *The environment*. www.emecs.or.jp/setoinfo97/1/1-4.htm. Accessed on 12 Nov 2012.

Karkut, J. (2010). Under the volcano—can sustainable tourism development be balanced with risk management? In P. Erfurt-Cooper & M. Cooper (Eds.), *Volcano and Geothermal Tourism: Sustainable Geo-Resources for Leisure and Recreation* (Chapter 16, pp 233–246). London: Earthscan.

Kaye, G., Finnis, K. K., Johnston, D. M. & Paton, D. (2008). *Volcanic hazard awareness in the tourism sector in Mammoth Lakes, California, USA*. GNS Science Report 2008/35, 13 pages.

Knipe, R. (1982). Pele: Volcano goddess of Hawaii. *Psychological Perspectives: A Quarterly Journal of Jungian Thought, 13*(2), 114–126.

Lockwood, J. P., & Hazlett, R. W. (2010). *Volcanoes: Global perspectives*. Chicester: Wiley-Blackwell.

Marshall, A. (2008, January). The Gods must be restless: Living in the shadow of Indonesia's volcanoes. *National Geographic*, pp. 34–57.

Meyers, L. (2004). *Pierce County—Region 5 hazard mitigation plan. Breaking the disaster cycle: Planning for disaster-resistant communities*. www.co.pierce.wa.us/xml/abtus/ourorg/dem/EMDiv/MP/PC%20Cover-Table%20of%20Contents.pdf. Accessed on 13 Nov 2012.

Montalvo, J. (2010). Hazard assessment and risk mitigation for tourists at Hekla Volcano, South Iceland. Master's Thesis, Faculty of Earth Sciences, University of Iceland.

Moriarty, G. (2012). Giant's causeway visitor centre opens to public. www.irishtimes.com/newspaper/ireland/2012/0704/1224319342448.html. Accessed on 12 Nov 2012.

Munro, A. & Parkin, D. (1999). *Volcanic risk mitigation plan*. Environment Waikato Policy Series 1999/10. www.waikatoregion.govt.nz/PageFiles/4192/Volcano.pdf. Accessed on 1 June 2012.

National Geographic (2010, 2 April) Iceland volcano: Eruption sparks tourist boom. *National Geographic News*. http://news.nationalgeographic.com/news/2010/04/photogalleries/100402-iceland-volcano-tourism-pictures/. Accessed on 21 May 2012.

Neal, C. A., Murray, T. L., Power, J. A., Adleman, J. N., Whitmore, P. M. & Osiensky, J. M. (2010). Hazard information management, interagency coordination, and impacts of the 2005–2006 eruption of Augustine Volcano. In J. A. Power, M. L. Coombs & J. Freymueller (Eds.), *The 2006 eruption of Augustine Volcano, Alaska* (Chapter 28, pp. 645–667). U.S. Geological Survey Professional Paper 1769, Alaska. http://pubs.usgs.gov/pp/1769/chapters/p1769_chapter28.pdf. Accessed on 2 June 2012.

Newhall, C. G., & Self, S. (1982). The volcanic explosivity index (VEI): An estimate of explosive magnitude for historical volcanism. *Journal of Geophysical Research, 87*(C2), 1231–1238.

Nimmo, H. A. (1986). Pele, ancient goddess of contemporary Hawaii. *Pacific Studies, 9*(2), 121–179.

NPS Public Use Statistics Office. (2014a). *Hawaii Volcanoes NP*. https://irma.nps.gov/Stats/SSRSReports/Park%20Specific%20Reports/Annual%20Park%20Recreation%20Visitation%20(1904%20-%20Last%20Calendar%20Year)?Park=HAVO. Accessed 4 Apr 2014.

NPS Public Use Statistics Office. (2014b). *Sunset Crater NM*. https://irma.nps.gov/Stats/SSRSReports/Park%20Specific%20Reports/Annual%20Park%20Recreation%20Visitation%20(1904%20-%20Last%20Calendar%20Year)?Park=HAVO. Accessed 4 Apr 2014.

Plunket, P. & Uruñuela, G. (2000). The archaeology of a Plinian eruption of the Popocatépetl Volcano. *Geological Society, London*. Special Publications (vol. 171, pp. 195–203).

Rhinehart, J. S. (1980). *Geysers and geothermal energy*. New York: Springer-Verlag.

Rosi, M., Papale, P., Lupi, L. & Stoppato, M. (2003). *Volcanoes: A Firefly Guide*. Buffalo, New York: Firefly Books Ltd

Schmincke, H. U. (2006). *Vulkanism*. Berlin: Springer Verlag.

Sigurdsson, H. & Lopes-Gautier, R. (2000). Volcanoes and tourism. In H. Sigurdsson, B. Houghton, S. McNutt, H. Rhymer & J. Stix (Eds.), *Encyclopedia of volcanoes*. San Diego: Academic Press.

Sigurdsson, H., Houghton, B., McNutt, S., Rhymer, H., & Stix, J. (Eds.). (2014). *Encyclopedia of volcanoes* (2nd ed.). San Diego: Academic Press. (An Imprint of Elsevier).

SMH—The Sydney Morning Herald (2011, 22 July) After the ash comes volcano tourist cash. *SMH*. www.smh.com.au/travel/after-the-ash-comes-volcano-tourist-cash-20110722-1hrry.html. Accessed on 21 May 2012.

Struck, A. (2010). Volcano tourism and its influence on the territory of Mount Etna (Italy)—Explored with digressions to Stromboli (Italy). In P. Erfurt-Cooper & M. Cooper (Eds.), *Volcano and geothermal tourism: Sustainable geo-resources for leisure and recreation* (Chapter 15, pp. 221–232). London: Earthscan.

SVE—Société Volcanologique Européenne (2012) Volcanoes and risk assessment. *SVE*. www.sveurop.org/gb/articles/articles/volchazards.htm. Accessed on 3 June 2012.

Theodorou, C. (2010, 4 April) Erupting volcano boosts tourism in Iceland. *CNN News*. http://articles.cnn.com/2010-04-04/travel/iceland.volcano.tourism_1_volcano-lava-plumes?_s=PM:TRAVEL. Accessed on 21 May 2012.

The total Yellowstone page (2012). www.yellowstone-natl-park.com/stats.htm. Accessed on 3 June 2012.

Tilling, R. I. (2005) Volcano hazards. In J. Marti & G. G. J. Ernst (Eds.), *Volcanoes and the environment* (Chapter 2). Cambridge: Cambridge University Press.

TM Tenerife Magazine (2012) Mount Teide National Park. *TM Tenerife Magazine*. www.tenerifemagazine.com/tag/mount-teide-national-park. Accessed on 12 Nov 2012.

U.S. Geological Survey (2000) *Viewing Hawaii's lava safely: Common sense is not enough*. U.S. Geological Survey Fact Sheet 152-00, version 1.1. http://pubs.usgs.gov/fs/2000/fs152-00/. Accessed on 12 Nov 2012.

USGS Fact Sheet 030-97 (2004) *Volcanic ash—Danger to aircraft in the North Pacific*. http://pubs.usgs.gov/fs/fs030-97/. Accessed on 3 June 2012.

USGS Volcano Hazards Program (2012) *Yellowstone volcano observatory—Hydrothermal explosions*. http://volcanoes.usgs.gov/volcanoes/yellowstone/yellowstone_hazard_43.html. Accessed on 20 Oct 2012.

Verne, J. (1871) *A journey to the center of the Earth*. France: Pierre-Jules Hetzel (Republished 1965. London: Penguin Group).

Viramonte, J. G., & Incer-Barquero, J. (2008). Masaya, the 'mouth of hell', Nicaragua: Volcanological interpretation of the myths, legends and anecdotes. *Journal of Volcanology and Geothermal Research, 176*, 419–426.

Visitor Centre Þorvaldseyri (2011) *Eyjafjallajökull erupts*. www.icelanderupts.is/. Accessed on 24 Aug 2011.

White, D. E., Hutchinson, R. A., & Keith, T. E. C. (1988). *The geology and remarkable thermal activity of Norris Geyser Basin, Yellowstone National Park, Wyoming*. U.S. Geological Survey Professional Paper 1456. Washington: USGS.

Whittlesey, L. H. (1995). *Death in Yellowstone: Accidents and foolhardiness in the first national park*. Lanham: Roberts Rinehart Publishers.

Winchester, S. (2003) *Krakatoa: The day the world exploded: August 27, 1883*. New York: Harper Collins.

Yellowstone up close and personal (2014) *Yellowstone visitor statistics page*. www.yellowstone.co/stats.htm. Accessed on 3 June 2014.

Part II

European Volcanoes

Part II (Chapters 2–11) of this book includes a number of European destinations based on active and dormant volcanic geosites. In Chap. 2 Deanne Bird and Guðrún Gísladóttir share their research into Iceland's volcano tourism under special consideration of the eruptive events of Eyjafjallajökull volcano as well as recent developments in risk management for visitors of active environments in Iceland.

In Chap. 3 Jasmine Cardozo Moreira, Roberto Rendeiro Martín-Cejas and Guillermo Melendez-Hevia take a closer look at tourism based on active volcanism on the Canary Islands while in Chap. 4 João Carlos Nunes reviews the unique Archipelago of the Azores and their volcanic heritage. Tourism on two volcanic islands in Greece, Santorini and Milos, has been explored and assessed by Kalliopi Gaki-Papanastassiou and Dimitrios Papanastassiou in Chap. 5.

Turkey's rich volcanic heritage is discussed in detail by Gülpınar Akbulut in Chap. 6 with a focus on geotourism. Chapter 7 is contributed by Szabolcs Harangi and presents an opportunity to learn about volcanic geosites in Hungary. Chapter 8 by Francesco Stoppa is introducing the Italian San Venanzo Volcano Park to the reader while in Chap. 9 Nathalie Cayla examines the history and the development of volcanic geotourism in the French Massif Central. Several key aspects of the Garrotxa Volcanic Zone in Spain are the focus of Chap. 10 by Jose M. Prats and Emili Bassols and Chap. 11 by Malcolm Cooper is assessing the famous Giant's Causeway and Fingals Cave and their tourism potential.

For those readers who wonder about the absence of Mount Vesuvius and Mount Etna, these volcanoes were included in a previous volume about volcano tourism. Therefore alternative destinations were given preference, including several dormant volcanic regions, which to date have not been documented in detail in the tourism literature.

Southern Iceland: Volcanoes, Tourism and Volcanic Risk Reduction

2

Deanne K. Bird and Guðrún Gísladóttir

2.1 Introduction

Iceland is well known for its active volcanoes with small eruptions (<0.1 km^3 Dense Rock Equivalent—DRE) occurring approximately every 4–5 years, and large flood-basalt eruptions (>10 km^3 DRE) occurring every 500–1,000 years (Guðmundsson et al. 2008). The exceptionally high rate of volcanism is due to the interaction of the Mid-Atlantic Ridge (MAR) with a mantle plume centred beneath Iceland (Sigmundsson, 2006). Offshore, ridges to the north and southwest of Iceland represent the MAR. Onshore, the MAR is comprised of a series of seismic and volcanic zones (Einarsson and Saemundsson 1987). Holocene volcanism is contained within these discrete 15–50 km wide zones that jointly cover approximately one-third of the country (Fig. 2.1).

Iceland boasts nearly all known volcano types and eruption styles on Earth with volcanoes located in subaerial, subglacial and submarine environments (Thordarson and Larsen 2007). Volcanic hazards are therefore diverse and often catastrophic. The Icelandic Meteorological Office (Veðurstofa Íslands—www.vedur.is) monitors volcanic activity through the South Iceland Lowland (SIL) national seismic network, various volumetric strain meters and continuous geodetic GPS stations. The three most significant volcanic hazards produced during eruptive events are jökulhlaups (a sudden burst of meltwater from a glacier, often caused by a subglacial eruption), tephra fallout and lightning. Other hazards include earthquakes, lava flows, poisonous gases, pyroclastic density currents and tsunami (Guðmundsson et al. 2008).

Located in a major tourist region, the infamous Hekla, Katla and Grímsvötn volcanoes, and their associated fissure systems, have the highest eruption frequency and greatest volcanic productivity in Iceland (Thordarson and Larsen 2007). In 2010, however, Eyjafjallajökull became Iceland's most famous volcano when tephra circulated through international airspace causing widespread chaos to the aviation industry. While Eyjafjallajökull remained quiet in 2011, its neighbours Hekla, Katla and Grímsvötn did not. In the next sections we describe the volcanic systems of Hekla, Eyjafjallajökull, Katla and Grímsvötn and their recent activity.

2.2 Hekla

The ridge-shaped stratovolcano, Hekla is Iceland's second most active volcano with 23 confirmed eruptions (Table 2.1) since its first historic eruption in 1104 AD. Although the 1104 AD eruption was a purely explosive Plinian eruption, Hekla is typically characterised by mixed eruptions. Usually beginning with a vigorous, but short-lived (0.5–2 h long) Plinian or sub-Plinian phase, the eruption transitions into a several-hour long phase, with simultaneous sustained emissions of moderately widespread tephra fall and fountain-fed lava flows (Thordarson and Larsen 2007). During these initial stages, the eruption plume can reach heights of 36 km above sea level (Höskuldsson et al. 2007). As the lava/tephra ratio increases dramatically, the eruption intensity equally decreases. The final phase of a Hekla eruption typically consists of discrete Strombolian explosions with very low magma discharge (Thordarson and Larsen 2007, and references within).

Prior to the 20th century, Hekla erupted one to three times per century. Since the 1947 eruption, however, the volcano has experienced regular eruptions in 1970, 1980–1981, 1991

D. K. Bird (✉)
Risk Frontiers, Macquarie University, North Ryde, NSW 2109, Australia
e-mail: deanne.bird@gmail.com

G. Gísladóttir
Department of Geography and Tourism, Institute of Life and Environmental Sciences and Earth Science Institute, University of Iceland, 101 Reykjavík, Iceland
e-mail: ggisla@hi.is

P. Erfurt-Cooper (ed.), *Volcanic Tourist Destinations*, Geoheritage, Geoparks and Geotourism,
DOI: 10.1007/978-3-642-16191-9_2, © Springer-Verlag Berlin Heidelberg 2014

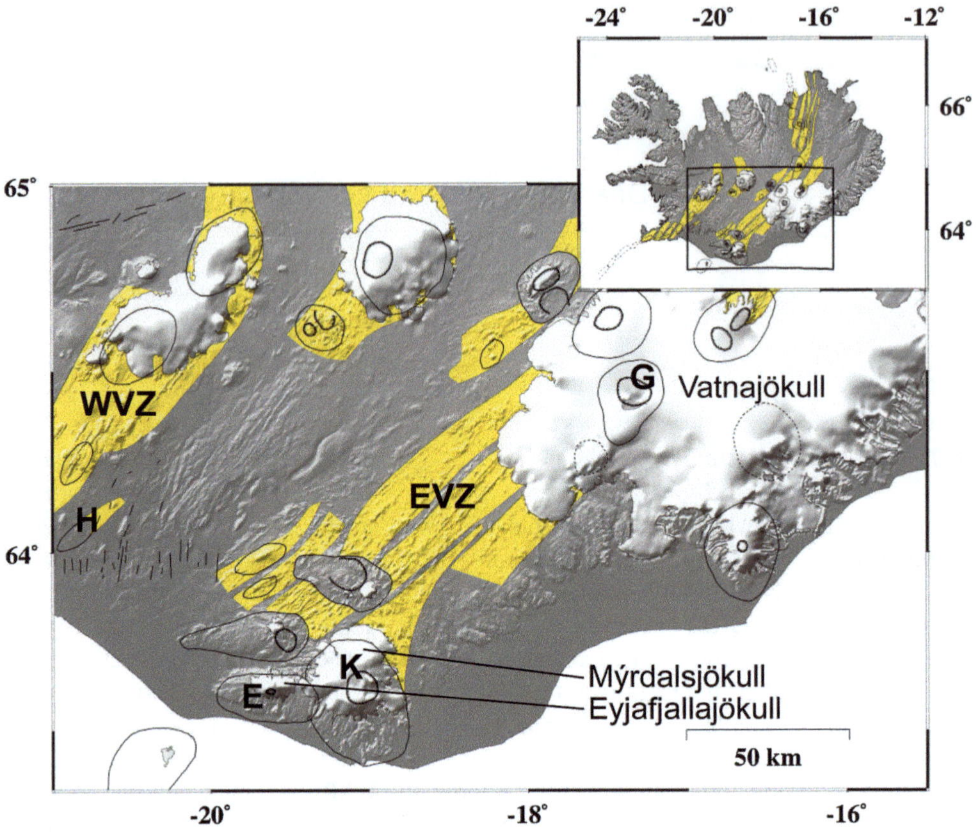

Fig. 2.1 The Western Volcanic Zone (*WVZ*) and Eastern Volcanic Zone (*EVZ*), and the volcanoes Hekla (*H*), Eyjafjallajökull (*E*), Katla (*K*) and Grímsvötn (*G*). The inserted map shows all the volcanic zones across Iceland. *Source* Einarsson and Saemundsson (1987)

Table 2.1 The Hekla, Eyjafjallajökull, Katla and Grímsvötn volcanic systems (from Thordarson and Larsen 2007, unless otherwise stated)

Volcano	Maximum elevation above sea level (m)	Length (km)	Width (km)	Area (km^2)	Confirmed eruptions since ~870 AD[a]	Last confirmed eruption[b]	Largest explosive eruption[c]	Largest effusive eruption[d]
Hekla	1491	60	19	720	23	2000	1104 VEI 5	1766–1768 1.3 km^3
Eyjafjallajökull	1666	30	14	300	4	2010	2010 VEI 4[e]	~920 0.05 km^3
Katla	1480	110	30	1300	21	1918	934 VEI 5	934 18 km^3
Grímsvötn	1722	100	23	1350	~71	2011	1783 VEI 4	1783 14 km^3

[a] From M. T. Guðmundsson et al. (2008) and Veðurstofa Íslands (2011a)
[b] From Veðurstofa Íslands (2011a)
[c] From M. T. Guðmundsson et al. (2008)
[d] From M. T. Guðmundsson et al. (2008)
[e] From Sammonds et al. (2010)

and 2000 (Gronvold et al. 1983; Veðurstofa Íslands 2011a), with each consecutive eruption decreasing in magnitude. Nevertheless, Hekla has produced significant quantities of tephra and lava during historic times, which equates to an estimated total volume of 13–14 km^3 of erupted magma (Thordarson and Larsen 2007).

Due to the high quantity of fluorine in Hekla's magma, nearly all eruptions have led to fluoride poisoning (fluorosis) in grazing livestock. In addition, extensive tephra fallout during the 1693 eruption caused mass trout deaths in lakes up to 110 km from the source (Guðmundsson et al. 2008). Hekla eruptions are also known to produce small pyroclastic flows (Höskuldsson et al. 2007), but seismicity has only been recorded in association with its eruptions (Einarsson 1991) making eruption predictions difficult. However, the distinct pattern of crustal deformation (measured by continuous borehole strain meters), and a small swarm of earthquakes recorded half an hour before the 1991 eruption, led to a short-term prediction of the 2000 eruption (Soosalu et al. 2005).

The Earth Science Institute at the University of Iceland and the Icelandic Meteorological Office recorded small but significant seismicity at Hekla volcano at 16:55 and 17:07 on 26 February 2000, respectively (Stefánsson 2000). A subsequent warning, stating that an eruption was imminent within an hour, was issued at 17:20. Based on strain meter data, the eruption began at 18:17 and an eruption column was observed at 18:20, which extended to 11 km above sea level by 18:25 (Stefánsson 2000). This eruption resulted in a 6.6 km fissure running SW to NE along the length of the Hekla volcano, slightly south of the summit (Höskuldsson et al. 2007). Considered the smallest and shortest of all Hekla eruptions, the 2000 eruption also followed the shortest repose period (Lacasse et al. 2004). Deformation around Hekla continued throughout 2011. However, media reports that an eruption was imminent were unfounded. Nevertheless, the Icelandic Department of Civil Protection and Emergency Management advised that hikers should take adequate precaution when climbing Hekla.

2.3 Eyjafjallajökull

Considered one of the oldest active volcanoes in Iceland, Eyjafjallajökull typically generates relatively small amounts of material (Sturkell et al. 2009). The shield-like central volcano consists of an elliptical 2.5 km wide caldera that is covered by a 200 m thick ice-cap of the same name. Due to the icecap, Eyjafjallajökull typically yields subglacial phreatomagmatic eruptions with very fine-grained ash deposits (Larsen et al. 1999) and small to medium jökulhlaup (Table 2.2) (Guðmundsson et al. 2005). Eyjafjallajökull is far less active than Hekla, Katla and Grímsvötn with only four eruptions since ~870 AD: in ~920, 1612, from 1821–1823, and the recent 2010 events. The first event of the 2010 eruptions occurred on the flank of the volcano along a fissure at Fimmvörðuháls in between the Mýrdalsjökull and Eyjafjallajökull icecaps (Sigmundsson et al. 2010) and, despite the sophisticated volcanic monitoring system, it was local residents who initially spotted the first signs of the eruption (Bird et al. 2011). After this uncertain start, the Icelandic emergency management organisations worked together with the local population to effectively mitigate risk.

The initial phase of the eruption produced spectacular fire-fountain activity and lava flows, which attracted many domestic and international onlookers. As the "Fimmvörðuháls" flank eruption was reduced to a 'hazard phase' on 13 April 2010, the Icelandic Meteorological Office recorded increased seismicity in Eyjafjallajökull. This activity was a precursor to the subglacial eruption of Eyjafjallajökull that commenced on 14 April 2010. The explosive eruption resulted in medium-sized jökulhlaups to the north, small jökulhlaups to the south and considerable ash fall to the east and east-southeast of the volcano (Almannavarnadeild Ríkislögreglustjórinn, 2010). The tephra that was ejected during the first few days of the 2010 subglacial eruption had a high proportion of very fine-grained ash that reached high altitudes in the atmosphere and, due to the prevailing north-westerly to westerly winds, air traffic was grounded in many countries throughout Europe. In comparison to the three largest producers of tephra—Hekla, Katla and Grímsvötn –Eyjafjallajökull volcano is considered to be a small tephra producer and the 2010 eruption was nothing out of the ordinary (Davies et al. 2010). The main factor that facilitated chaos within the aviation industry was the prevailing wind.

Table 2.2 Categorisation of Icelandic jökulhlaup (from Guðmundsson et al. 2005)

Category		Peak discharge (m^3s^{-1})
1	Very small	<3,000
2	Small	3,000–10,000
3	Medium	10,000–30,000
4	Big	30,000–100,000
5	Catastrophic	>100,000

The continuing eruption also produced lightning, gas emissions and lava flows within close proximity of the crater and heavy sound blasts were heard especially to the south and east of the volcano (Veðurstofa Íslands 2010). While the Fimmvörðuháls eruption had a positive economic impact in Iceland, the subglacial Eyjafjallajökull eruption did not and many communities are still dealing with ash impacts. Despite the risk that Eyjafjallajökull poses to farms and tourist accommodation at the base of the volcano, prior to 2010, much more interest in emergency management was given to its very active and more hazardous neighbour, Katla.

2.4 Katla

The Katla central volcano consists of a 5 km wide magma chamber sitting at a shallow depth of 1.5 km below sea level (Sturkell et al. 2003). The caldera is 14 km long, 600–750 m deep and is overlain by the Mýrdalsjökull icecap (Fig. 2.2), which has a maximum thickness of ~740 m (Björnsson et al. 2000). Historically, Katla is characterised by subglacial phreatomagmatic eruptions and has erupted on average, one to three times per century with the last confirmed eruption in 1918 (Thordarson and Larsen 2007). Small, sudden jökulhlaup in 1955 and 1999 are believed to have been the result of minor eruptions that did not break the glacier surface (Gudmundsson 2005). However, eruptions under the icecap typically melt large volumes of ice

Fig. 2.2 Katla volcano is blanketed by the Mýrdalsjökull icecap in the central southern region of Iceland. *Photo* Guðrún Gísladóttir

within a few hours with some resulting in catastrophic jökulhlaup that attain a peak discharge of 100,000–300,000 m^3s^{-1} (Björnsson et al. 2000).

The Katla jökulhlaup, produced during the eruption that began on 12 October 1918, reached a peak discharge of over 300,000 m^3s^{-1} and transported vast amounts of sediment and ice. The jökulhlaup carved its way through the glacier creating a gorge 1,460–1,830 m in length, 366–550 m in width and more than 145 m in height. This caused the detachment of a segment of glacier, which was transported in blocks estimated to be 40–60 m high onto the floodplain (Tómasson 1996). This flood was considered to be the world's largest known historic flood caused by volcanism (O'Connor and Costa 2004). Although, resident's descriptions recorded in annals suggest that the jökulhlaup produced during the 1755 Katla eruption was most likely larger (Guðmundsson and Högnadóttir 2006).

However, not all Katla eruptions have been subglacial. The 934–938 AD Eldgjá flood lava eruption occurred along a ~75 km discontinuous fissure that began under the Mýrdalsjökull icecap in the southwest and continued for 60 km through subaerial mountainous terrain to the northeast of the icecap (Larsen 2000). This eruption produced jökulhlaups, a hyaloclastite flow and two major lava fields. Pahoehoe lavas dominate Eldgjá lava fields although these are interspersed with rootless cones (pseudo-craters), which indicate that the lava was emplaced over shallow lakes or moist ground. Not only is this eruption the largest volcanic event in Iceland's history but it is also the largest of its kind on Earth in the last 2000 years (Thordarson and Larsen 2007). It is therefore not surprising that Eldgjá caused the most extensive environmental changes created by volcanism during the last 11 centuries in Iceland (Larsen 2000).

In addition to jökulhlaups and lava, Katla eruptions have produced heavy tephra fall approximately 20 cm thick to distances of 30 km from source and lightning in areas within 30–40 km of the eruption site (Guðmundsson et al. 2008; Larsen 2000). Coupling these hazards with its proximity to inhabited regions on the south coast, Katla has earned the reputation of being one of Iceland's most hazardous volcanoes (Guðmundsson et al. 2007). High levels of seismicity and crustal deformation, particularly in 2011, suggest that Katla is in a heightened state of activity. Earthquake activity within and around the Katla caldera intensified prior to a small jökulhlaup flood from the southeast corner of Mýrdalsjökull on 8 July 2011. Local residents and tourists were evacuated from the area and the main highway was cut as the flood destroyed one bridge. While several emergency centres were established as a precaution, an eruption column did not penetrate the glacier and the event was declared over within two days.

2.5 Grímsvötn

Due to its position underneath Europe's largest glacier, Vatnajökull, the Grímsvötn central volcano, also typically generates subglacial phreatomagmatic eruptions. Since settlement, Grímsvötn has been the most active volcanic system in Iceland (Guðmundsson et al. 2004) with recent eruptions in 1902, 1922, 1933, 1934, 1938, 1983, 1998, 2004 and 2011 (Veðurstofa Íslands 2011a). Within the Grímsvötn caldera lies a subglacial lake that drained in periodic jökulhlaups every 4–6 years prior to the 1996 fissure eruption in Gjálp. Occurring to the north of Grímsvötn, the 1996 Gjálp eruption caused glacial meltwater to accumulate in the

Fig. 2.3 Twisted metal debris from a bridge destroyed by the jökulhlaup produced from the 1996 Gjálp eruption near Grímsvötn. *Photo* Deanne Bird

Grímsvötn subglacial lake for a month before it flooded to the south of Vatnajökull between 4 and 7 November 1996. This jökulhlaup reached a peak discharge rate of approximately 55,000 m^3s^{-1} (Stefánsdóttir and Gíslason 2005), draining 3.2 km^3 of meltwater from the Grímsvötn lake in about 40 h (Bjornsson 1998).

Emerging across the alluvial floodplain, the jökulhlaup impacted 750 km^2 of land, extended the coastline by 800 m and created approximately 7 km^2 of new land (Smellie 2000). Carrying massive icebergs weighing up to 1000 tons, the floodwaters damaged and destroyed several bridges and roads, including the main highway. A section of a destroyed bridge is displayed on the main highway and serves as a permanent reminder to passing motorists of the shear catastrophic force of Iceland's volcanoes (Fig. 2.3). However, not all jökulhlaups are triggered by eruptions. For example, the reverse occurred in 2004 when an eruption was triggered by the pressure release of draining meltwater (jökulhlaup) that had accumulated in the Grímsvötn caldera due to geothermal activity (Sturkell et al. 2006).

Similar to Katla, the largest eruption from the Grímsvötn system was a subaerial fissure event along a mixed cone row that extended continuously for 27 km (Thordarson and Larsen 2007). This flood lava eruption, known as the Laki eruption, occurred over eight months from 1783 to 1784 AD and is the second largest event of its kind in Iceland (Sigurdsson 1982). Emitting 122 megatons of SO_2 into the atmosphere and generating a sulphuric aerosol veil that was suspended over the Northern Hemisphere for more than five months, the Laki eruption resulted in a reduction of the annual mean surface temperature in Europe and North America by 1.3° for two to three years. Locally, the volcanic haze or dry fog caused plants to wither and killed more than 60 % of Iceland's livestock, mostly through fluoride poisoning (fluorosis). This led to widespread famine which lasted from 1783–1786 and caused severe malnutrition, scurvy and the death of ~20 % of the Icelandic human population (Thordarson and Self 2003). Further afield, the volcanic haze spread westward across Europe to the Altai Mountains in China (Sigurdsson 1982), causing considerable damage to vegetation and crops throughout Europe and stunted tree-growth in Scandinavia and Alaska (Thordarson and Hoskuldsson 2002).

The most recent volcanic activity in Grímsvötn, however, was nowhere near severe. On 21 May 2011 a subglacial eruption created an ash plume that reached a maximum altitude of 20 km (Veðurstofa Íslands 2011b). Ash fallout was experienced locally to the south and east of the volcano and 60–70 lightning strikes were recorded per hour on 22 May 2011 (Jakobsdóttir et al. 2011). Since this eruption occurred at the same location as the 2004 eruption, ice-melt was not significant enough to produce a jökulhlaup (Guðmundsson et al. 2011). This eruption ended on 28 May 2011 (Veðurstofa Íslands 2011b), only 7 days after it began.

Despite, or because of, these infamous volcanoes, many visitors flock to Iceland every year to experience its natural beauty and witness firsthand, the geological processes taking place. Importantly, visitors can feel at ease with the

Fig. 2.4 Hiking trails scattered around the landscape surrounding the Mýrdalsjökull icecap and Katla volcano. *Photo* Deanne Bird

knowledge that, in Iceland, only two people perished in volcanic eruptions during the 20th century and none have died during the 21st century. However, due to the rising number of tourists on and around Iceland's volcanoes, the possibility for fatalities in moderate-sized explosive eruptions is increasing (Guðmundsson et al. 2008). In the next section we describe tourism in southern Iceland and what government agencies are doing to reduce risk in these volcanic environments.

2.6 Tourism in Southern Iceland

Iceland is an accessible and popular destination with many local and international tourists exploring the dramatic volcanic landscape of the south (Benediktsson et al. 2011; Bird et al. 2010b). Located within a short drive from the capital Reykjavík and the international airport at Keflavík, the southern region offers a myriad of activities including canyoning, ice and mountain hiking (Fig. 2.4), horse riding, kayaking, mountain climbing, quad biking, snowmobiling, 4WD super jeep trips and white water rafting. There are plenty of organised single or multiday trips from Reykjavík or, alternatively, visitors can hire a car and discover the south coast themselves.

Obviously, Iceland's volcanic landscape and geothermal activity are significant features of many organised tours. In particular, super jeep and hiking tours explore the lava fields and summit of Hekla. Organised ice-hiking, snowmobiling and super jeep tours also traverse the glaciers covering Katla and Grímsvötn while special tours can be arranged to drive or hike over Eyjafjallajökull. However, the most popular hiking path in Iceland is the Laugavegur route, which traverses 55 km of glacial and volcanic landscapes from Landmannalaugar to Þórsmörk. For the more adventurous traveller, this hike continues along the Fimmvörðuháls hiking track that negotiates the rough mountain pass between Eyjafjallajökull and Mýrdalsjökull (Fig. 2.5) and encompasses the region affected by the recent 2010 flank eruption.

Figure 2.6 highlights a steady increase in international tourist numbers staying overnight in hotels and guest houses in southern Iceland since 2002, with a somewhat small decrease in 2010. The total number of overnight stays by Icelandic and international tourists in wilderness lodges and camping grounds has been less predictable during the same period. Nevertheless, there was a definite decrease in 2010 from 2009 in both international (29.1 %) and Icelandic tourists (38.0 %) (Hagstofa Íslands, 2011), which might be attributed to the subglacial Eyjafjallajökull eruption, but this was followed by an overall increase in 2011 of 12.9 % and 40.5 %, respectively (Hagstofa Íslands 2012). Although wilderness lodge accommodation in 2012 remained constant, hotel and guesthouse accommodation increased by 15.6 % in 2012 from 2011 (Hagstofa Íslands 2013). Due to the increasing risk of another eruption in southern Iceland, government agencies and tourism operators are implementing risk management strategies.

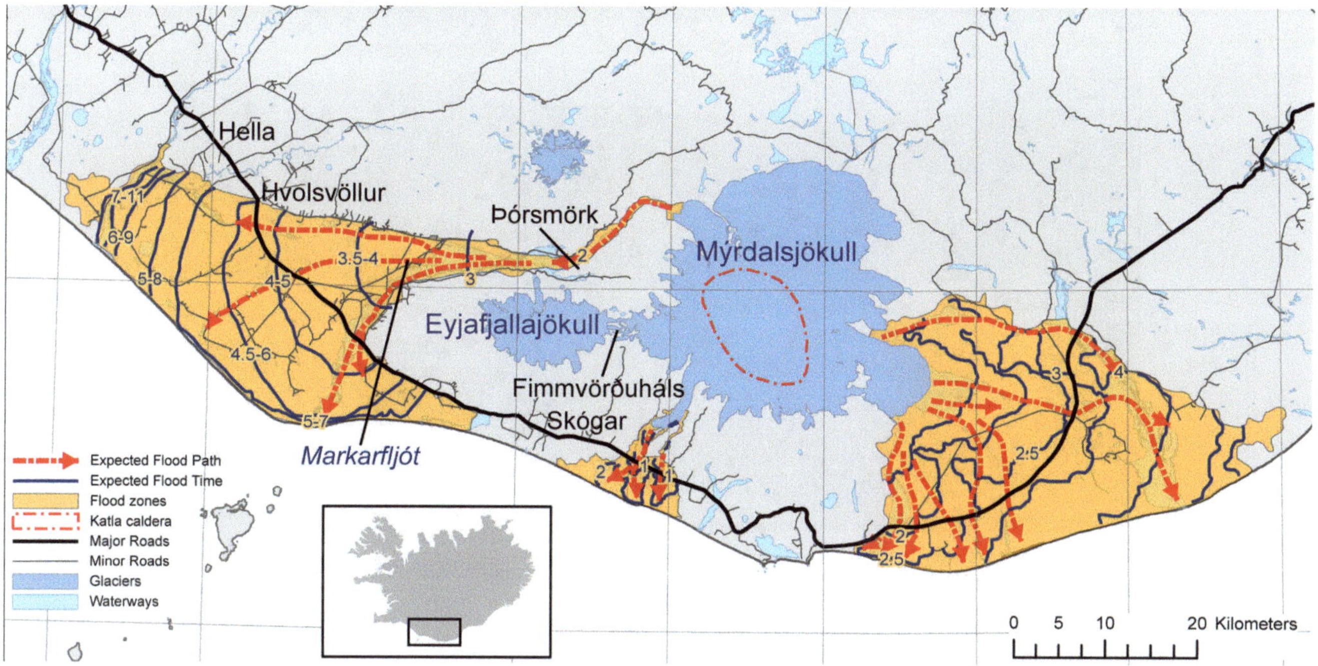

Fig. 2.5 The popular tourist destination of Þórsmörk, situated to the west of Mýrdalsjökull, and the eastern, southern and western jökulhlaup hazards zones around Katla. *Source* (from Ríkislögreglustjóri almannavarnadeild 2008a, b). *Map* produced by James O'Brien, Risk Frontiers

Fig. 2.6 **a** The total number of overnight stays by Icelandic and international tourists in wilderness camping grounds and lodges in southern Iceland from 1998 to 2012; **b** The total number of overnight stays by Icelandic and international tourists in hotels and guesthouses in southern Iceland from 1998 to 2012. *Source* (Hagstofa Íslands 2004, 2005, 2006, 2007, 2008, 2009, 2010, 2011, 2012, 2013)

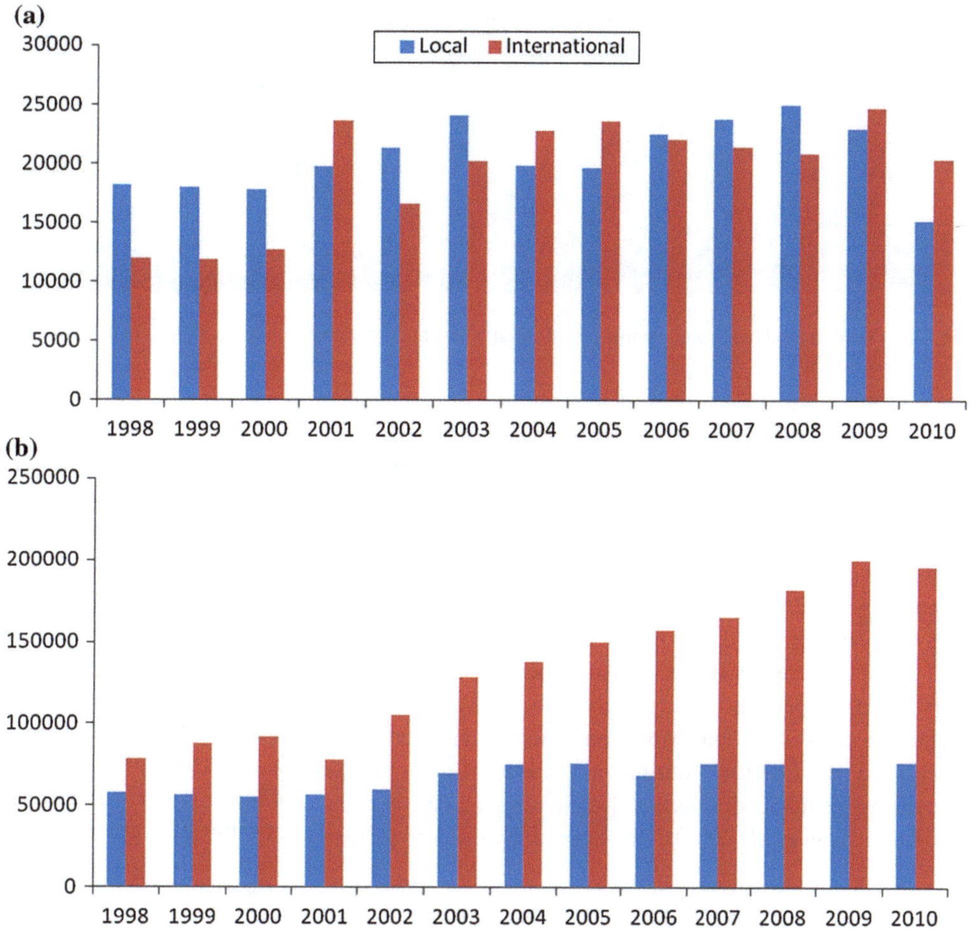

Fig. 2.7 A jeep crossing the river Krossá, which flows into the Markarfljót near Þórsmörk. *Photo* Deanne Bird

Fig. 2.8 Wilderness lodges and camping grounds at a mountain hut area on the floodplain in Þórsmörk. Mýrdalsjökull is seen in the background. *Photo* Deanne Bird

2.7 Volcanic Risk Management in Southern Iceland

Post-eruption famine is highly unlikely to occur in contemporary Iceland (Guðmundsson et al. 2008). However, loss of livestock and damage to property and infrastructure will cause severe economic consequences for a nation of only 320,000 people, as experienced following the 2010 Eyjafjallajökull eruption. Moreover, there is an increasing likelihood of fatalities during an eruption due to the increasing number of tourists frequenting active regions (Guðmundsson et al. 2008). This potential would be further exacerbated if a Katla eruption was to produce a catastrophic jökulhlaup.

Government agencies have therefore developed regional volcanic risk management strategies. In particular, emergency management officials have been preparing for a Katla eruption (Sigthorsson et al. 2006) and emergency response procedures have recently been developed or revised for communities located in the eastern, southern and western jökulhlaup hazard zones (see Fig. 2.5) (Almannavarnir 2006; Bird 2010; Bird et al. 2009, 2011). If an eruption is imminent, officials will enforce road closures preventing people from crossing glacial rivers that are vulnerable to jökulhlaup. In

Fig. 2.9 Travellers reading the hazard and emergency response information sign entitled 'Katla-Mýrdalsjökull' on a hiking track in Þórsmörk. *Photo* Damian Gore

the popular tourist destination of Þórsmörk (see Fig. 2.5), this is considered a necessary precaution since the Þórsmörk access road traverses the Markarfljót floodplain, crossing dangerous rivers and tributaries (Fig. 2.7). It is estimated that a catastrophic jökulhlaup with a discharge rate greater than 100,000 m^3s^{-1} would produce a flood height across the plain in excess of 20 m and that it would reach Þórsmörk around 2 h after an eruption commences (Guðmundsson and Gylfason 2005). It is therefore essential to close roads susceptible to flooding until the threat has passed.

Each of the mountain hut communities in Þórsmörk: Húsadalur, Langidalur and Básar, are either on, or adjacent to, floodplains (Fig. 2.8). During a Katla eruption, people located in the Þórsmörk region will be instructed to go to higher ground and avoid low-lying areas, especially near glacial rivers.

During the 2008 summer, the Icelandic Department of Civil Protection and Emergency Management together with local police and rescue teams, scheduled hazard education and emergency response training campaigns with the Þórsmörk tourism sector. These consisted of information meetings with tourism companies, onsite training at each mountain hut community, distribution of brochures entitled 'Eruption Emergency Guidelines' available in six languages, and the erection of hazard and emergency response information signs entitled 'Katla-Mýrdalsjökull' in mountain huts and in prominent positions along hiking trails (Fig. 2.9).

Similar precautions have been developed surrounding Hekla, Eyjafjallajökull and Grímsvötn with local information centres (Table 2.3) providing essential hazard and emergency response information for travellers.

2.8 How Prepared are Tourists for a Volcanic Eruption in Southern Iceland?

To assess the vulnerability of the tourism sector in southern Iceland, surveys were carried out from July to September in 2007 and 2009. In total, 139 face-to-face questionnaire interviews were conducted in 2007 and 124 in 2009. For information about the methodology used see Bird et al. (2010a, b). The results of the survey show that education and training campaigns implemented in 2008 were well accepted by tourists and tourism employees. However, they were not entirely successful at increasing knowledge or preparedness (Table 2.4). Respondents believed the map was inappropriately scaled for the Þórsmörk region and found it confusing. For example, information within the brochure refers to 'Mt. Katla' but 'Mt Katla' is not marked on the map. Additionally, the first instruction in the brochure states:

> Examine the map to see where glacial floods have flowed in the past. If an eruption warning signal is given, keep to the upper slopes. The main escape routes are shown by red arrows on the map. No entry signs indicate road closures.

However, many respondents commented that it is very difficult to determine their location on the map, the extent of previous floods or designated escape routes. The only instruction for tephra is: '*It is advisable not to drive where tephra is falling, as it can damage the engine.*' Although the section on tephra states that it can be harmful to the eyes and respiratory system, it provides no advice on how to protect oneself. In comparison, the sections entitled 'Risk of lightning' and 'Toxic gases' provide useful, detailed information about these hazards and how to reduce personal vulnerability.

2.9 Conclusion

The results of these surveys suggest that emergency management agencies are making progress to reduce the risk within the tourism sector but improvements can be made by:

- Improving the "Eruption Emergency Guidelines" brochure including the hazard and risk map to ensure that they are comprehendible and in a user-friendly format. This can be achieved through public consultation and workshops;
- Exploring alternate methods for educating tourists e.g. travel guides (Lonely Planet, Rough Guide, etc.), social media (Facebook, Twitter, etc.), tourist websites, and educational films on local and tourist buses; and,

Table 2.3 Information centres located in southern Iceland that provide tourists with details of local volcanic areas and emergency response procedures

Centre	Address	Telephone	Email
Hveragerði regional information centre	Sunnumörk 2–4 810 Hveragerði	+354 483-4601	tourinfo@hveragerdi.is
Hekla centre	Leirubakki 851, Hella	+354 487-8700	bookings@leirubakki.is
Hvolsvöllur district information centre	Austurvegur 8 860 Hvolsvöllur	+354 487-8043	tourinfo@hvolsvollur.is
Eyjafjallajökull visitor centre	Þorvaldseyri 861 Hvolsvöllur	+354 487 5757	info@icelanderupts.is
Vík district information centre—including the Katla centre	Víkurbraut 28 870 Vík	+354 487-1395	info@vik.is
Skaftafell visitors centre	Vatnajökull National Park Skaftafell 785 Öraefi	+354 470-8300	skaftafell@vjp.is

Table 2.4 Levels of knowledge and preparedness in 2007 and 2009 among tourists in Þórsmörk. Some sections do not equal 100 % due to rounding

	2007	2009
% of participants who were travelling with a guide	9	11
% of participants who carried a mobile phone	71	91
% of participants who had informed someone of their location in this region	78	53
% of participants who stated they knew there was an early warning system	22	57
% of participants who stated they knew the emergency procedures	4	28
If a jökulhlaup warning is issued, % of participants who would:		
go to the highest point	40	39
escape Þórsmörk	18	21
report to wardens	15	23
follow procedures	12	7
other	14	10
If there was a volcanic eruption, % of participants who would:		
report to wardens or guide	54	32
call an emergency number (e.g. 112) or friend	19	7
listen to radio	9	2
evacuate Þórsmörk	5	20
go to higher ground	–	28
other	13	10
% of participants who had:		
Seen an 'Eruption Emergency Guidelines' brochure	–	26
Read an 'Eruption Emergency Guidelines' brochure	–	17
Seen a 'Katla-Mýrdalsjökull' sign	–	44
Read a 'Katla-Mýrdalsjökull' sign	–	42

- Adopting an all-hazards approach to education, i.e. include information on other volcanic systems (e.g. Eyjafjallajökull, Hekla and Grímsvötn) and provide additional detail in relation to glacial melt associated with geothermal activity and, information on seasonal melt in addition to earthquakes, extreme weather conditions, etc.

For more information on Iceland's volcanic hazards and related preparedness measures see:

- http://www.safetravel.is/
- http://en.vedur.is/
- http://www.earthice.hi.is/page/ies_volcanoes
- http://www.vegagerdin.is/english/road-conditions-and-weather/

For more information on the National and Geo-Parks that encompass southern Iceland's volcanoes see:

- http://katlageopark.is/
- http://en.south.is/
- http://www.vatnajokulsthjodgardur.is/english

References

Almannavarnadeild Ríkislögreglustjórinn. (2010). Eruption at the Eyjafjallajökull volcanic system. Retrieved April 15, 2010, from http://www.almannavarnir.is/displayer.asp?cat_id=413

Almannavarnir. (2006). Bergrisinn 2006. Retrieved February 18, 2008, from http://www.almannavarnir.is/upload/files/Bergrisinn%20lokask%C3%BDrsla%200%204%20%282%29.pdf

Benediktsson, K., Lund, K. A., & Huijbens, E. (2011). Inspired by eruptions? Eyjafjallajökull and Icelandic tourism. *Mobilities, 6*(1), 77–84.

Bird, D. K. (2010). *Social dimensions of volcanic hazards, risk and emergency response procedures in southern Iceland.* PhD thesis, University of Iceland/Macquarie University, Reykjavik/Sydney. (ISBN 978-9979-9967-5-0).

Bird, D. K., Gísladóttir, G., & Dominey-Howes, D. (2009). Resident perception of volcanic hazards and evacuation procedures. *Natural Hazards and Earth Systems Sciences, 9*(1), 251–266.

Bird, D. K., Gísladóttir, G., & Dominey-Howes, D. (2010a). *Tourists, volcanoes and education—are hazard, risk and emergency response procedures effectively communicated in southern Iceland?* Paper presented at the Cities on Volcanoes Conference Abstract, Tenerife.

Bird, D. K., Gísladóttir, G., & Dominey-Howes, D. (2010b). Volcanic risk and tourism in southern Iceland: Implications for hazard, risk and emergency response education and training. *Journal of Volcanology and Geothermal Research, 189*(1–2), 33–48.

Bird, D. K., Gísladóttir, G., & Dominey-Howes, D. (2011). Different communities, different perspectives: Issues affecting residents' response to a volcanic eruption in southern Iceland. *Bulletin of Volcanology, 73*(9), 1209–1227. doi:10.1007/s00445-011-0464-1.

Bjornsson, H. (1998). Hydrological characteristics of the drainage system beneath a surging glacier [10.1038/27384]. *Nature, 395*(6704), 771–774.

Björnsson, H., Pálsson, F., & Guðmundsson, M. T. (2000). Surface and bedrock topography of the Mýrdalsjökull ice cap, Iceland: The Katla caldera, eruption sites and routes of jökulhlaups. *Jökull, 49*, 29–46.

Davies, S. M., Larsen, G., Wastegård, S., Turney, C. S. M., Hall, V. A., Coyle, L., et al. (2010). Widespread dispersal of Icelandic tephra: how does the Eyjafjöll eruption of 2010 compare to past Icelandic events? *Journal of Quaternary Science, 25*(5), 605–611. doi:10.1002/jqs.1421.

Einarsson, P. (1991). Earthquakes and present-day tectonism in Iceland. *Tectonophysics, 189*(1–4), 261–279.

Einarsson, P., & Saemundsson, K. (1987). Earthquake epicenters 1982–1985 and volcanic systems in Iceland. In T. Sigfússon (Ed.), *Í hlutarins eðli: Festschrift for Þorbjörn Sigurgeirsson.* Menningarsjóður: Reykjavík.

Gronvold, K., Larsen, G., Einarsson, P., Thorarinsson, S., & Saemundsson, K. (1983). The Hekla eruption 1980–1981. *Bulletin of volcanology, 46*(4), 349–363. doi:10.1007/bf02597770.

Guðmundsson, G. B., Óskarsson, N., Einarsson, E. M., Sigurðsson, Á., Thorbjarnardóttir, B. S., Roberts, M. J. et al. (2011). Grímsvötn volcano: Status report: 17:00 GMT. Retrieved May 23, 2011, from http://en.vedur.is/media/jar/Grimsvotn_status_2011-05-23_IES_IMO.pdf

Gudmundsson, M. T. (2005). Subglacial volcanic activity in Iceland. In C. J. Caseldine, A. Russell, J. Hardardóttir & Ó. Knudsen (Eds.), *Iceland: Modern processes, past environments* (pp. 127–151). Amsterdam: Elsevier.

Guðmundsson, M. T., Elíasson, J., Larsen, G., Gylfason, Á. G., Einarsson, P., Jóhanesson, T., et al. (2005). Yfirlit um hættu vegna eldgosa og hlaupa frá vesturhluta Mýrdalsjökuls og Eyjafjallajökli. In M. T. Guðmundsson & Á. G. Gylfason (Eds.), *Hættumat vegna eldgosa og hlaupa frá vestanverðum Mýrdalsjökli og Eyjafjallajökli* (pp. 11–44). Reykjavík: Ríkislögreglustjórinn and Háskólaútgáfan.

Guðmundsson, M. T., & Gylfason, Á. G. (2005). *Hættumat vegna eldgosa og hlaupa frá vestanverðum Mýrdalsjökli og Eyjafjallajökli.* Reykjavík: Ríkislögreglustjórinn and Háskólaútgáfan.

Guðmundsson, M. T., Högnadóttir, P., Kristinsson, A. B., & Guðbjörnsson, S. (2007). Geothermal activity in the subglacial Katla caldera, Iceland, 1999–2005, studied with radar altimetry. *Annals of Glaciology, 45*(1), 66–72.

Guðmundsson, M. T., & Högnadóttir, P. (2006). Ísbráðnun og upptakarennsli jökulhlaupa vegna eldgosa í Kötluöskju og austanverðum Mýrdalsjökli, Skýrsla unnin fyrir Almannavarnadeild Ríkislögreglustjóra, RH-02-2006 (p. 33). Reykjavík: Jarðvísindastofnun Háskólans.

Guðmundsson, M. T., Larsen, G., Höskuldsson, Á., & Gylfason, Á. G. (2008). Volcanic hazards in Iceland. *Jökull, 58*, 251–268.

Guðmundsson, M. T., Sigmundsson, F., Björnsson, H., & Högnadóttir, T. (2004). The 1996 eruption at Gjálp, Vatnajökull ice cap, Iceland: efficiency of heat transfer, ice deformation and subglacial water pressure. *Bulletin of Volcanology, 66*(1), 46–65.

Hagstofa Íslands. (2004). Hagtíðindi: Ferðamál og samgöngur/Statistical series: Tourism and transport. Tourist accommodation 2003. Reykjavík: Hagstofa Íslands. ISSN 1670-4584.

Hagstofa Íslands. (2005). Hagtíðindi: Ferðamál og samgöngur/Statistical series: Tourism and transport. Tourist accommodation 2004. Reykjavík: Hagstofa Íslands. ISSN 1670-4584.

Hagstofa Íslands. (2006). Hagtíðindi: Ferðamál og samgöngur/Statistical series: Tourism and transport. Tourist accommodation 2005. Reykjavík: Hagstofa Íslands. ISSN 1670-4584.

Hagstofa Íslands. (2007). Hagtíðindi: Ferðamál og samgöngur/Statistical series: Tourism and transport. Tourist accommodation 2006. Reykjavík: Hagstofa Íslands. ISSN 1670-4584.

Hagstofa Íslands. (2008). Hagtíðindi: Ferðamál og samgöngur/Statistical series: Tourism and transport. Tourist accommodation 2007. Reykjavík: Hagstofa Íslands. ISSN 1670-4584.

Hagstofa Íslands. (2009). Hagtíðindi: Ferðamál og samgöngur/Statistical series: Tourism and transport. Tourist accommodation 2008. Reykjavík: Hagstofa Íslands. ISSN 1670-4584.

Hagstofa Íslands. (2010). Hagtíðindi: Ferðamál og samgöngur/Statistical series: Tourism and transport. Tourist accommodation 2009. Reykjavík: Hagstofa Íslands. ISSN 1670-4584.

Hagstofa Íslands. (2011). Hagtíðindi: Ferðamál og samgöngur/Statistical series: Tourism and transport. Tourist accommodation 2010. Reykjavík: Hagstofa Íslands. ISSN 1670-4584.

Hagstofa Íslands. (2012). Hagtíðindi: Ferðamál, samgöngur og upplýsingatækni/Statistic Series: Tourism, transport and IT. Tourist accommodation 2011. Reykjavik: Hagstofa Islands. ISSN 1670-4584.

Hagstofa Íslands. (2013). Hagtíðindi: Ferðamál, samgöngur og upplýsingatækni/Statistic Series: Tourism, transport and IT. Tourist accommodation 2012. Reykjavík: Hagstofa Íslands. ISSN 1670-4584.

Höskuldsson, Á., Óskarsson, N., Pedersen, R., Grönvold, K., Vogfjörð, K., & Ólafsdóttir, R. (2007). The millennium eruption of Hekla in February 2000. *Bulletin of volcanology, 70*(2), 169–182. doi:10.1007/s00445-007-0128-3.

Jakobsdóttir, S. S., Sigurðsson, G., Árnason, Ó., Guðmundsson, M. T., Jónsdóttir, I., Högnadóttir, T. et al. (2011). Grímsvötn volcano, Status Report: 17:00 GMT, 22 May 2011. Retrieved September 15, 2011, from http://en.vedur.is/media/jar/Grimsvotn_status_2011-05-22_IES_IMO.pdf

Lacasse, C., Karlsdóttir, S., Larsen, G., Soosalu, H., Rose, W. I., & Ernst, G. G. J. (2004). Weather radar observations of the Hekla 2000 eruption cloud, Iceland. *Bulletin of Volcanology, 66*(5), 457–473. doi:10.1007/s00445-003-0329-3.

Larsen, G. (2000). Holocene eruptions within the Katla volcanic system, south Iceland: Characteristics and environmental impact. *Jökull, 49*, 1–28.

Larsen, G., Dugmore, A., & Newton, A. (1999). Geochemistry of historical-age silicic tephras in Iceland. *The Holocene, 9*(4), 463–471. doi:10.1191/095968399669624108.

O'Connor, J. E., & Costa, J. E. (2004). The world's largest floods, past and present—their causes and magnitudes (p. 13). Reston: U.S. Geological Survey.

Ríkislögreglustjóri almannavarnadeild. (2008a). Katla—Austur hluti. Retrieved December 15, 2009, from www.almannavarnir.is/default.asp?cat_id=193

Ríkislögreglustjóri almannavarnadeild. (2008b). Katla—Vestur hluti. Retrieved December 15, 2009, from www.almannavarnir.is/default.asp?cat_id=195

Sammonds, P., McGuire, B., & Edwards, S. (2010). Volcanic hazard from Iceland: analysis and implications of the Eyjafjallajökull eruption (p. 21). London: UCL Institute for Risk and Disaster Reduction.

Sigmundsson, F. (2006). *Iceland geodynamics, crustal deformation and divergent plate tectonics*. Chichester: Praxis Publishing Ltd.

Sigmundsson, F., Hreinsdóttir, S., Hooper, A., Árnadóttir, T., Pedersen, R., Roberts, M. J. et al. (2010). Intrusion triggering of the 2010 Eyjafjallajökull explosive eruption. *Nature, 468*(18 November 2010), 426–432.

Sigthorsson, I. H., Gudmundsson, A. T., Marteinsdottir, F., & Nielsen, T. (2006). Evacuation plan for volcanic eruptions in the Katla caldera in Iceland. *Geophysical Research Abstracts, 8*(05254).

Sigurdsson, H. (1982). Volcanic pollution and climate: The 1783 Laki eruption. *Eos, Transactions American Geophysical Union, 63*(32), 601–602. doi:10.1029/EO063i032p00601.

Smellie, J. L. (2000). Subglacial eruptions. In H. Sigurdsson (Ed.), *Encyclopedia of volcanoes* (pp. 403–418). San Diego: Academic Press.

Soosalu, H., Einarsson, P., & Þorbjarnardóttir, B. ó. (2005). Seismic activity related to the 2000 eruption of the Hekla volcano, Iceland. *Bulletin of Volcanology, 68*(1), 21–36. doi:10.1007/s00445-005-0417-7.

Stefánsdóttir, M. B., & Gíslason, S. R. (2005). The erosion and suspended matter/seawater interaction during and after the 1996 outburst flood from the Vatnajökull Glacier, Iceland. *Earth Planetary Science Letter, 237*, 433–452.

Stefánsson, R. (2000). A volcanic eruption in Hekla, February 26, 2000. Retrieved September 23, 2011, from http://hraun.vedur.is/ja/englishweb/heklanews.html#strain

Sturkell, E., Einarsson, P., Sigmundsson, F., Geirsson, H., Ólafsson, H., Pedersen, R., et al. (2006). Volcano geodesy and magma dynamics in Iceland. *Journal of Volcanology and Geothermal Research, 150*(1–3), 14–34.

Sturkell, E., Einarsson, P., Sigmundsson, F., Hooper, A., Ófeigsson, B. G., Geirsson, H., et al. (2009). Katla and Eyjafjallajökull Volcanoes. In J. J. M. van der Meer (Ed.), *The Mýrdalsjökull ice cap, Iceland: glacial processes, sediments and landforms on an active volcano* (Vol. 13, pp. 5–21). Amsterdam: Elsevier Science.

Sturkell, E., Sigmundsson, F., & Einarsson, P. (2003). Recent unrest and magma movements at Eyjafjallajökull and Katla volcanoes, Iceland. *Journal of Geophysical Research, 108*(B8), 2369.

Thordarson, T., & Hoskuldsson, A. (2002). *Iceland*. Hertfordshire: Terra Publishing.

Thordarson, T., & Larsen, G. (2007). Volcanism in Iceland in historical time: Volcano types, eruption styles and eruptive history. *Journal of Geodynamics, 43*(1), 118–152.

Thordarson, T., & Self, S. (2003). Atmospheric and environmental effects of the 1783-1784 Laki eruption: A review and reassessment. *Journal of Geophysical Research, 108*(D1), 4011. doi:10.1029/2001jd002042.

Tómasson, H. (1996). The jökulhlaup from Katla in 1918. *Annals of Glaciology, 22*, 249–254.

Veðurstofa Íslands. (2010). Update on activity: Eruption in Eyjafjallajökull, Iceland. Retrieved July 13, 2011, from http://en.vedur.is/earthquakes-and-volcanism/articles/nr/1884

Veðurstofa Íslands. (2011a). List of recent volcanic eruptions in Iceland: Eruptions in Iceland in since 1902. Retrieved September 14, 2011, from http://en.vedur.is/earthquakes-and-volcanism/articles/nr/1874

Veðurstofa Íslands. (2011b). Update on volcanic activity in Grímsvötn: Status reports, photos and facts. Retrieved September 15, 2011, from http://en.vedur.is/earthquakes-and-volcanism/articles/nr/2180

3 Tourism and Volcanism in the Canary Islands, Spain

Jasmine Cardozo Moreira, Roberto Rendeiro Martín-Cejas, and Guillermo Melendez-Hevia

3.1 Introduction

The Canary Islands, which are located in the North Atlantic Ocean near Europe and Northern Africa, consist of seven volcanic islands and many islets, with a spring climate that prevails year round. Their vertiginous contrasting landscapes and gentle climates have created an ideal environment for a rich diversity of species and habitats. These environmental features of the Canary Islands, together with their proximity to the most important departure points for European tourism, enhance their tourist appeal.

The Canarian volcanoes are important socioeconomic resources for the inhabitants of the islands. Pyroclastic materials and lava flows have been consistently used (Dóniz-Páez et al. 2011), and since 2010, aspects related to volcanism are part of the "Competitive Touristic Plan of Canary Islands—A volcanic experience". One of the objectives of this plan is to improve the tourism in the Autonomous Community through promoting programs and policies for the modernisation and upgrading to meet the demand trends of the tourism industry. The Plan involves a number of innovative activities to diversify the tourist offer of the four participant Islands Gran Canaria, Tenerife, Lanzarote and La Gomera to ensure regular fluxes of tourists throughout the year in consolidated tourist areas as well as at other locations under development (Boletin Oficial del Estado 2010).

This way, some features of geological heritage, where specific geological problems or elements appear particularly well exposed, can play a relevant role as educational tools. These are distinct components of the landscape with an outstanding geological, geomorphological or palaeontological value. Volcanism-related features are the most conspicuous and important geological tourist attractions of the Canary Islands. The tourist routes include all relevant points of geological heritage which are accessible to visitors. Some of the most significant geosites of the islands are geological features such as the picturesque volcanic cones in Lanzarote, El Hierro and La Palma; the impressive volcanoes of Teide, Teneguía and Taburiente; the volcanic tunnels or lava caves as well as the striking rock formations of Las Cañadas in Tenerife.

However, despite the relevance of the geological features and the large number of visitors, this constant tourist flow unfortunately exerts great pressure on the need for mobility and may have a substantial environmental impact. The relationship between tourism and transportation goes beyond the mere objective of travelling from departure point to destination, as it also continues throughout the whole tourism experience. Surface transport for tourists now becomes crucial in tourism services and, at the same time, could be an important factor due to its unsustainability (Martin-Cejas and Sanchez 2010).

3.2 General Aspects of the Canary Islands

3.2.1 Volcanic Aspects

The Canary Islands are an oceanic archipelago with volcanic origins, composed of seven major islands and several islets of different geological ages, ranging from 22 to 0.7 million years (Carracedo et al. 1998) (Fig. 3.1).

J. C. Moreira (✉)
Departamento de Turismo, Unversidade Estadual de Ponta Grossa, Ponta Grossa, Brazil
e-mail: jasmine@uepg.br

R. R. Martín-Cejas
Facultad de Economía, Empresa y Turismo, Departamento de Análisis Económico Aplicado, Universidad de Las Palmas de Gran Canaria, Las Palmas, Spain
e-mail: rrendeiro@daea.ulpgc.es

G. Melendez-Hevia
Depto. Geología (Paleontología), Universidad de Zaragoza, Zaragoza, Spain
e-mail: gmelende@unizar.es

P. Erfurt-Cooper (ed.), *Volcanic Tourist Destinations*, Geoheritage, Geoparks and Geotourism,
DOI: 10.1007/978-3-642-16191-9_3, © Springer-Verlag Berlin Heidelberg 2014

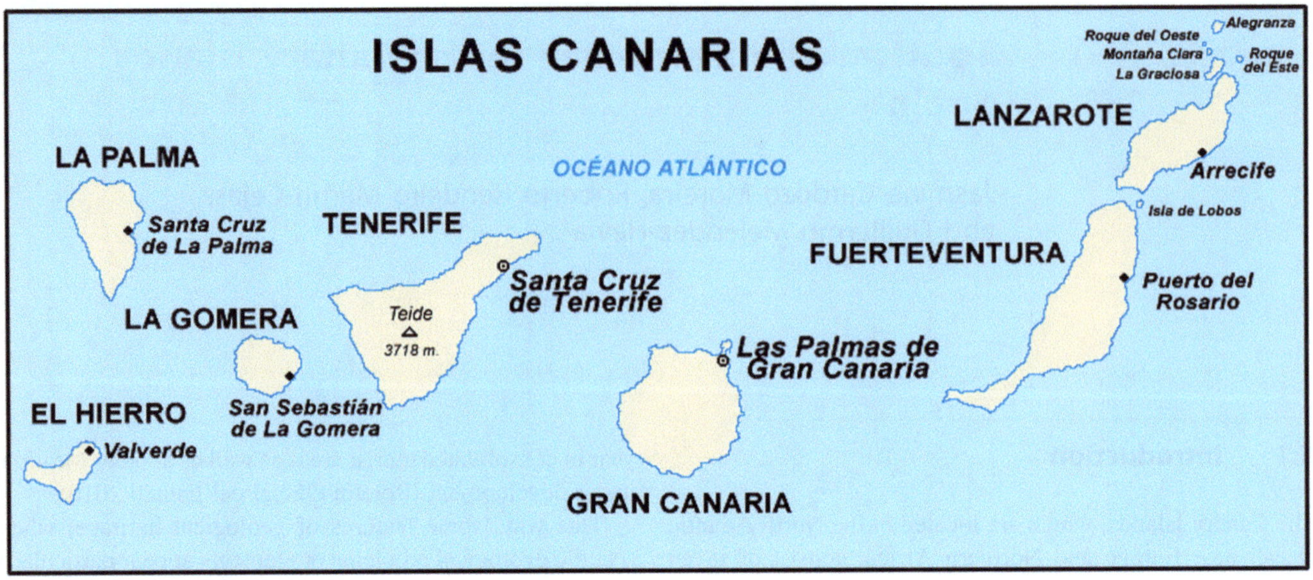

Fig. 3.1 Map of the Canary Islands. *Source* Wesisnay (2005)

The volcanically active alignments of these islands are situated in a band 200 by 500 km off the African continental margin opposite Cape Juby. The islands can be separated into three categories according to their eruptive activity (Carracedo 1994, 226):

1. The islands of Tenerife, La Palma, Lanzarote and probably El Hierro (Hernández-Pacheco 1982) which have had eruptions in historic time (<500 years) and are thus, by definition, volcanically active;
2. The islands of Fuerteventura and Gran Canaria with Quaternary volcanism;
3. The island of Gomera, where no evidence of Quaternary eruptions has been found (Cantagrel et al. 1984), although longer periods without eruptions on Lanzarote, Gran Canaria and Fuerteventura were followed by renewed activity.

Taking into account the risks, Carracedo (1994) says that one of the main factors invoked as hindering volcanic hazards assessment in the Canarian Archipelago is the apparent spatial dispersion of volcanism, in which an eruption may happen with the same statistical probability at any point in the archipelago. Volcanic hazards associated with the basaltic fissural-type eruptions of rift zones are generally of small magnitude. Damage is usually caused by tephra fall within a radius of a few hundred meters around the vent and by small volume and low velocity lava flows, which have courses that are closely controlled by the topography. An exception is the eruption that occurred in 1730 in Lanzarote, in wich 3–5 km^3 volcanic products were emitted over 6 years, covering 23 % of the island with pyroclastic ejecta and lava flows (Carracedo et al. 1992, Carracedo 1994). Thus, according to the characteristics of the volcanism of the Canary Islands and more recent manifestations, it can be said that the volcanic risk in the archipelago is relatively low (Araña and Ortiz 1989 *apud* Araña 2000).

3.2.2 Tourism and Mobility in the Canary Islands

Today, Canary's economy depends almost exclusively on tourism; more than 70 % of the jobs are directly and indirectly related to this sector. The average tourist stays on the islands for around 10 days, and the expenditure per tourist are about €40.40 per day. The main source countries are Great Britain (45 %), Germany (20.7 %) and Ireland (10.3 %) (Lanzarote 2010). Most of the island's terrestrial surface is considered a protected area. This is paradoxical because while the local government really tries to conserve the environment, it also allows continuous tourism development of coastal areas.

The rapid growth of urbanized areas in the Canary Islands is seriously degrading the natural ecosystem. The moratorium on tourism[1] has not been effective at controlling the development of tourism, and the environmental impact is still increasing. Sustainable development requires the economic and social needs of current generations to be met, without sacrificing the ability of future generations to achieve an acceptable quality of life. However, at present this principle is far from being met by the Canaries' tourism

[1] Tourism moratorium is merely a normative law to regulate the growth of tourism infrastructure in the region. A moratorium on the building of new hotels was declared in the Canary Islands at the end of the 1990s.

Table 3.1 National Parks of the Canary Islands and the total number of visitors in 2012

National park	Canary Island	Visitors in 2012	Percentage of total visitors to Spanish National Parks
Teide	Tenerife	2.660.854	27.90
Timanfaya	Lanzarote	1.546.003	15.46
Garajonay	La Gomera	744.304	7.80
Caldera de Taburiente	La Palma	424.832	3.72
Total		5.375.993	54.88

Source Red de Parques Nacionales 2014

development. The rapid growth in tourism and collateral services like surface transport makes evaluating the impact of the Canaries' transportation network on its environment crucial, in order to formulate alternative integrated tourism and transport planning in accordance with the principles of sustainability.

The constant flow of tourists exerts a great pressure on the need for mobility, and must substantially impact upon the environment. The relationship between tourism and transport goes beyond the mere objective of travelling from the departure point to the destination, as it also continues throughout all the tourism experience. Surface transport becomes crucial in tourism services, and at the same time is an important factor in its unsustainability.

3.2.3 Protected Areas: The Most Significant Geological Heritage Linked with Tourism

Based on the data in Table 3.1, it becomes apparent that the parks located in the Canary Islands in 2012 amounted to 5.375.993 visitors, compared to the combined total of 9.535.808 visitors of Spanish parks. The four parks Teide, Timanfaya, Garajonay and Caldera de Taburiente are therefore responsible for more than 50 % of the total number of tourists visiting the 14 National Parks of Spain, further confirming the visitor's interest in volcanic aspects.

According to Carcavilla et al. (2011), the analysis of the number of visitors reflects the power of attraction exercised by certain sites of geological interest. The most significant example is the Teide National Park (in Tenerife), a protected natural area for its high environmental significance and geological protagonism, that received three and a half million visitors in 2006 was more visited than any other Spanish monument, surpassing even the values of the Prado Museum (Madrid) and Guggenheim Museum (Bilbao) together. Thus, we present briefly some of the characteristics of Protected Areas and aspects relating to volcanism in the Canary Islands, which has over the years attracting numerous tourists.

3.3 Tenerife

Situated on the island of Tenerife, Teide National Park features the Teide-Pico Viejo stratovolcano that, at 3718 m, is the highest peak on Spanish soil (Fig. 3.2). Rising 7500 m above the ocean floor, it is regarded as the world's third-tallest volcanic structure and stands in a spectacular environment (UNESCO 2011a). The Park was created in 1954 and is one of the oldest parks in Spain. It was declared World Heritage Site in 2007. The Caldera de Las Cañadas is a depression 17 km across, bounded to the north-east, east and south by steep escarpments rising as high as 700 m. Above this, Mt Teide rises up to over 2000 m, beyond the influence of the moist trade winds. In this place, the climatic conditions are very different from those in the rest of the archipelago.

The Park is essentially configured by geographic elements that have very defined morphological and geological elements. Mt Teide is still active today and is a main attraction for tourism. The high fumarolic activity is evident in the areas near the crater. Remnants of historical eruptions from 1492, 1706, 1798 and 1909 are evident on Teide's slopes and from its central crater. The National Park is a paradigmatic enclave for geology and volcanology, both for its creation and history and for the great variety of volcanic materials that can be found, which allows for the observation and study of a broad range of processes and structures within a sheltered and clearly delimited space (Villalonga and Gomez 2006).

Other activities that are very popular are:

- Trails: There is an extensive network of trails and to climb up to Teide's crater (Telesforo Bravo trail, Fig. 3.3), permission is needed. The smell of sulfur is strong and there are numerous fumaroles;

Fig. 3.2 The Teide volcano is located within the Teide National Park in Tenerife and is one of the most visited National Parks in Spain with 2,407,480 visitors in 2010. *Photos* Jasmine Moreira

Fig. 3.3 Telesforo Bravo Trail can be used with special permission to reach the crater of Mount Teide. The Teide Cable Car is a modern ropeway that climbs up to the summit of Teide in 8 min. *Photos* Jasmine Moreira

- Teide Cable Car: This modern cable car climbs up to the Teide Cableway Terminal in 8 min (from 2,356 to 3,555 m), with a cabin that accommodates 35 people.

There are two visitor centres, one of them focused in the geological history of the region and the relations with environment life that inhabit it.

The Garajonay National Park is special because of the protected Laurel forest (a fossil forest), a subtropical flora that covered the lands of the Mediterranean through most of the Tertiary period. Garajonay's forest is a relic of this previously widespread floral type (IUCN 1985). The area was also declared a World Heritage Site in 1986 (Fig. 3.4).

3.4 La Gomera

As the only island in the Canaries La Gomera has not experienced an eruption in recent times. Thus, ash and lava fields have been eroded away leaving mature soils formed from horizontal basalts cut by a series of ravines. The local landscape is further characterized by volcanic dykes and domes.

3.5 Furteventura

Since 2009, the entire island has been declared a World Biosphere Reserve. Some of the protected areas are the Corralejo Natural Park, a large area of dunes, a solidified lava stream and the Montaña Roja volcanic cone. Isla de Lobos Natural Park is a small islet (around 5 sq. km),

Fig. 3.4 One of the outstanding landmarks of the Garajonay National Park is an extrusive dome, the 'Roque de Agando'. *Photo* Jasmine Moreira

situated 2 km northeast of Fuerteventura. With a beautiful seabed surrounding the isle, this is a Submarine Natural Reserve. In Malpaís de La Arena volcanic features can be observed, which are related to past eruptions around 10,000 years ago.

3.6 Las Palmas

This island was declared a World Biosphere Reserve in 2002, and contains a wide range of representative habitats with all the diversity provided by the transition from the coast up to the mountain peaks (UNESCO 2012). In the Caldera de Taburiente National Park the El Paso Visitor Centre offers exhibitions, audiovisual information, a library and a bookshop as well as a botanical garden. At La Cumbrecita, Lomo de Los Caballos and Roque de Los Muchachos there are additional information points. The Taburiente Service Centre is located in the heart of the Caldera and has an exhibition room with a scale model (Ministerio de Agricultura, Alimentación y Medio Ambiente 2013a).

3.7 El Hierro

El Hierro has some 500 volcanic cones, with another 300 covered by more recent deposits. According to Doniz-Paez et al. (2011), El Hierro receives a type of sustainable tourism related to its ecotourist offerings based on products like diving or hiking. However, the volcanoes on the island are not appreciated enough as a tourist attraction, even though El Hierro offers a different experience from other islands of the archipelago, which construct their offer on the sunshine, sea and sand model. The profile of El Hierro's visitors reveals a great sensitivity to nature and respect for the local population. Not only will this island be the first of the Canaries to base its ordinary power supply on renewable energy, El Hierro will probably also be the first Spanish island to join the Global Geoparks Network.

La Corona del Lajial Volcano is close to the main touristic centre on the island La Restinga, where the principal offer is diving in the Mar de Las Calmas. The recent submarine eruption (in progress from 10 October 2011 to March 2012, Fig. 3.5) near La Restinga village has impeded the main tourist attraction in El Hierro, diving (*id.* 2011).

Fig. 3.5 Aerial views of the submarine eruption at El Hierro in January 2012. *Source* Involcan (Instituto Volcanologico de Canárias)—Guardia Civil Helicópteros

3.8 Gran Canaria

The Gran Canaria Biosphere Reserve covers approximately a third of the island on the south-western side. From a geological standpoint, this Reserve is of exceptional value in understanding the process of formation and evolution of ocean islands. This Reserve includes all the catchment areas from the mountain tops down to the beaches and marine areas, through the valleys and farms. It includes also a considerable heritage of material assets (agricultural, architectural and craft structures, etc.) and intangible assets such as oral tradition and working songs (UNESCO 2011b).

3.9 Lanzarote

The processes involved in the formation of Lanzarote were complex and involved several periods of volcanism, followed by subsequent periods of erosion and decreased volcanic activity (Carracedo et al. 2003). The Timanfaya National Park was created in 1974 and the whole island was declared a Biosphere Reserve in 1993 with the Park as an example of recent volcanic activity in the Canary Islands. It contains geological features of considerable scientific value, including *'a'ā* and *pahoehoe* lava flows interspersed by islets of land that survived historic eruptions. Other features consist of volcanic cones, lava tubes, tumuli, domes or hornitos as well as geothermal phenomena (Ministerio de Agricultura, Alimentación y Medio Ambiente 2013b; Timanfaya National Park 2013). This Park also contains the Mancha Blanca Visitor Centre which offers a panoramic balcony, walkway facilities for the disabled, gift and bookshop, displays, a screening room with four different audiovisuals in several languages and a special effects room where visitors can see and hear the simulation of a volcanic eruption.

At Echadero de los Camellos the Museum Information Centre offers exhibitions about the traditional use of dromedaries and offers camel rides, which are equipped for physically and sensorially disabled visitors. The Termesana Route is a guided walking tour offered to only 8 people per group, on Mondays, Wednesday and Fridays (Fig. 3.6). The route, about 3 km long, illustrates how humans adapted their lifestyle after a large magnitude catastrophe. Traditional crops blend with a landscape dominated by the beauty

Fig. 3.6 The Termesana Route is limited to only two groups of visitors per day. While some tourists prefer hiking others are using camel rides at Echadero de los Camellos. *Photos* Jasmine Moreira

Fig. 3.7 Visitors entering in Cueva de Los Verdes (*left*) and a sculpture in Jameos del Agua by the famous local artist Cesar Manrique. *Photos* Jasmine Moreira

of its volcanoes, the lava flows and man's effort to survive after the eruptions. Another route is the Coastal Route, which can be used only one Wednesday per month (Ministerio de Agricultura, Alimentación y Medio Ambiente 2013b).

Located within the Timanfaya National Park is Montañas del Fuego (Mountains of Fire), the 'Centre for Art, Culture and Tourism' of Lanzarote's Insular Municipality. Here the visitors can take the 'Route of the Volcanoes' (a tour by bus only, 14.5 km), which fits perfectly into its surroundings and runs along the main centre of the eruptions, where concentrations of unusual elements are of great geological and geomorphic interest. This area also offers geothermal manifestations and the option of eating in a restaurant that uses residual heat from geothermal features to grill the food.

Two other important tourist attractions in Lanzarote are the Cueva de los Verdes and Jameos Del Agua, also managed by the Centre for Art, Culture and Tourism (Fig. 3.7). The Cueva is a volcanic tube over 6 km in length, which stretches down to the sea, and is the result of the volcanic eruption of the Volcán de La Corona (around 4000 years ago). The cave was used as a refuge by the natives of the island in times of invasion which were frequent in the 18th century. It was the first centre to be opened to the general public in 1964 and receives now 370,000 visitors per year (Fernandez and Fajardo 2008). The Jameos del Agua is part of the complex of cave galleries and was created, when the roof of the volcanic tube collapsed naturally, producing various holes or hollows (Jameos) connected by lava channels. There is a natural lagoon which is fed by the tidal seawater where a tiny endemic species of albino, blind crab exists (www.centrosturisticos.com/). The Jameos del Agua number of visitors per year is around 700,000 (Fernandez and Fajardo 2008).

3.10 Conclusion

Policy makers have a wide range of measures to intervene in the transport market to stimulate a more sustainable pattern of transport use; i.e. the weak sustainability principle. For instance, the government could introduce subsidies for cleaner modes of transport and tax more polluting forms. By considering the existing trade off between economical gains and environmental impacts, the government may, at least in theory, be able to reach the social optimum. The main issues in integrating transport planning into tourism policy are to maintain the objective of making the transport sector more sustainable, and to simultaneously allow growth in tourism (Martin-Cejas and Sanchez 2010).

Therefore alternative transport strategies become necessary to conciliate sustainable tourism development with the need of leisure mobility. The efforts to reduce transport intensity in the leisure segment can be made without major changes in the level of transport provision. For instance, the regulation of a minimal level of car occupancy for cars travelling to Timanfaya Park probably would perform better than price intervention. Public transport improvements, such as transferring leisure trips from private car to public bus service may be a suitable "soft measure". The advantage of promoting a modal shift to public transport, without reducing the numbers of tourists, is that it can be implemented at local government level. In addition, it will bring benefits to residents, visitors and the local ecosystem by reducing congestion, pollution, energy use and CO_2 emissions.

On the other hand, the relevance of the geological heritage of the Canary Islands as a touristic attraction seems to be beyond doubt, as is demonstrated by the visitors they receive annually and the contribution to regional economic development. Similarly, the interest of geological features and volcanic events as educational tools appears also clear, as every new geological event becomes a first page topic in the media. To further the knowledge of the local geological heritage and in its recognition as a tourism and economic resource, it seems to be necessary to foster the public concern about the need to protect this heritage through legal definition and a rational, sustainable use.

In conclusion, the possibility of establishing Geoparks on some of the smallest and protected islands could play an excellent role in building a holistic attitude towards nature conservation by encompassing attention to biotic (biodiversity) and abiotic (geodiversity) elements. Together with the concept of sustainable tourism the conservation of not only the natural aspects, but also the protection of cultural aspects of the Canary Islands would be ensured.

References

Araña, V. (2000). Volcanismo de las islas canarias. In M. Astiz, A. Garcia (Eds.), Curso Internacional de Volcanolgia y Geofísica. Lanzarote: Servicio de publicaciones del Exmo. Cabildo Insular de Lanzarote.

Araña, V., Ortiz, R. (1989). Riesgo volcánico en el archipelago canario. ESF Meeting on Canarian Volcanism. CSIC. Madrid, 247–253.

Boletin Oficial del Estado. (2010). Otras disposiciones. Resolucion de 20 de enero. 18-02-2010. 43(III), 15934. Retrieved August 19, 2013, from www.boe.es/boe/dias/2010/02/18/pdfs/BOE-A-2010-2677.pdf

Cantagrel, A., Cendrero, J. M., Fuster, J. M., Ibarrola, E., & Jamond, C. (1984). K-Ar chronology of the volcanic eruptions in the Canarian Archipelago: Island of Gomera. *Bulletin Volcanologique, 47*(3), 597–609.

Carcavilla, L., Belmonte, A., Durán, J. J., & Hilario, A. (2011). Geoturismo: Concepto Y Perspectivas em España. *Ensenanza de lãs Ciencias de La Tierra, 19*(1), 79–92.

Carracedo, J. C., Rodriguez B. E., & Soler, V. (1992). The 1730-1736 eruption of Lanzarote: an unusually long, high magnitude fissural basaltic eruption in the recent volcanism of the Canary Islands. *Journal of Volcanology and Geothermal Research, 53*, 239–250.

Carracedo, J. C. (1994). The Canary Islands: An example of structural control on the growth of large oceanic-island volcanoes. *Journal of Volcanology and Geothermal Research, 60*, 225–241.

Carracedo, J. C., Day, S., Guillou, H., Rodríguez Badiola, E., Canas, J. A, & Pérez-Torrado, F. J. (1998). Hotspot volcanism close to a passive continental margin: The Canary Islands. *Marine Geology, 135*(5), 591–604.

Carracedo, J. C., Singer, B., Jicha, B., Guillou, H., Rodríguez Badiola, E., Meco, J., et al. (2003). La erupción y el tubo volcánico del Volcán Corona (Lanzarote, Islas Canarias). *Estudios Geológicos, 59*, 277–302.

Doniz-Paez, J., Becerra-Ramirez, R., Gonzalez-Cárdenas, E., Guillén-Martín, C., & Escobar-Lahoz, E. (2011). Geomorphosites and geotourism in volcanic landscape: The example of La Corona de Lajial Cinder Cone (El Hierro, Canary Islands, Spain). *Geojournal of Tourism and Geosites, 8*(2), 185–197.

Fernandez, H., & Fajardo, V. (2008). The impact of museum exhibitions and heritage sites in Lanzarote: Visitor communication. In H. Fernandez-Betancourt (Ed.), *Tourism, heritage and informal learning: Museums as E-labs (emotions and education)*. Lanzarote: Escuela Universitaria de Turismo de Lanzarote.

Hernández-Pacheco, A. (1982). Sobre una posible erupción en 1793 en la isla del Hierro (Canarias). *Estudios Geológicos, 38* (1–2): 15–25.

IUCN. (1985). *World Heritage Nomination—IUCN Technical Report—Garajonay National Park.* Retrieved August 19, 2013, from http://whc.unesco.org/archive/advisory_body_evaluation/380.pdf

Lanzarote. (2010). *Anuario Estadístico de Lanzarote*. Edited by the data centre in Cabildo de Lanzarote.

Martin-Cejas, R., & Sanchez, P. P. R. (2010). Ecological footprint analysis of road transport related to tourism activity: The case for Lanzarote Island. *Tourism Management, 31*, 98–103.

Ministerio de Agricultura, Alimentación y Medio Ambiente (2013a). *Parque Nacional Caldera de Taburiente.* Retrieved August 19, 2013, www.reservasparquesnacionales.es/real/parquesnac/usu/html/listado-actividades-oapn.aspx?cen=4

Ministerio de Agricultura, Alimentación y Medio Ambiente (2013b). *Parque Nacional Timanfaya*. Retrieved August 19, 2013, from www.reservasparquesnacionales.es/real/parquesnac/usu/html/listado-actividades-oapn.aspx?cen=6

Red de Parques Nacionales: Visitantes (2014) Datos de visitantes a los Parques Nacionales. Retrieved April 14, 2014, from www.magrama.gob.es/es/red-parques-nacionales/la-red/gestion/visitantes_tcm7-280140.pdf

Timanfaya National Park. (2013). Retrieved August 19, 2013, from www.discoverlanzarote.com/timanfaya.asp

UNESCO (2011a). *Teide National Park*. Retrieved August 19, 2013, from http://whc.unesco.org/en/list/1258/

UNESCO (2012). La Palma. Retrieved August 19, 2013, from www.unesco.org/new/en/natural-sciences/environment/ecological-sciences/biosphere-reserves/europe-north-america/spain/la-palma/

UNESCO (2011b). *Gran Canaria*. Retrieved August 19, 2013, from www.unesco.org/new/en/natural-sciences/environment/ecological-sciences/biosphere-reserves/europe-north-america/spain/gran-canaria/

Villalonga, M.D. & Gomez, J.R. (2006). *Proposal to inscribe Teide National Park on the world heritage list*. Gobierno de Canarias. Retrieved August 19, 2013, from http://whc.unesco.org/uploads/nominations/1258.pdf

Wesisnay. (2005). *Mapa de las Islas Canarias*. Retrieved August 19, 2013, from http://commons.wikimedia.org/wiki/File:Mapa_de_Canarias.png

The Azores Archipelago: Islands of Geodiversity

4

João Carlos Nunes

4.1 Introduction

The Azores Archipelago (a Portuguese Autonomous Region) is located in the North Atlantic at the distance of 1,815 km from the Mainland Portugal, and is formed by nine islands and several islets, which are dispersed along a strip with 600 km length and with a WNW-ESE trend. The archipelago lies on the triple junction between the North American, Eurasian and African (or Nubian) plates (Laughton and Whitmarsh 1974; Searle 1980; Miranda et al. 1998; Vogt and Jung 2004) and, in general terms, the intervening plate boundaries are the Mid-Atlantic Rift (MAR), separating the American plate from the Eurasian and African plates, and the Azores-Gibraltar Fault Zone (AGFZ), bounding the latter two plates (Fig. 4.1).

In the Azores area, the MAR trends N-S to N20E and is divided by transform faults into seven short segments (Luis et al. 1994). The western AGFZ segment in the Azores region includes the east-west GLORIA Fault and a WNW-ESE general trend area (Fig. 4.1) oblique to the spreading direction and allowing magmatic intrusion along faults, feeding the volcanism that built the islands (Lourenço et al. 1998). Volcanic and tectonic activities are well displayed in the geomorphology of the islands, the former includes 26 eruptions since the settlement of the islands in the early fifteenth century (Zbyszewski 1963; Weston 1963/64; Queiroz et al. 1995; Madeira and Brum da Silveira 2003). Earthquakes reaching magnitude 7 that caused about 6,350 deaths, and successive pre-historic surface fault ruptures that produced well-developed fault scarps, represent recent seismotectonic and neotectonic activity (Madeira and Ribeiro 1990; Nunes et al. 2001; Madeira and Brum da Silveira 2003; Nunes 2008).

All the Azores islands are of volcanic origin and are all oceanic islands that emerged from the surrounding seafloor due to the progressive build-up of submarine volcanic products, a deep-sea process that could have started about 36 million years ago. Thus, the Azores islands emerge from the Azores *Plateau* (or Azores Platform) defined by the 2,000 m bathymetric line (Fig. 4.1) and which makes the transition to the surrounding abyssal seafloor. The oldest terrestrial volcanism (8.12 million years) is evident in outcrops on Santa Maria Island while Pico is the youngest island of the archipelago that emerged about 300,000 years ago.

There are 27 main volcanic systems in the Azores Islands with 16 major central volcanoes (most of them silicic and with summit subsidence calderas) and 11 volcanic ridges associated with basaltic fissure volcanism. Among these, 9 major polygenetic volcanoes and 7 areas of basaltic fissure volcanism are classed as active (although currently dormant) and are located on the islands of São Miguel, Terceira, Graciosa, São Jorge, Pico and Faial and the Dom João de Castro Bank, a large submarine volcano. Located offshore near the islands are located important active submarine volcanic ridges including the Monaco Bank (south of São Miguel), the Princes Alice Bank (SW of Faial) and the Serreta Ridge (W of Terceira), where in 1998–2001 a 'Serretian' type eruption took place (Forjaz et al. 2001), the last one in the Azores region.

Moreover, there are about 1750 monogenetic volcanoes in the archipelago, either dispersed along the flanks and inside the summit depression of the polygenetic volcanoes, or belonging to the 11 basaltic fissure volcanic systems located in different islands (Nunes and Lima 2008). These monogenetic eruptive centres include domes and *coulées*, tuff rings and tuff cones, *maars*, scoria and spatter cones, and eruptive fissures. Many of those features may be considered as geosites, together with other volcanic structures (such as historical eruptive centres and products, hydrothermal fields, pillow lava and prismatic jointing outcrops, volcanic caves

J. C. Nunes (✉)
Department of Geosciences, University of the Azores,
Azores Geopark Scientific Coordinator, Ponta Delgada, Portugal
e-mail: jcnunes@uac.pt

P. Erfurt-Cooper (ed.), *Volcanic Tourist Destinations*, Geoheritage, Geoparks and Geotourism,
DOI: 10.1007/978-3-642-16191-9_4,

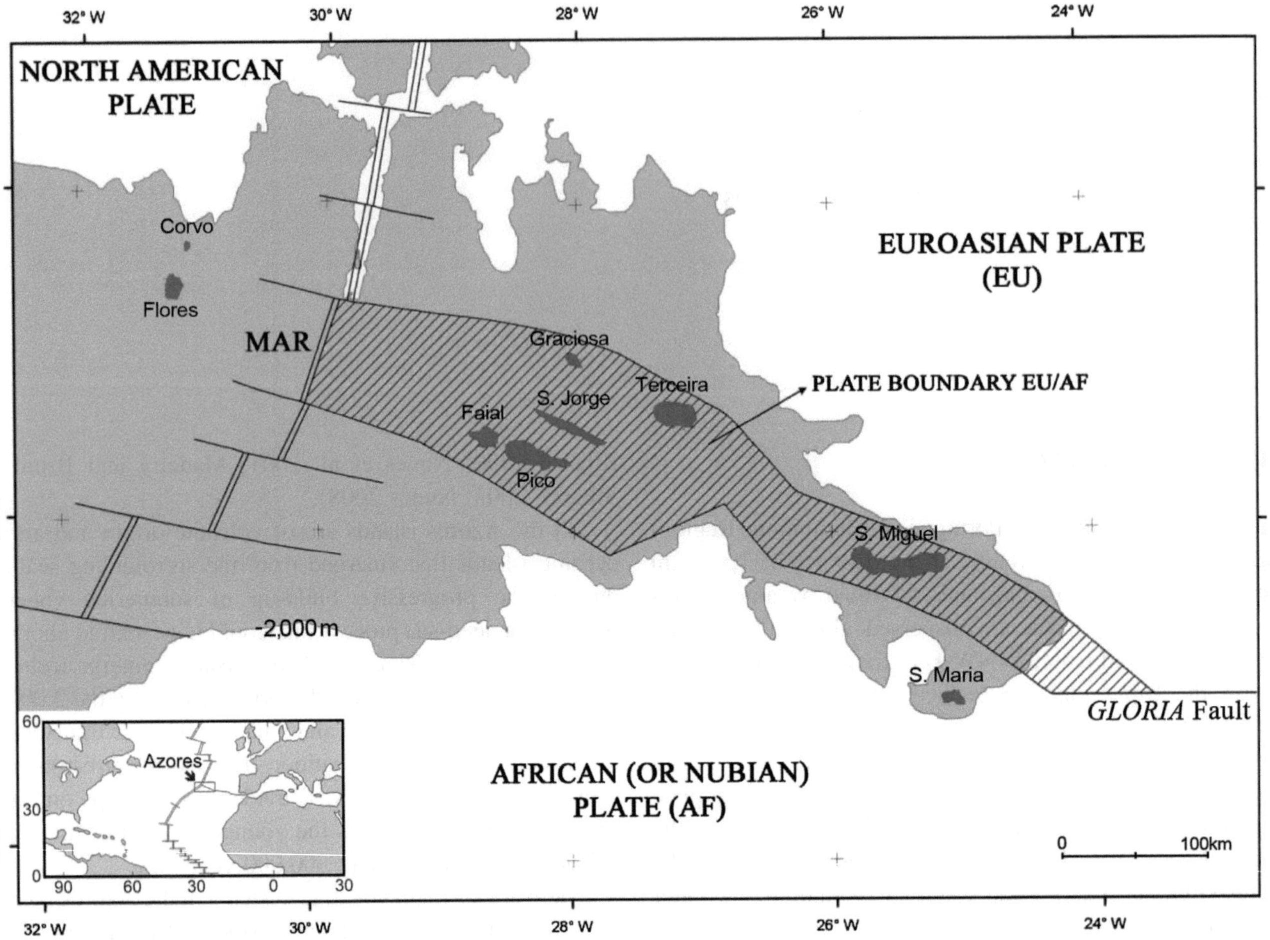

Fig. 4.1 General geotectonic framework of Azores Archipelago. MAR: Mid-Atlantic Rift. © J. C. Nunes

and primary pyroclastic deposit exposures), tectonic structures (fault scarps, sag ponds), sedimentary deposits (fossiliferous marine deposits of Miocene to Quaternary age, flood deposits, secondary lahars), and littoral features (e.g. littoral platforms of volcanic or landslide origin—locally called 'fajãs'). Additionally, some offshore sites are also worth mentioning, such as the Lucky Strike and Menez Gwen black smokers associated with this deep-sea hydrothermal fields and the Dom João de Castro Bank seamount volcano that erupted in 1720 A.D. and whose summit (at a depth of 12 m) presents an impressive field of fumaroles (Nunes et al. 2003).

Thus, the Azores may be considered a natural laboratory of international relevance with regard to active volcanism, volcanic and tectonic landforms, global plate tectonics, and neotectonics (Figs. 4.2 and 4.3). The archipelago displays varied and abundant geological features of scientific, educational, scenic, socio-cultural and economic (touristic) interest, both on the islands and at sea, whose intrinsic value was evaluated during the preparation of the Azores Geopark project (Lima 2007; Nunes et al. 2009, 2011; Nunes and Lima 2013).

4.2 The Azores Islands Geology: A Brief Summary

4.2.1 Santa Maria Island

Santa Maria Island has an area of 97 km^2 and is part of the Eastern Group of the archipelago, together with São Miguel Island and the Formigas Islets, the latter located about 37 km northeast from Santa Maria Island.

Santa Maria is the southern and easternmost island of the archipelago, was the first one to be discovered and settled and was also the first to be formed. Therefore the island can be considered the geological birth place of the archipelago which emerged about 10 million years ago at the Baia dos Cabrestantes area, where the older rocks of the Azores are still visible as outcrops (Fig. 4.4).

Showing a volcanic origin like the other Azorean islands, in spite of not presenting active volcanism, Santa Maria includes several volcanic complexes, almost all of basaltic rocks, either associated with submarine or terrestrial volcanism. Besides the oldest geological formations of the archipelago, Santa

Fig. 4.2 Examples of volcanic features and landforms from the Azores Islands. **a** Lava delta and tuff cone (Velas, São Jorge Island). **b** Prismatic jointining (Ribeira do Maloás, Santa Maria Island). **c** Maar (Caldeira Negra, Flores Island). **d** Subsidence caldera (Caldeirão, Corvo Island). **e** Stratovolcano (Pico Mountain, Pico Island). **f** Boiling water-type fumarole (Furnas, São Miguel Island). **g** Silica stalactites (Algar do Carvão volcanic pit, Terceira Island). **h** Slumping marks (Monte da Guia, Faial Island). **i** Volcanic neck (Ilhéu da Baleia, Graciosa Island). © *Photos* J. C. Nunes

Fig. 4.3 Examples of tectonic features of the Azores Islands. **a** Normal faulting in a plinian pumice type sequence (Fogo Volcano, São Miguel Island). **b** Fault scarps of Pedro Miguel graben (Faial Island). **c** Lagoa do Capitão fault scarp and sag pond (Pico Island). © *Photos* J. C. Nunes

Fig. 4.4 The Surtseyan tuffs at the Baia dos Cabestantes on Santa Maria Island are the oldest rocks of the Azores archipelago and are covered by basaltic lava flows (*top left*). © J. C. Nunes

Maria island exhibits unique and distinctive characteristics, such as (i) major outcrops of pillow lavas, (ii) important areas with consolidated sedimentary rocks, including limestone, sandstone, claystone and conglomerates, and (iii) a remarkable fossil content in many of the sedimentary rocks.

4.2.2 São Miguel Island

São Miguel Island is part of the Eastern Group and is the biggest island of the archipelago with a 747 km^2 surface area and about 56 % of the Azores population.

In general terms São Miguel Island's volcanism is associated with four main silicic polygenetic volcanoes, each with a summit caldera (Povoação, Furnas, Sete Cidades and Fogo central volcanoes) and two areas of exclusively basaltic volcanism (Fig. 4.5). The latter display different ages: the Nordeste volcano complex is the oldest on the island, and the Picos complex is the most recent, occupying the area between Relva-Lagoa (on the South coast) and Capelas-Ribeira Grande (on the North coast).

About 500 monogenetic volcanoes, 35 lakes of different size and a wide variety of mineral and thermal waters and hydrothermal fields are some testimonies of the geodiversity of São Miguel Island. Since the beginning of human settlement several eruptions have occurred, like the 1563 eruption (Fogo Volcano), the 1630 eruption (Furnas Volcano), and the Pico do Fogo, 1652 eruption, at the Picos Volcanic Complex. Besides these, submarine eruptions have occurred offshore São Miguel Island, the most famous being the Sabrina Island eruption, which took place offshore Ponta da Ferraria in 1811.

4.2.3 Terceira Island

Terceira Island is part of the islands of the Central Group, and with a 400 km^2 area is the second most populated of the archipelago. Terceira is composed of four polygenetic volcanoes with caldera (Cinco Picos, Guilherme Moniz, Santa Bárbara and Pico Alto central volcanoes) and by a basaltic fissure volcanism area, split in two main sectors of the island. The geology of the island is marked by impressive silicic effusive volcanism products, occurring as numerous domes and thick *coulées*, often with obsidian (Fig. 4.6).

The historical volcanism dates back to the eighteenth century, with terrestrial eruptions in 1761 in the Mistérios Negros and in the Basaltic Fissural Zone, and submarine eruptions in 1867 and 1998–2001, the latter on the Serreta submarine ridge, northwest of Terceira Island (França et al. 2003). The submarine volcano of Serreta, which was active from 1998 to 2001, was the last volcanic episode of the archipelago and an important milestone for the international volcanology community, as it allowed the first observations and interpretation of Serretian-type eruptions.

The Dom João de Castro Bank is an important active seamount volcano, located about 63 km southeast of Terceira Island, with an impressive submarine fumarolic field, and was an ephemeral island following the 1720 A.D. eruption.

4.2.4 Graciosa Island

Graciosa is the northernmost island of the Central Group and the second smallest island of the archipelago (with an area of 61 km^2). It is also the one with the lowest altitude

Fig. 4.5 Furnas volcano caldera (São Miguel Island) is the most relevant terrestrial geosite of the Azores archipelago. © J. C. Nunes

Fig. 4.6 Santa Barbara volcano at Terceira Island with its summit nested caldera and flank domes and coulées. © M. P. Costa

of the archipelago, displaying only 405 m at its highest point.

Graciosa Island includes the smallest polygenetic volcano of the Azores (the Caldeira Volcano) which takes up most of its southeast side. In contrast, the northwest sector of the island is dominated by a field of 32 basaltic scoria cones and their associated lava flows.

After discovery and settlement of the island no volcanic eruptions were reported, and the last volcanic eruption was associated with the Pico Timão scoria cone, less than

Fig. 4.7 Furna do Enxofre volcanic cave at Graciosa Island. © GESPEA and J. Góis

2,000 years ago. Besides this Holocene eruption, the active volcanism of the island is well expressed in its fumarolic fields, the most important one being located inside the Furna do Enxofre volcanic cave (Fig. 4.7).

4.2.5 São Jorge Island

São Jorge is the most central island of the archipelago with an area of 244 km^2 and is characterized by its elongated shape, being the second biggest island in length after São Miguel.

The distinctive geological features of the island are: (i) its exclusive basaltic volcanism, (ii) the absence of a central volcano, and (iii) being an island with a long volcanic ridge, composed of about 350 cones (mostly scoria cones) and associated basaltic lava flows (Fig. 4.8). The smooth and flatten morphology east of the Ribeira Seca stream shows the older age of this sector of the island, where 1.3 million years old lava flows outcrop.

Besides the regional tectonic control that influences its shape, the volcanic landscape of the island is dominated by the several 'fajãs' that are emplaced at the foothills of the high sea cliffs of the island (either as detritic slope deposits or as lava deltas), which are the main touristic exhibits of the island.

The volcanism of the island includes also some submarine episodes, such as those responsible for the genesis of the Morro Grande de Velas and Morro de Lemos Surtseyan tuff cones. Historical eruptions took place on this island in 1580 (at three different eruptive centres) and in 1808. More recently in 1964 a submarine eruption took place off the coast of Velas.

4.2.6 Pico Island

Pico is the biggest island of the Central Group, the second largest of the archipelago (445 km^2) and exhibits the highest point of Portugal (2,350 m), the Pico Mountain volcano, which is the 3rd highest volcano of the North Atlantic Ocean. In relation to the surrounding sea-floor Pico Mountain is a 3,500 m high stratovolcano, characterized by dominant effusive basaltic volcanism episodes of Hawaiian type eruptions and extensive pahoehoe lava fields (Fig. 4.9).

The main geological features of the island are related to the fact that Pico is the youngest island of the archipelago (about 300,000 years old) and its volcanism is essentially basaltic, evident either in the Pico Mountain volcano, the Topo shield volcano or in the volcanic ridge of the Planalto da Achada. This latter area has a length of 30 km and is composed of about 190 scoria and spatter cones and eruptive fissures.

Since the island's settlement there have been historical eruptions in 1562/64, in 1718 and in1720. It is important to mention that the 1718 eruption had two subaerial eruptive centres and a third submarine vent offshore the southern coast of the island. The last eruption took place in 1963 off the northern coast of the island, north of Cachorro, being a serretian-type submarine eruption.

4.2.7 Faial Island

Faial, with an area of 173 km^2, is the westernmost island of the Central Group and the one closest to the Mid-Atlantic Ridge, located about 120 km east of the MAR (Fig. 4.10).

Fig. 4.8 Volcano-tectonic alignment of scoria cones (*foreground*) and Morro Grande de Velas Surtseyan tuff cone peninsula (*top left*) at São Jorge Island. Pico Island is in the background. © J. C. Nunes

Fig. 4.9 Basaltic pahoehoe lava field at the UNESCO's site 'Landscape of the Pico Island Vineyard Culture'. Note the 'rilheiras' excavated by the wheels of vehicles pulled by animals. © J. C. Nunes

Generally speaking, the islands' volcanism is associated with the presence of two main central volcanoes (the Caldeira and Ribeirinha volcanoes) and two areas of markedly fissural basaltic volcanism (the Horta Basaltic Zone and the Peninsula of Capelo).

The polygenetic volcano of Caldeira dominates all of the central part of the island and in recent times is characterized by explosive eruptions of trachytic nature, with the extrusion of large volumes of pumice. The summit of the volcano is occupied by a 2 km diameter, 470 m deep and 10,000 years old caldera.

Besides the Ribeirinha shield volcano, the eastern part of the island of Faial is dominated by an important tectonic structure (the *Graben* of Pedro Miguel), with WNW-ESE

Fig. 4.10 Costado da Nau fossil sea cliff, the westernmost Faial Island shoreline prior to the Capelinhos eruption (1957–58 A.D.). © J. C. Nunes

trending active faults that deeply shape the volcanic landscape of the island.

Faial Island has seen two historical eruptions, in 1672/73 and in 1957/58, the latter at Capelinhos and inside the Caldeira. The Capelinhos eruption increased the size of the island by 2.4 km^2 (nowadays only 0.6 km^2), and represents a historical mark in the World volcanology and for Faial's population and society.

4.2.8 Flores Island

Flores Island, with 141 km^2, is part of the Western Group of the archipelago and is the westernmost territory of Europe (Fig. 4.11). While it does not exhibit a major central volcano of large dimensions (like on the neighbouring island of Corvo), the most significant characteristic of this island is the presence of several explosion craters associated with hydromagmatic eruptions, which caused the formation of maars (e.g. Lagoa Funda and Lagoa Seca) and tuff rings (e.g. Caldeira Branca).

Also, the important drainage basins existing on this island (e.g. Badanela and Santa Cruz creeks basins) and the presence of several ancient volcanic cones, have shaped a landscape marked by several volcanic necks and outcropping dikes, the latter as walls that intersect the nearby landscape.

Along the shoreline are several sea erosion caves and well preserved prismatic and spheroidal jointing of lava flows. The Rocha dos Bordões is an excellent example of prismatic jointing in the 570,000 years old mugearitic lava flows.

4.2.9 Corvo Island

Corvo, with only 17 km^2 in area, is the smallest island of the Azores and together with Flores forms the Western Group of the archipelago. In spite of its small area, the island displays an interesting geodiversity, expressed namely by a wide diversity of rocks that includes basalts, trachytes, pumice, scoria, ignimbrites, etc.

The entire island is consistent with a polygenetic volcano with a summit caldera (named Caldeirão) and can therefore be considered an island-volcano. This volcanic depression has a 2.1 km average diameter and is occupied by a lake with a maximum depth of about 2 m. This central volcano has several secondary cones, either on its flanks like the Cova Vermelha scoria cone, or inside the caldera, like the Montinho do Queijo spatter cone.

Due to the marine erosion the island undergoes, the nature of its volcanic products and the fact that this island does neither show historic volcanism nor any recent volcanic activity (e.g. during the last thousands of years), the shoreline of the island of Corvo is high and steep. A single exception is the Vila do Corvo lava delta, where the only community of the island is emplaced and where took place the last eruption that occurred in the island, at the Pão do Açucar-Moinhos area (Fig. 4.12).

Fig. 4.11 The western coastline of Flores Island, which is the westernmost European territory. © J. C. Nunes

Fig. 4.12 Vila do Corvo lava delta at Corvo Island. Flores Island is in the background (*top left*). © E. A. Lima

4.3 The Azores Geopark and Geotourism

The international relevance of the Azorean geodiversity, the high number and quality of its geosites and the undoubted importance of its geological heritage (Brilha et al. 2005; Lima 2007; Nunes et al. 2011; Nunes and Lima 2013), together with a rich biological and cultural heritage, supported a major effort of the Azores Government to implement Geoconservation and Environmental Education policies. This resulted in the creation of the Azores Geopark and its application to the UNESCO's Global Geopark Network.

The Azores Geopark, the 53rd geopark of the European Geoparks Network after March 2013, is also supported on strategic decisions of the regional and local authorities to develop Nature Tourism policies based on the most effective tourism icon of the Azores: its volcanoes and volcanic landscape.

Despite its small land surface (2,324 km^2), the archipelago offers a wide diversity of landforms, together with different types of rocks, structures and features, which derive namely from the type of eruption they originated from, its dynamics, the nature of the magmas, and the subsequent actions of weathering and erosion processes, and thus constitutes a natural laboratory of volcanic geodiversity.

The Azores archipelago includes 121 geosites, dispersed over the nine islands and the surrounding seafloor, with relevant scientific, educational and touristic value. Thus, volcanoes, calderas, lakes, lava fields, fumaroles, hot springs and thermal waters, volcanic caves, 'fajãs', fault scarps and marine fossil deposits, among many others, belong to this network of geosites.

Among those, 57 are considered as priority geosites for the Azores Geopark, and include 6 geosites of international relevance: the Mid-Atlantic Rift and associated deep-sea hydrothermal fields, the Furnas Volcano caldera, the Pico Mountain Volcano, the Graciosa Caldeira and Furna do Enxofre volcanic cave, the Capelinhos Volcano and the Algar do Carvão volcanic pit.

Given the insular nature of the region, the Azores Geopark is supported on a geosites network dispersed over the nine islands and the surrounding seafloor, (i) that ensures the representativeness of the geodiversity that characterizes the Azorean territory, (ii) that reflects its geological and eruptive histories, (iii) with common conservation and promotion strategies, and (iv) is based on a decentralized management structure with support of all the islands.

Besides the geological heritage of the Azores Islands, there are other values of reference in the archipelago, such as its rich biodiversity and the architectural, cultural, ethnographic and immaterial heritage of undeniable value. The Azorean geo-landscapes are thus the main *ex-libris* of the archipelago, with huge potential for tourism due to its attributes of great attractiveness and good opportunities for sustainable use.

Following the "Strategic Plan of Marketing for Tourism in the Azores—2008/2010" coordinated by the Azores Tourism Association (ATA) and the Azores Government, a geotouristic strategy was implemented based on two main principles: (i) the development of intra and inter-islands routes, and (ii) the implementation of dissemination actions accompanied by monitoring actions (Nunes et al. 2010a; Nunes et al. 2010b). The main routes proposed for the development of the Azores Islands geotourism include:

(a) The Volcanic Caves Route—to discover the subterranean world of the islands, valuing the volcanic caves and associated visitor centers, such as 'Gruta das Torres' (Pico island), 'Gruta do Carvão' (São Miguel island), 'Furna do Enxofre' (Graciosa island) and 'Algar do Carvão' and 'Gruta do Natal' (Terceira island);
(b) The Belvederes Route—to discover, by car, the Azores volcanic landscapes, valuing the numerous belvederes ('Miradouros' in Portuguese), and viewpoints that exist on all islands of the archipelago, often superbly located, with supporting infrastructure (e.g. benches, fences, parking areas, restrooms) and in general impeccably maintained;
(c) The Walking Trails Route—to discover, by foot, the Azores geosites, enhancing the Regional Network of Walks and other trails in Azores, with more than 100 different options;
(d) The Thermal Route—to discover and enjoy the power of Azores volcanism, taking advantage of the benefits in terms of health, leisure and well-being of the Azorean thermal waters and mud, as in Ferraria, Furnas, Caldeiras da Ribeira Grande, Caldeira Velha and Poça da D. Beija (São Miguel island), Carapacho (Graciosa island), Varadouro (Faial island), and Furnas do Enxofre (Terceira island);
(e) The Science and Interpretation Centers Route—to learn and interpret the Azorean volcanic phenomena, valuing the 25 science, interpretation and visitors centers existing on all the islands, such as the Capelinhos Volcano Interpretation Center (Faial island), the Mountain House (Pico island) or the Astronomical Observatory and the ExpoLAB (São Miguel island);
(f) The Urban Georoutes—to discover the geology of my village/town, that allow local population and visitors to see and learn about the rocks in buildings and monuments and to understand about the geology of urban areas.

The promotion and development of this geoturistic strategy and all these routes is a key issue and a priority goal for the Azores Geopark, taking into account the rich Azorean geodiversity, the high number and quality of its geosites and the undoubted importance of its geological heritage.

Thus, visitors and tourists are welcomed to use the walking trails, visit viewing points, belvederes and tea plantations, go whale-watching, dive in the blue ocean, go bird-watching, enjoy the regional handicraft, taste the geothermal Furnas stew (cooked in the ground at a fumarolic field), taste the regional sweets and wines or take a bath in warm waters… while visiting and being part of a fascinating tour in an active volcanic landscape.

The slogan: "Come to meet the Azorean volcanoes *and enjoy an eruption… of Flavours, Smells and Experiences"*, says it all.

References

Brilha, J., Andrade, C., Azerêdo, A., Barriga, F. J. A. S., Cachão, M., Couto, H., et al. (2005). Definition of the Portuguese frameworks with international relevance as an input for the European geological heritage characterisation. *Episodes, 28*(3), 177–186.

Forjaz, V. H., França, Z. & Nunes, J. C. (2001). Serretian, a new type of submarine eruptions. In *Cities on Volcanoes 2—Abstracts Book* (p. 39). Poster: Auckland, New Zealand.

França, Z., Cruz, J. V., Nunes, J. C., & Forjaz, V. H. (2003). Geologia dos Açores: uma perspectiva actual. *Açoreana, 10*(1), 11–140.

Laughton, A. S., & Whitmarsh, R. B. (1974). The Azores-Gibraltar plate boundary. In L. Kristjansson (Ed.), *Geodynamics of Iceland and the North Atlantic Area* (pp. 63–81). Dordrecht: D. Reidel.

Lima, E. A. (2007). *Património geológico dos Açores: Valorização de Locais com Interesse Geológico das Áreas Ambientais, Contributo para o Ordenamento do Território*. Tese de Mestrado em Ordenamento do Território e Planeamento Ambiental. Departamento de Biologia. Universidade dos Açores.

Lourenço, N., Miranda, J. M., Luis, J. F., Ribeiro, A., Mendes-Victor, L. A., Madeira, J., et al. (1998). Morpho-tectonic analysis of the Azores Volcanic Plateau from a new bathymetric compilation of the area. *Marine Geophysical Researches, 20*, 141–156.

Luis, J. F., Miranda, J. M., Galdeano, A., Patriat, P., Rossignol, J. C., & Mendes-Victor, L. A. (1994). The Azores triple junction evolution since 10 Ma from an aeromagnetic survey of the Mid-Atlantic Ridge. *Earth and Planetary Science Letters, 125*, 439–459.

Madeira, J. & Brum Da Silveira, A. (2003). Active tectonics and first paleoseismological results in Faial, Pico and S. Jorge islands (Azores, Portugal). In D. Pantosti, K. Berryman, R. S. Yeats, & Kinugasa, Y. (Eds.), Ten Years of Paleoseismology in the ILP: Progress and Prospects, *Annals of Geophysics, 46* (5), 733–761.

Madeira, J., & Ribeiro, A. (1990). Geodynamic models for the Azores triple junction: a contribution from tectonics. *Tectonophysics, 184*(3/4), 405–415.

Miranda, J. M., Mendes-Victor, L. A., Simões, J. Z., Luis, J. F., Matias, L., Shimamura, H., et al. (1998). Tectonic setting of the Azores Plateau deduced from a OBS survey. *Marine Geophysical Researches, 20*, 171–182.

Nunes, J. C. (2008). Caracterização sumária da sismicidade da Região dos Açores. In C. S. Oliveira, A. Costa & J. C. Nunes (Eds.), *Sismo 1998—Açores. Uma Década Depois* (pp. 59–72). Orion: Governo dos Açores/SPRHI, SA.

Nunes, J. C. & Lima, E. A. (2008). Paisagens Vulcânicas dos Açores: Valor Intrínseco Enquanto Recurso Natural e Património Geológico. *Livro de Resumos—IV Congresso Nacional de Geomorfologia* (p. 31). Braga: Outubro.

Nunes, J. C. & Lima, E. A. (2013). Ranking and classifying the Azores Islands geosites. In *Rendiconti Online della Società Geologica Italiana, Special Issue for the 12th European Geoparks conference* (Vol. XX, p. 4), Ascea, Italy. (In Press).

Nunes, J. C., Forjaz, V. H. & França, Z. (2001) Principais sismos destrutivos no arquipélago dos Açores—uma revisão. In M. R. Fragoso (Ed.), *5° Encontro Nacional de Sismologia e Engenharia Sísmica.* Ponta Delgada (Açores) (pp. 119–131). Funchal: Laboratório Regional de Engenharia Civil.

Nunes, J. C., Forjaz, V. H., Alves, J. L. & Bernardes, A. C. (2003) Caracterização vulcanológica do Banco D. João de Castro (Açores): novos dados. *Ciências da Terra (UNL), n° esp.* V, CD-ROM: D55–D58.

Nunes, J. C., Madeira, J. & Brilha, J. (2009). The Azores archipelago in the America-Eurasia-Africa triple junction: A geological framework as the scientific background for the creation of the Azores Geopark. In C. N. Carvalho & J. Rodrigues (Eds), *Proceedings 8th European Geoparks conference* (pp. 176–179). Idanha-a-Nova, Geopark Naturtejo: Setembro. (14–16).

Nunes, J. C., Lima, E. A., Costa, M. & Porteiro, A. (2010a). Azores Islands geodversity and geological heritage: contribution to tourism development in the Autonomous Region of The Azores. In M. L. Rodrigues & M. E. Freire (Eds.), *Proceedings of the international conference on Geoheritage and Geotourism* (pp. 63–64). Lisbon: APGeotur & CEG-University.

Nunes, J. C., Lima, E. A., Costa, M. P. & Porteiro, A. (2010b). Azores Islands volcanism and volcanic landscapes: its contribution to the geotourism and the Azores Geopark Project. *Geociences On-line Journal, 18*(16), 4.

Nunes, J. C., Lima, E. A., Ponte, D., Costa, M. P. & Castro, R. (2011). Azores Geopark Application. Available at www.azoresgeopark.com, p. 50.

Queiroz, G., Gaspar, J. L., Cole, P. D., Guest, J. E., Wallenstein, N., Duncan, A. M., et al. (1995). Erupções vulcânicas no vale das Furnas (ilha de São Miguel, Açores) na primeira metade do século XV. *Açoreana, 8*(1), 159–165.

Searle, R. (1980). Tectonic pattern of the Azores spreading centre and triple junction. *Earth and Planetary Science Letters, 51*, 415–434.

Vogt, P. R., & Jung, W. Y. (2004). The Terceira Rift as hyper-slow, hotspot-dominated oblique spreading axis: a comparison with other slow spreading plate boundaries. *Earth and Planetary Science Letters, 218*, 77–90.

Weston, F. S. (1963/64). List of recorded volcanic eruptions in the Azores with brief reports. *Boletim do Museu e Laboratório Mineralógico e Geológico da Faculdade de Ciências de Lisboa, 10*(1), 3–18.

Zbyszewski, G. (1963). Les phénomènes volcaniques modernes dans l'archipel des Açores. *Comunicações dos Serviços Geológicos de Portugal, 47*, 231.

Volcano Tourism in Greece: Two Case Studies of Volcanic Islands

5

Kalliopi Gaki-Papanastassiou and Dimitrios Papanastassiou

Santorini Island Volcano, Greece

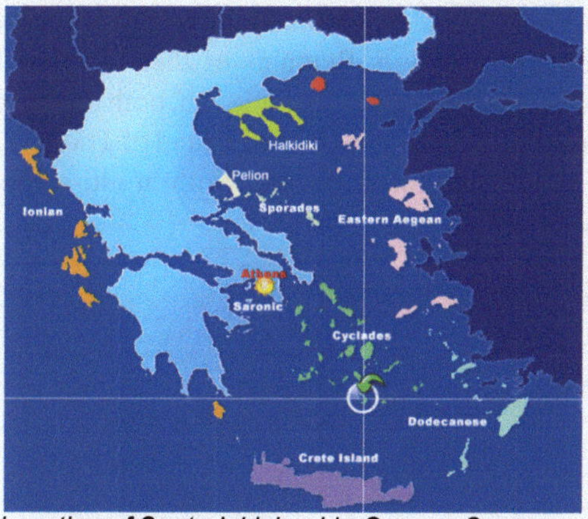

Location of Santorini Island in Greece. Source: Drawn by the Authors

Typical eruption style: Typical eruption style: Dominantly explosive, with effusive activity during intervals. Powerful explosive, so-called Plinian eruptions appear to occur every 10-30,000 years. Construction of lava shields and minor to irregularly spaced subplinian explosive activity during interval times.

Current status: Dormant (1 of 5)

Geographical location including map co-ordinates:
Cyclades Islands, Aegean sea, Greece
Latitude: 36° 25' N
Longitude: 25° 25' E

Type of volcano:
Caldera type (complex stratovolcano)

Last known eruption: 1613 BC ± 13 years (the so-called "Minoan eruption", one of the largest Plinian eruptions during the past 10,000 years on earth), 197 BC, 46/47 AD, 726, 1570-1573, 1707-1711, 1866-1870, 1925-1928, 1939-1941, 1950.

Summit Elevation: The height of the caldera rim is 329 m, the top of most recent (1950) lava shield, Nea Kameni is 128 m, while to highest altitude of the island is 564m.

Major attractions:
Spectacular sea-filled volcanic caldera surrounded by steep colourful cliffs, beautiful villages and fields, volcanic sand beaches. The important archaeological sites at Akrotiri

Status of protection:
Existing National laws provide the necessary protection

Visitor numbers: About 1 million per year

5.1 Santorini Island Volcano

5.1.1 Introduction

The Greek volcanoes are part of the Aegean Volcanic Arc, which is caused by the subduction of the African tectonic plate under the Eurasian plate. Santorini is the most famous volcano in Greece, and was the site of the most recent volcanic eruption in Greece in 1950; the others are Methana, Mílos, Nisyros, Yali and Kos (Table 5.1).

K. Gaki-Papanastassiou (✉)
Department of Geography and Climatology, Faculty of Geology and Geoenvironment, National University of Athens, Panepistimiopolis, 15784 Athens, Greece
e-mail: gaki@geol.uoa.gr

D. Papanastassiou
Institute of Geodynamics, National Observatory of Athens, 11810 Thissio, Athens, Greece
e-mail: d.papan@noa.gr

P. Erfurt-Cooper (ed.), *Volcanic Tourist Destinations*, Geoheritage, Geoparks and Geotourism,
DOI: 10.1007/978-3-642-16191-9_5,

Table 5.1 A list of active and dormant volcanoes in Greece. *Source* Siebert and Simkin (2002)

Name	Elevation in metres	Last eruption
Kos	430	Pleistocene
Methana	760	258 BCE
Mílos	751	140 CE
Nisyros	698	1888
Poros	80	Pliocene
Santorini (Kolumbo)	−18	1650
Santorini (Nea Kameni)	130	1950

Greece also has 752 hot springs which are popular tourist destinations. The first part of this chapter concentrates on the island of Santorini, which is described as follows:

Santorini is a volcanic Island group at the central south end of the Cyclades in the Aegean Sea, Greece. As a popular tourist destination in Greece, it is renowned for its spectacular sea-filled volcanic caldera surrounded by steep colourful cliffs, the beauty of its villages and fields and the sandy beaches. The caldera has a diameter of 11 km NS and 7.5 km EW, with a depth of 390 m in the north (Fig. 5.1). Santorini archipelago consists actually of a group of islands: the main island Thira (75.8 km^2, ca. 10,000 inhabitants in 13 villages), Thirasia (9.3 km^2, ca. 350 inhabitants), Aspronisi (White Island), Palea Kameni (Old Burned Island) and Nea Kameni (New Burned Island).

5.1.2 The History of Santorini

The Minoans were an ancient people who set up a civilization on Crete and the surrounding islands, including Santorini about 3000 BC and they were living on Santorini at the time of the massive eruption. Remains of their villages have been discovered buried under volcanic ash, both on the main island of Santorini and on the adjoining island of Thirasia. Evidently they were a cultured people, with multi-storied houses,

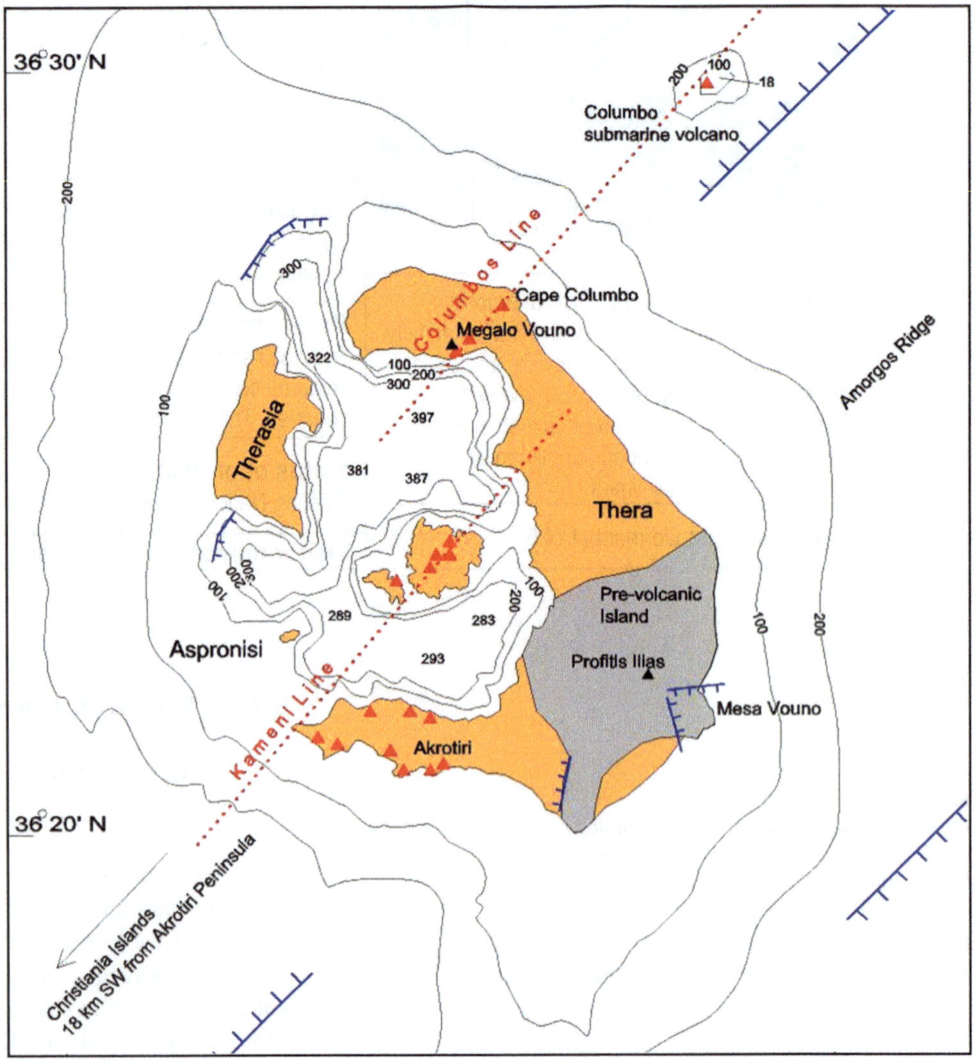

Fig. 5.1 The caldera of Santorini Volcano (based on Heiken and McCoy 1984)

plumbing and painted frescoes. At this time the island of Thira was called *Strongili*, which means Round Island. After the first historical eruption of the volcano that resulted in the island's present shape, Thira was named *Kallisti*, meaning the Fairest One. During the tenth century BC, the Dorians invaded the island which they named Thira after their ruler Thiras from Sparta, and established seven villages.

During the Roman period, the island was used as a place for exiles, but later they helped in building up the island. The Christianization of Santorini took place between the second and fifth century. The island was often ravaged and even destroyed by pirates and in 1204 it was conquered by the Venetians. It was about then the island got its current name. The islands patron saint was Agia Irini (St Irene) and the foreign sailors called her St Irini—thus Santorini.

The Venetians fortified five different places on the island with Imerovigli as the capital. They chose Imerovigli because it is the highest point of the Caldera. There is also a strange rock formation in the front of the village, called Skaros, where they built the strongest of five castles, which was destroyed in 1956 during a strong earthquake. The island was devastated by the Venetians in 1354 and once again in 1397, this time by the conquering Turks. In 1821 Santorini joined forces with Greek revolutionists and the island was subsequently freed from Turkish rule. On 9 July 1956 there was a terrible earthquake which ruined many buildings. The present population of Kamari for example is almost all from a village that was totally destroyed back then. The island's buildings have been reconstructed after the earthquake.

5.1.3 Geology of Santorini

Up to about two million years ago, Santorini was a small non-volcanic island. The non volcanic basement of the island, consists of Mesozoic-Cenozoic schists and marbles, are exposed at the Profitis Ilias Mountain, Mesa Vouno, the Gavrillos ridge, Pirgos, Monolithos and the inner side of the caldera wall between Cape Plaka and Athinios. They represent a former non-volcanic island of about 9 km × 6 km extension with rocks similar to neighbouring Cycladic islands (Fig. 5.2).

The first volcanic activity occurred in early Quaternary, about 2 million years ago. These volcanic rocks lie on top of the basement. The oldest volcanic deposits outcrop in the south-west part of the island, produced by volcanic centres located in the actual southern submarine area.

Eventually (around 500,000 years ago) there were two giant 'shield volcanoes', which are mountains in the shape of flat cones. These mountains united then with the non-volcanic island to form a single large island. Although neither of these mountains exists any longer, geologists have given them names. The northern mountain is called Mount Peristeria while the southern one is called Mount Thira.

About 200,000 years ago, things warmed up when Mount Thira started to produce vast amounts of magma and ash, eventually completely emptying the magma chamber under the mountain. The structure of the mountain was not able to support itself and it went crashing downwards into the empty magma chamber, leaving a caldera—a wide, deep hole in the ground. This process was repeated in a whole series of eruptions over the next 200,000 years, with both mountains producing magma, collapsing, re-growing and collapsing again, each time deepening the caldera.

The products of these centres are pyroclastics (tuffs, tuffites, hyaloclastites, scoriae, pumices and ashes), domes, lava flows and pillow lavas of andesitic-dacitic composition. A great part of these series is formed by dacitic submarine tuffs and tuffites with a thickness greater than 200 m. At the upper part of the pyroclastic series, up to 200 m above sea level, there are sedimentary levels with marine fossils which testify the intense tectonic activity of the area in the last 2 million years.

Volcanic activity in Santorini continued with Megalo Vouno, Thira, Mikros Profitis Ilias, Thirasia and Skaros volcanic centres alternating effusive, explosive and extrusive activity. The last paroxysmal eruption was the Minoan phreato-plinian explosion, which formed the pumice fall, and flow and surge deposits that cover the entire Santorini area. The first phase of the Minoan eruption was when Mount Thira blew its top and churned out pumice, a white rock filled with bubbles of gas which is lighter than water. A column about 35 km high of pumice and ash was blown into the stratosphere raining back down on the island. Where it has not been removed, Santorini is still covered in a layer of pumice, in places as much as 5 m thick. In the next phase, large amounts of magma were given off in explosive blasts by the mountain. This would have produced 'mushroom clouds' similar to those produced by atom bombs. In total, between the pumice and the magma, 30–40 km^3 of rock were ejected from the volcano. In the final stage of the eruption, the magma chamber beneath the surface of the mountain was empty and the surface of the volcano collapsed, further deepening the existing caldera and leaving the vertical cliffs that can be seen today. The new caldera was deeper than the sea around it, so the sea rushed in and engulfed the centre of the island.

Mount Thira did not die as a result of the eruption. It started to reform, eventually its tip emerged from the water in 197 BC. A number of islands have formed and reformed in the centre of the caldera. At present there are two, known as Palea Kameni (Old Burnt Island) and Nea Kameni (New Burnt Island), which have erupted sporadically over the

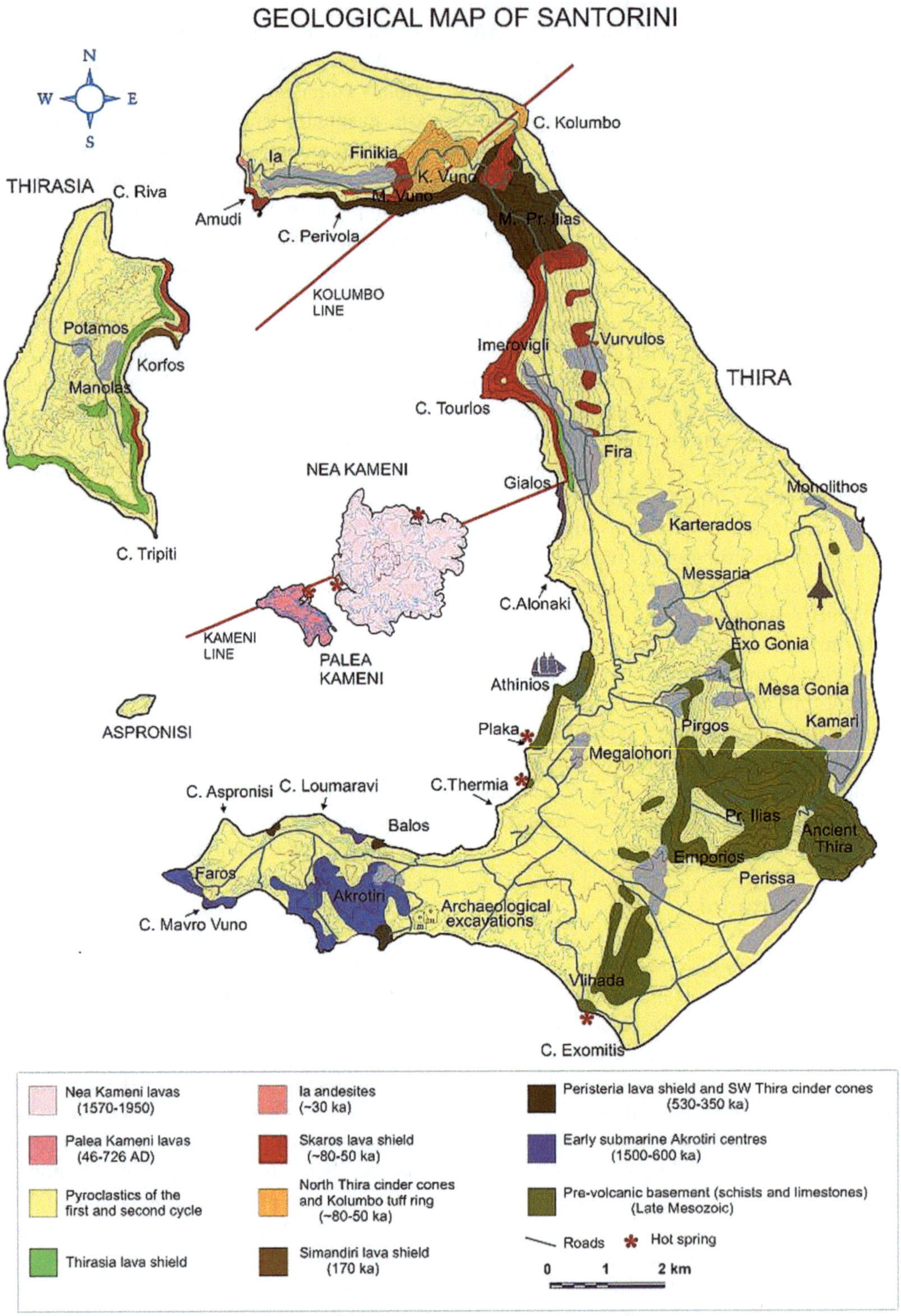

Fig. 5.2 Geological map of Santorini Island (Modified from the Geological map of IGME, 1994)

centuries. Most dacitic lava flows of Nea Kameni erupted during the last four centuries and four eruptive periods occurred between 1925 and 1950 (Fig. 5.3). Nea Kameni is still emitting smoke even today, although these are minor affairs, compared to the mind-boggling scale of the previous eruptions. The Peristeria volcano seems now to be extinct with no trace of where it once stood, at the north end of the present caldera.

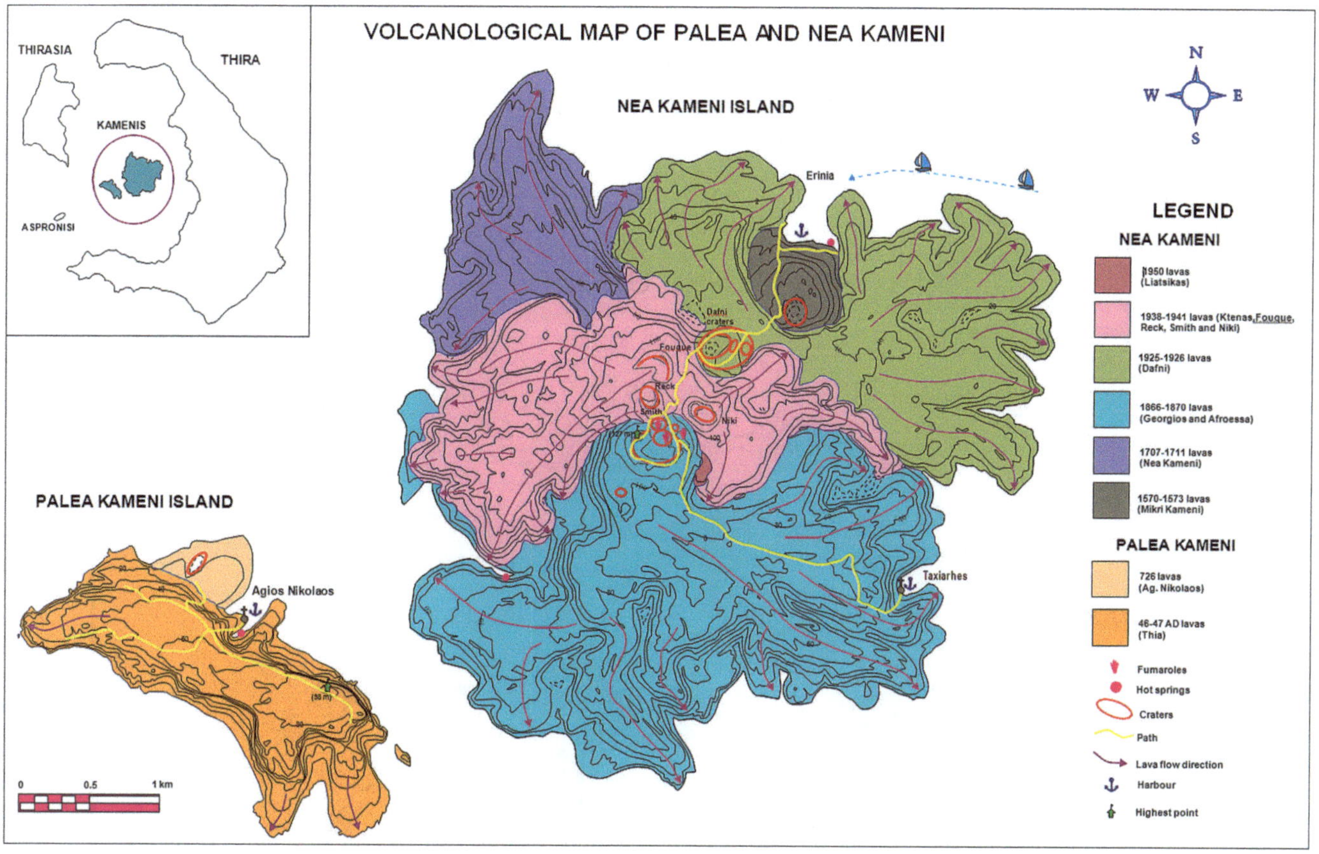

Fig. 5.3 Volcanological map of Palea and Nea Kameni. *Source* Institute for the study of Santorini Volcano

5.1.4 Volcanic Hazards

Santorini is the most active volcanic field of the South Aegean volcanic arc, and classed as one of the world's most violent caldera volcanoes. It comprises two of the three active volcanic Aegean centres, Kameni and Kolumbo. The group of the islands of Santorini is heavily populated (14,000 inhabitants on ca. 90 km^2). A continuously increasing amount of people (about 1,000,000 persons) visit Santorini yearly. More than 1/3 of them visit Nea Kameni islet and walk on the historic craters.

The maximum expected event on Santorini is a catastrophic paroxysmal eruption like the Minoan event. The probability of occurrence of such an event in the near future is very low, as geological and radio dating indicate that major explosive events on Santorini are separated by more than 20,000 years time gap. The second magmatic cycle of Santorini lasted about 200 ka and comprises seven major events. The repose time between the two late major explosive events (Cape Riva and Minoan eruptions) is 17.4 ka. Therefore it is improbable, that a major event occurs after only 3.6 ka. Furthermore, recent seismic data does not support the presence of a big magma chamber in the Santorini area.

The most probable event to occur on Santorini is an eruption of post-Minoan type, similar to those that built up Nea Kameni and Kolumbo. In the case of the repetition of such an event, the potential radius of the hazardous effects can be traced with relative precision, based on the already known effects of this type of activity. Based on this and on other probable scenarios, a volcanic hazard zonation map was plotted (Fig. 5.4) which reports the probable position of the future eruptive centres, the phreatic explosions hazard zones, the ballistic ejecta hazard zones, the tsunami hazard zones, the toxic gases and ash fallout exposed zones, the hydroclastic ejecta hazard zones, the lava flows and scoriae cones hazard zones as well as the landslides hazard zones (Vougioukalakis and Fytikas 2005).

The data available about the post-Minoan activity does not permit an accurate long-term forecasting of the next volcanic unrest. A short-term forecasting for the onset of such a type of activity is considered possible with an integral and efficiently operating monitoring network, using different parameters. For this purpose the Institute of Geodynamics operates a local seismological network in the island, which is connected in real time with the headquarters of the Institute in Athens. This network permits the

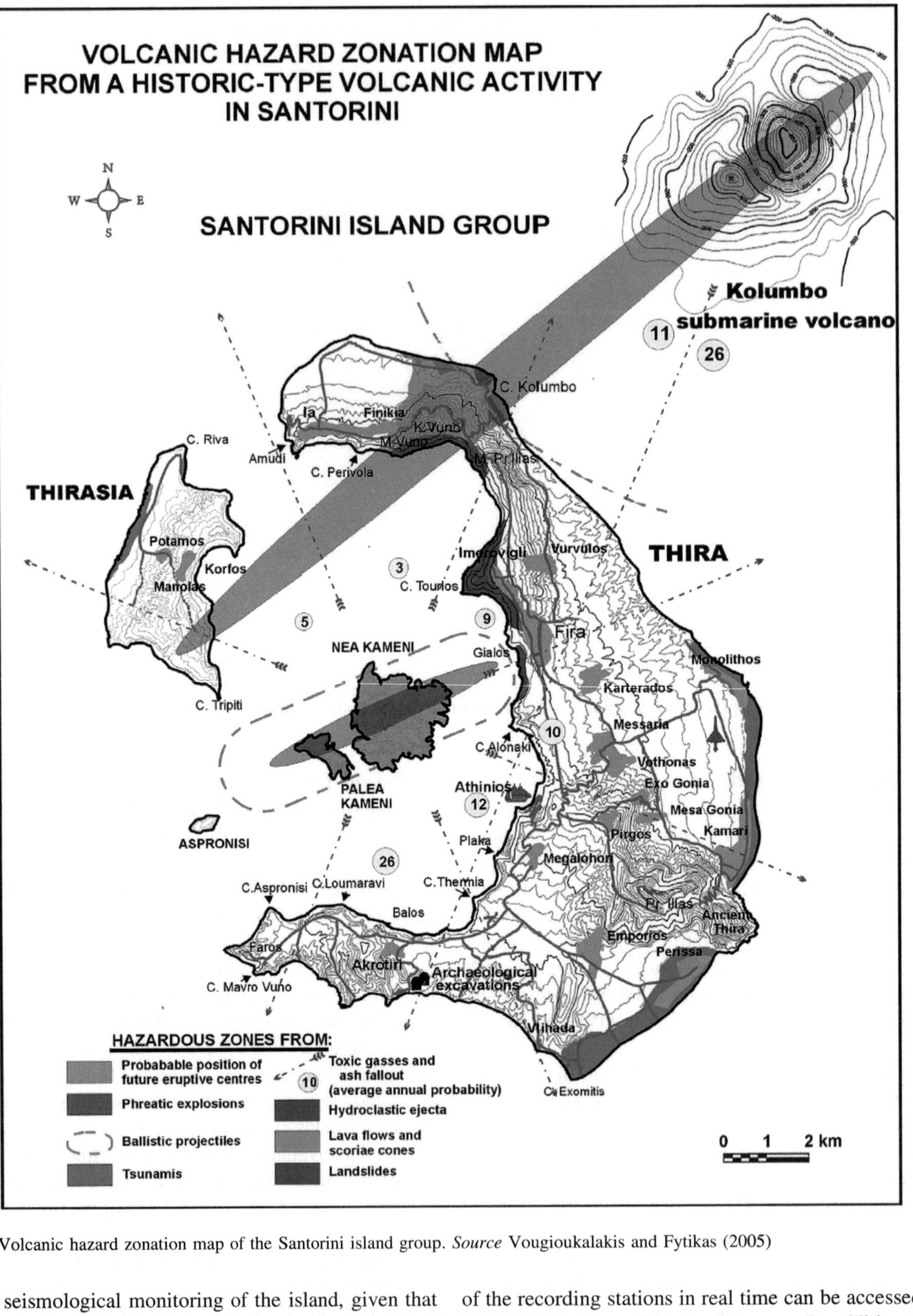

Fig. 5.4 Volcanic hazard zonation map of the Santorini island group. *Source* Vougioukalakis and Fytikas (2005)

accurate seismological monitoring of the island, given that it is capable to record micro earthquakes at magnitudes of the order of 1.0 on the Richter scale and to recognize anomalous seismic activity. Information and the waveforms of the recording stations in real time can be accessed at this URL address: http://bbnet.gein.noa.gr. In addition to these the Institute for the study and monitoring of the Santorini volcano operates a GPS network of three stations, while

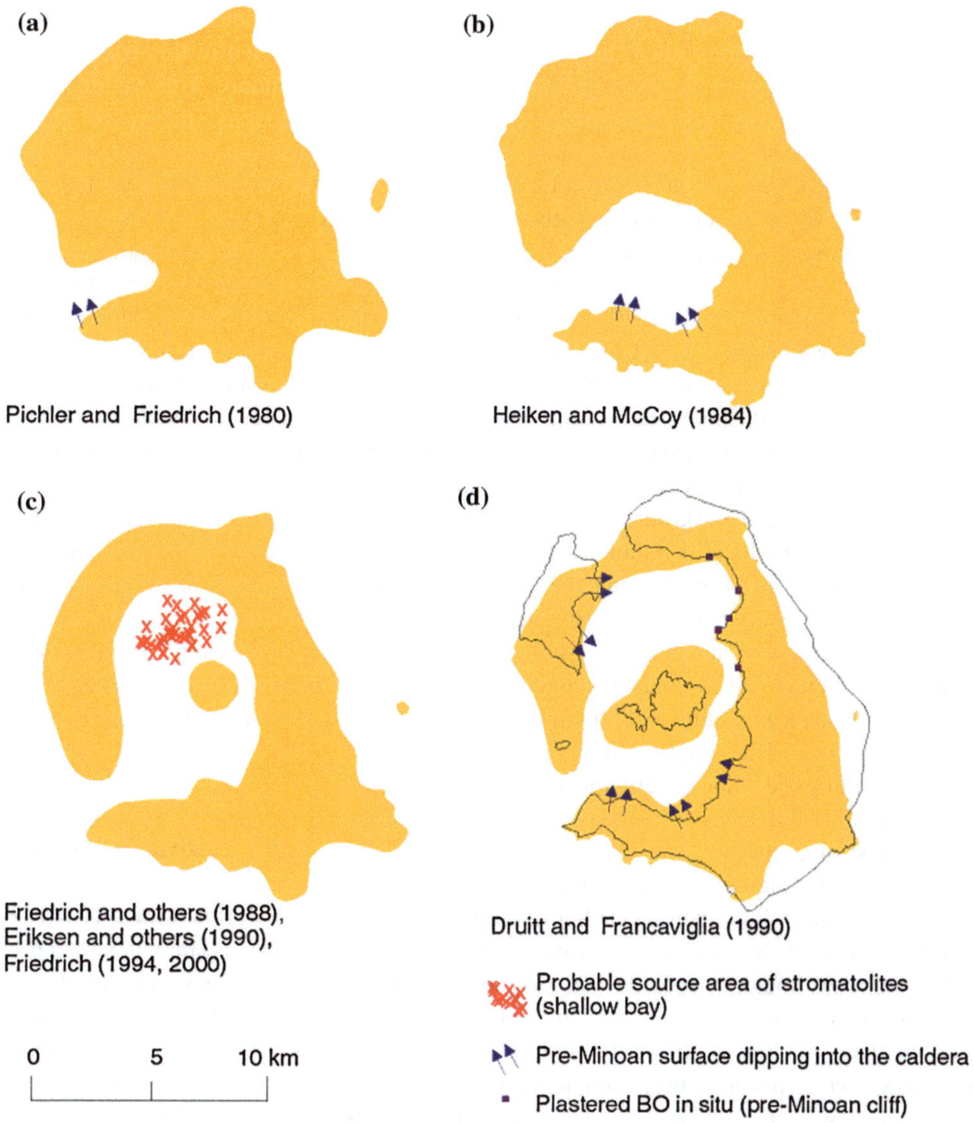

Fig. 5.5 Models of the Santorini Island before the Minoan eruption. Compiled by the Authors using the above mentioned sources

performing regularly high resolution GPS surveys and on a regular basis takes ground, water and gas temperature measurements (http://ismosav.santorini.net).

5.1.5 The Minoan Eruption

Until about 20 years ago it was thought that the shape of the island before the Minoan eruption was circular with a central cone or a system of several overlapping volcanic shields. This model however has been considerably modified over the last years (Fig. 5.5). Evidence has been found that the present-day caldera which consists of several basins, existed already before the eruption, at least in parts (Pichler and Friedrich 1980; Heiken and McCoy 1984; Eriksen et al. 1990; Pyle 1990; Druitt and Francaviglia 1990, 1992; Friedrich 1994).

The Minoan eruption provides a fixed point for aligning the entire chronology of the second millennium BC in the Aegean, because evidence of the eruption is found throughout the region. Despite this evidence, the exact date of the eruption has been difficult to determine, with current estimates based on radiocarbon dating indicate, that the eruption occurred between 1627 and 1600 BC [1613 BC $\pm$ 13 years] (Friedrich et al. 2006).

The Minoan eruption of Santorini, also referred to as the Thira eruption or Santorini eruption, was a major catastrophic volcanic eruption with a Volcanic Explosivity Index (VEI) of 6 or 7 (Dense-rock equivalent, DRE = 60 km^3) (Newhall and Self 1982; Decker 1990).

Therefore the Minoan eruption is considered as one of the biggest volcanic eruptions since the beginning of civilisation. Only the eruption of Tambora (Indonesia) in 1815 was bigger. For comparison, Santorini produced about 35 times as much rock and ash as the eruption of Mount St Helens in the USA in 1980 and had an explosivity rating about ten times that of Mt St Helens. The eruption of Krakatoa in Indonesia in 1883 was slightly smaller than that of Santorini. To give some idea of the scale, Krakatoa's eruption produced an ocean wave which killed 36,000 people and the sound of the eruption was heard over 4500 km away.

On Santorini, there is a 60 m thick layer of white tephra that overlies the soil clearly delineating the ground level prior to the eruption. This layer has three distinct bands that indicate the different phases of the eruption. Since no bodies have been found at the Akrotiri site, Heiken and McCoy (1990) note that the local population had advance warning of the impending eruption, leaving the island prior to its destruction. However, the thinness of the first ash layer, along with the lack of noticeable erosion of that layer by winter rains before the next layer was deposited; indicate that the volcano gave the local population only a few months warning. The eruption devastated the island of Thira, including the Minoan settlement at Akrotiri, as well as communities and agricultural areas on nearby islands and on the coast of Crete.

Since 1969, intense archaeological excavations have brought to light an important Cycladic/Minoan town which had been buried beneath the volcanic ash for almost 4,000 years. Although it appears that people had time to evacuate their island in time before the eruption, carrying most of their goods with them, the findings from Akrotiri are impressive: especially, they include well-preserved and magnificent wall paintings, ceramics and other objects. Thanks to the work of numerous archaeologists, a new light was thrown on an important prehistoric period and culture.

The Minoan eruption and the destruction of the city at Akrotiri have induced continuing speculations by many researchers that relate the destruction of Santorini to Plato's story of the sunken Atlantis. This is because Plato wrote about an island with a very advanced civilization that disappeared under the sea. Thira was also believed to have had a very advanced civilization and most of the island disappeared under the sea.

5.1.6 Natural Resources and the Local Economy

As a popular tourist destination in Greece, it is renowned for its spectacular sea-filled volcanic caldera surrounded by steep colourful cliffs (Fig. 5.6), the beauty of its villages and fields. Many people from abroad are coming to get married in this island and further more to spend their honeymoon. The majority are holidaymakers. According to statistics, there are 75 domestic and 85 charter flights arriving to Santorini per week, during summer time. During 2010 Santorini has been approached by 800 cruise ships.

Years ago, the main income of the islanders was the exportation of pumice stone. Two million tons of pumice was exported per year. The insulation of Suez Canal was made out of Thira's pumice stone mixed with cement.

The eastern slopes of the island are green and fertile, even in October. This is due to the abundant vineyards that grow in the fertile volcanic soil. The terraced slopes of the mountains use every available part of this fertile land. Santorini suffers from water scarcity because it has few natural water reserves, although the nature of the dry soil of the island produces grapes, which make one of the best wines of Greece. The climate is healthy and perfect for producing the famous wine (vinsando), fava beans and tomatoes.

However, tourism has become the main source of income for all islanders for the last 30 years, as tens of thousands tourists from Greece and around the world visit the island for unforgettable holidays.

5.1.7 What to See and What to Do

5.1.7.1 Ancient Thira (Phira)

Ancient Thira is situated on top of a high cliff just above Kamari, with an old church from 1100 AD as a wellknown landmark (Fig. 5.7). This was the Roman head quarters. Right at the top of the southeast peak of the highest mountain of Santorini in Profitis Helias is the ancient settlement of Thira.

The other most historic area of Santorini is located in Akrotiri, where excavations have brought to light the remains of the Minoan town (Figs. 5.8, 5.9, 5.10 and 5.11). Akrotiri is an ancient town that was buried when the volcano erupted 3,500 years ago. This area is considered to be 4,000 years old. The first excavation began in 1869, but not much work was done until 1969, when archaeologist Professor Marinatos started the excavations again. Today, the archaeologists have uncovered an area of over 10,000 m^2 and there still remains a lot more to be discovered. The quality of frescoes, the pottery and the magnificence of the buildings bear testimony of an important community that existed many centuries ago. The visitors of the excavations can walk around in the small street quarters and see the buildings almost like they were back then. You will be impressed by the modernity; the houses were two or three stories high, and they even had running water. The excavations have also revealed beautiful frescoes, tools and everyday life objects.

Fig. 5.6 Map showing the different places of interest of the Santorini Island. *Source* www.santorini-web.com

Fig. 5.7 View of Tira (Phira), the capital of Santorini. The caldera walls show the different layers of erupted volcanic material. *Photo* Henry Hartley

Fig. 5.8 Prehistoric ruins of Akrotiri with a staircase that was destroyed by earthquakes before the Minoan eruption covered everything in thick layers of tephra. *Photo* D. Papanastassiou

Fig. 5.9 The facade of a two story house is part of the excavated ruins of Akrotiri. *Photo* D. Papanastassiou

5.1.8 Cruising the Caldera

A slightly different, but equally enjoyable way to visit the Caldera and enjoy the stunning sunsets of Oia is a cruise. Many tour operators run these cruises and you will find many being advertised in Fira. The cruises last all day and leave from Fira around mid-morning. Typically, the ship visits Nea Kameni, the active volcano in the centre of the caldera and usually the visitors will be able to stop off here and climb the volcano (Fig. 5.12). Do not forget that you will need to be wearing suitable footwear for this climb. Ships usually spend 1 h at Nea Kameni before sailing on to Palea Kameni and its hot springs. After a refreshing bath in the spring water the ships continue on to Thirassia, or 'Little Thira' for lunch, a swim and perhaps a donkey ride up to the town of Manolas. From Thirassia, the ship continues on to Oia where there is time to explore the town and watch the sunset before returning to the ship for the cruise back to Fira.

Fig. 5.10 Fresco of a fisherman from the Akrotiri excavations, dating from around 1600 BC. *Source* www.santorini-art.com/akrotiri_paintings.htm

The Archaeological Museum and the Prehistoric Museum are the places where you can see items from the excavations in Akrotiri, as well as many artefacts and

Fig. 5.11 Fresco from the Akrotiri excavations. Panorama painting from the Bronze Age excavations at Akrotiri, Santorini. *Source* www.santorini-art.com/akrotiri_paintings.htm

Fig. 5.12 Nea Kameni Island in the Santorini caldera. *Photo* Bernard Gagnon

Fig. 5.13 Sunset at Oia, Santorini. *Photo* Pedro Szekely

sculptures from the rich historical past of Santorini including prehistoric times such as the Neolithic and Minoan era and the Hellenistic, Roman and Byzantine periods.

At the 'Thera Foundation' (www.idryma-theras.org.gr) is an exposition of reproductions of the frescos found in the prehistoric city of Akrotiri; a virtual tour of the Akrotiri archaeological site that allows visitors to gain an impression of the complexities and the size of Akrotiri. A photographic exhibition of Akrotiri presents a photographic tour of the prehistoric settlement of Akrotiri and is dedicated to the people, who have worked on the excavations for some 50 years. Moreover, there is also the Institute of the study and monitoring of Santorini Volcano.

Most people have read about the spectacular sunsets that occur on this island (Fig. 5.13). The sunsets at Santorini, viewed from the Caldera are breathtakingly beautiful when seen as a backdrop to the volcano. The colours that streak the sky change from lilac to deep purple, from yellow to orange to red, as the golden sun sinks and becomes blood red reflecting its light on to the sea and the surrounding little islands, an amazing scene for photography.

Maps, books with photos and touristic guides of the island can be found everywhere, in bookstores, kiosks, super markets as well as in the souvenir shops. The Institute of the study and monitoring of Santorini Volcano (http://ismosav.santorini.net) has published a book about the volcano as well as different brochures.

5.2 Mílos Island Volcano

5.2.1 Introduction

Mílos Island Volcano, Greece

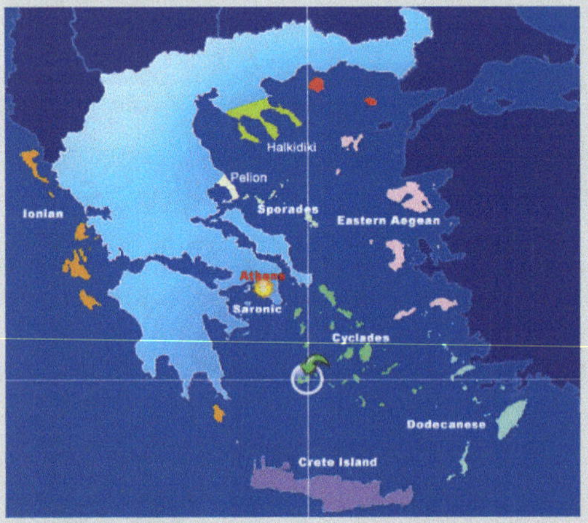

Location of Mílos Island in Greece. Source: Drawn by the Authors

Typical eruption style: Explosive, hydrothermal explosions

Current status: dormant (1 of 5)

Status of protection:
Existing National laws provide the necessary protection

Geographical location including map co-ordinates:
Cyclades Islands, Aegean sea, Greece
Latitude: 36°41'55"N
Longitude: 24°26'20"E

Type of volcano:
Stratovolcano

Last known eruption: 140 AD ± 60 years

Summit Elevation: 751m

Major attractions and Activities:
Beautiful beaches with unusual geological formations for swimming, some beaches have natural hot springs due to geothermal activity. Hiking and trekking through the extraordinary landscape of Milos, roundtrips by boat to visit the catacombs and other places of interest.

Visitor numbers:
About 100,000 per year

The Greek volcanoes are part of the *Aegean Volcanic Arc*, which is caused by the subduction of the African tectonic plate under the Eurasian plate. Santorini is the most famous volcano in Greece, and was the site of the most recent volcanic eruption in Greece in 1950; the others are Methana, Mílos, Nisyros, Yali and Kos (Table 5.1).

Greece also has 752 hot springs which are popular tourist destinations. This chapter concentrates on Mílos, and this volcano is described as follows: Mílos is located in the SW Cyclades, a large island group in the Aegean Sea (see previous part). It is 86 nautical miles from Piraeus and approximately half way between Piraeus and Crete. Its surface area is 151 km^2 and its varied coastline is 125 km long. In general terms it is a low island, whose highest peak is Prophet Elias in the north-west, rising to 751 m. The Small Mountain (Mikro Vouno) in the same area is 636 m high, while no other peak reaches over 400 m. The most characteristic feature of Mílos is its magnificent gulf, which provides a natural harbour and is possibly the largest in the Mediterranean. This, in combination with the island's vast mineral wealth, has played and is still playing a major role in the history, economy, culture and development of the island society throughout the last 7,000 years.

The population of Mílos is about 5,000, although it rises considerably every summer. Most of the inhabitants live in the eight villages Plaka (the capital), Plakes, Triovassalos, Pera Triovassalos, Tripiti, Pollonia, Zefiria, and Adamas, which are in the flat, northern part of the island. The locals live off fishing and exporting minerals from the island with tourism as an additional source of income during summer. One of the most famous sculptures of the ancient world, Venus de Milo (Aphrodite of Mílos) was found here. A copy can be seen in the archaeological museum, while the original is in the Louvre museum in Paris. It was given to the French during the period of Turkish rule. Everywhere on the island a tourist will see large or small copies of this work of art.

The island of Mílos is almost entirely a volcanic island. This formation is the reason for the island's lack of forests, but also for the rich vegetation and abundant natural water supplies and rivers, which have given Mílos its colour. This unique Mílos colour is dazzlingly evident in its multitude of beautiful beaches—more than 75—with their glistening silver sand and clear blue waters. Also in the chalky white, terracotta, ochre and jet black rocks, the likes of which can be seen nowhere else in the whole of Greece. Visitors, even without specialized knowledge, will surely be impressed by the images they see at every step. The colours and morphology of the geological phenomena… the coastlines, the gulfs, the cliffs, the gorges, the hills… are unique.

5.2.2 History

Nature gave to the island many bays, and the large natural harbour made the island a perfect place for some of the earliest settlers in Greece. There is evidence that the island was inhabited as far back as the sixth millennium BCE. Both the Minoans and Mycenaeans had settlements here in the second millennium BCE and the island flourished because of the good connections with other islands as well as the mainland. Mílos produced large quantities of obsidian, a volcanic rock well suited to make weapons and tools. Obsidian from Mílos has also been located in the Peloponnese, Crete, Cyprus and even in Egypt, which indicates that there was a flourishing export trade as well.

During the Bronze Age (2800–1100 BCE) the island was an extremely important center in the Cycladic world, especially the ancient city of Philakopi. During the Hellenistic period, around 1000 BCE, a new settlement was built in the area of modern Klima. This new town developed rapidly, in particular in the field of arts and craft. The so called Melian Vases of that period are greatly renowned. Later, the Melians refused to surrender to the Persians during their attempts to invade Greece during the fifth century BCE and fought with the rest of the Greeks in the battles of Salamis and Plataea. In their attempt to remain neutral during the Peloponnesian war, they were punished by the Athenians, who in 415 BCE put all the old people to death and sold the young men, women and children into slavery. The population of the island during the Hellenistic and Roman times is estimated to have been about 20.000 (pers. comm. with the archaeologists of the Cyclades Ephorate of Antiquity).

The history of the island during the following centuries was similar to that of the rest of the Cyclades Islands. Until 311 BCE Mílos was ruled first by Macedonia and then by Egypt. The powerful fleet of the Ptolemaids ensured the freedom and safety of the seas. As a result, the island of Mílos saw a phase of renewed economic growth which was also reflected culturally. Examples of this creative era are the famous statue of Venus (at the Louvre Museum in Paris), and the imposing 2.5 m tall statue of Poseidon (displayed in the National Museum in Athens). During the subsequent Roman conquest, a number of new buildings were constructed (the theatre is prominent), and Christianity made its appearance, probably during the first century. The Catacombs of the island (Fig. 5.14) are the most extensive in Greece and among the most significant in the whole Roman Empire; the skeletons of over 2,000 persons have been found there.

The most important event in the Byzantine era was the destruction of the ancient city of Klima (fifth–sixth

Fig. 5.14 The catacombs of Mílos Island date back to the first–fifth century and were used by the early Christians as places of worship as well as for burials. Chambers and graves were cut out of the relatively loose volcanic rock (pumice tuff). *Photo source* www.greecetravel.com

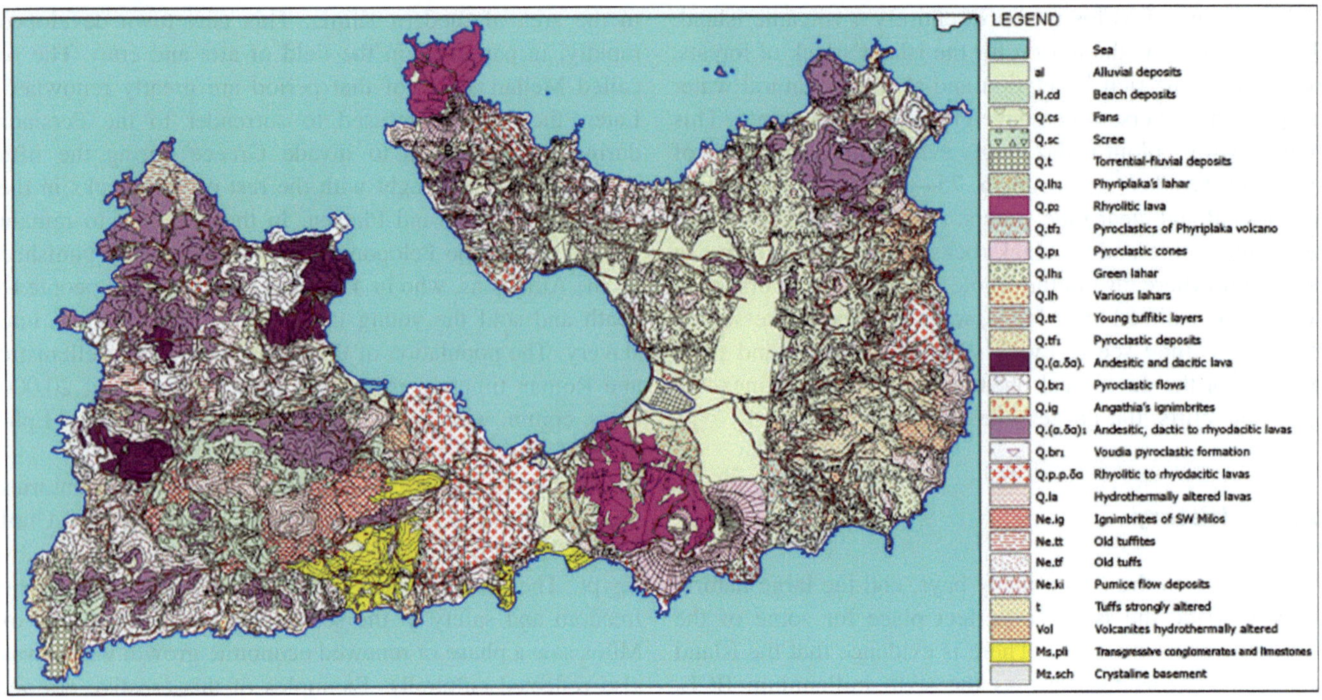

Fig. 5.15 Geological map of Mílos Island. *Source* Mílos Mining Museum

century), possibly as a result of an earthquake. Finally, during the Venetian and Ottoman rule, the inhabitants fought relentlessly for their freedom. Mílos was liberated in the nineteenth century after the War of Independence which had started in 1821.

5.2.3 Geology

Mílos and the adjacent small islands of Kimolos, Polyaigis, AntiMílos, Glaronisia and Akradia, have grown from submarine and subaerial volcanism that initially was dominantly andesitic and basaltic, but ended with predominantly rhyolitic eruptions (Fytikas et al. 1976, 1984, 1986 and Fytikas 1977, 1989). The oldest volcanic rocks are Neogene submarine rhyolitic pyroclastic-flow deposits formed 14 million years ago, overlying the metamorphic basement having an age of 33 and 64 million years (Fig. 5.15). The latest activity occurred during the late Pleistocene and consisted of two large explosive phreatomagmatic eruptions, 380,000 and 90,000 years ago. They are concentrated in the eastern half of the low, U-shaped Mílos Island and are forming the Trachilas and Fyriplaka lava domes and phreatic explosion craters. On Antimílos Island to the NW a composite volcano was constructed (Fig. 5.16). Phreatic explosions, commonly producing overlapping craters rarely more than 1 km in diameter, continued from late-Pleistocene to recent times. Historic magmatic eruptions have not occurred at Mílos Island, but a phreatic or explosive hydrothermal eruption of Mílos volcano occurred sometime between 80 and 200 CE, creating a lahar which buried the walls of a Roman town. Recent activity at Mílos consists of solfatara, fumaroles, and hot springs.

It is worth mentioning some geological features which are a direct result of the island's volcanic nature. Firstly, the two extinct volcanoes; the one at Firiplaka, on the south coast, has an impressive and well-preserved crater, with a diameter of 1700 m and a height of 220 m approximately. The other volcano is at Trachilas in the south of Mílos, although only a small part of it remains (Fig. 5.17). Another geological formation is that of the steep-sided hills such as Castro, which rises above Plaka, and Prophet Ilias. These volcanic plugs were formed when magma, from under the Earth's crust, failed to erupt due to low gas pressure. This molten rock solidified inside the volcanic pipe to form a hard rock called andesite. Subsequently, the softer surrounding rock was eroded away exposing the hard plug.

The Glaronisia or Seagull Islands, off the north coast, are a remarkable and rare example of andesite column formation (Fig. 5.18). These islets are entirely made up of vertical, pentagonal and hexagonal prisms, 20–30 cm in diameter, in majestic array. Also on the north coast is Sarakiniko where plenty of fossils of land and sea organisms exist. Here, layered variations of white volcanic rock interplay to create a dramatic land and sea vision (Fig. 5.19).

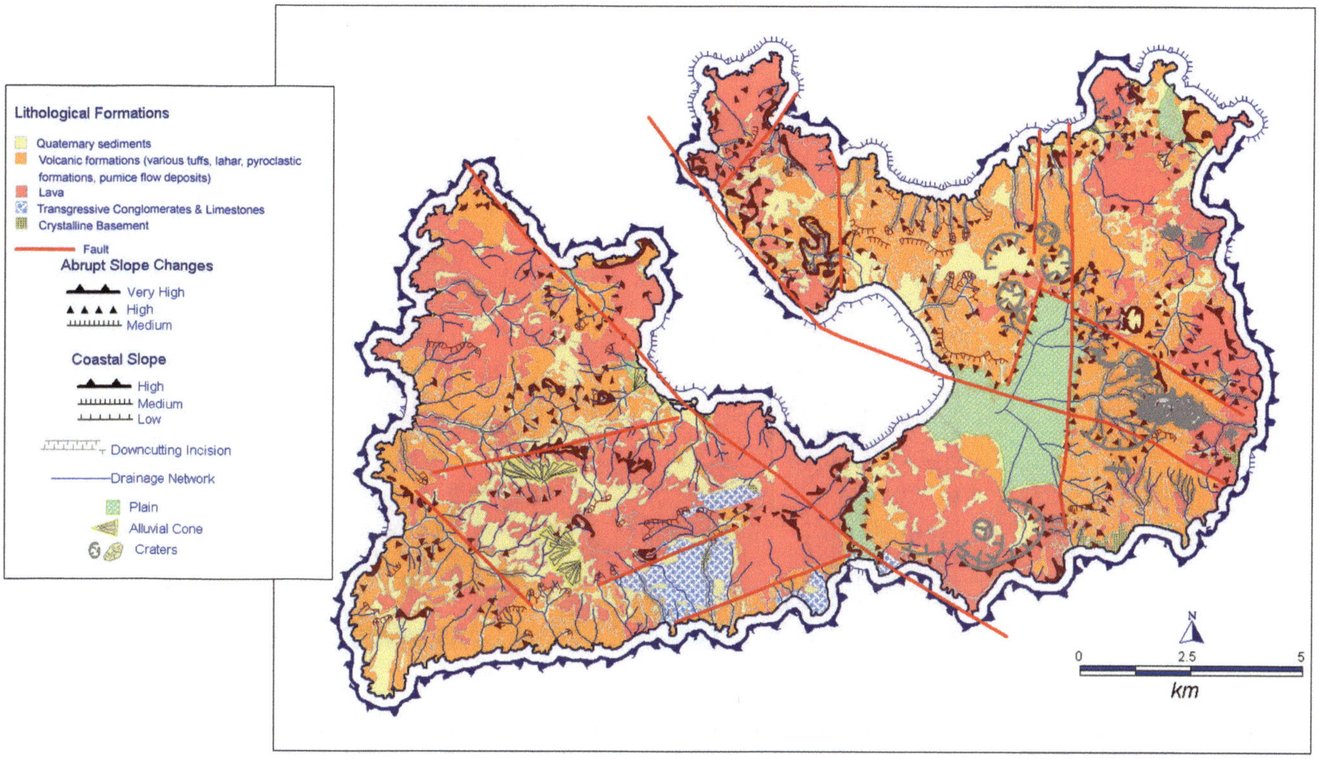

Fig. 5.16 Geomorphological map of Mílos Island. *Source* Kyriakopoulos et al. (2006)

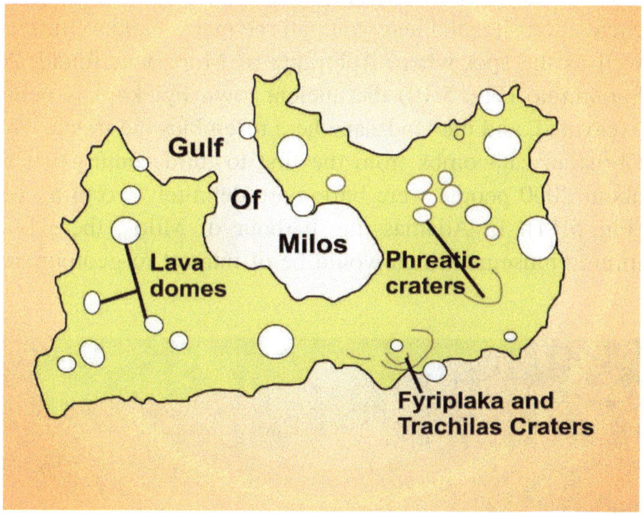

Fig. 5.17 The main domes of the island. *Source* Drawn by the authors

One of the volcanic characteristics associated with Mílos is the geysers, mainly in the east and south-east, where gases, mostly carbon dioxide, hydrogen sulphide and steam, issue through apertures in the rocks, often causing yellow coloration, the well-known sulphur smell and elevated surface temperature; e.g. Ag. Kyriaki 102 °C, Pyromenes 100 °C, Paliochori 101 °C, Vounalia 54 °C, Kastanas 86 °C, and Adamas 100 °C (Botz et al. 1996; Dando et al. 1995). At points where this occurs under the sea, bubbles of gas are visible on the surface and again the same characteristic smell and elevated temperature (Fontaine et al. 2003). Examples may be seen at Ag. Kyriaki, Paliochori, Kanava and Rivari. There are also a number of hot springs, some of which are spas and although they have never been adequately exploited, Hippocrates mentions them in his writings *On Epidemics*. We are told that Athenians with dermatological problems went to the spas on Mílos, in search of a cure.

These springs are at sea level in Adamas for example, where public baths have been built with temperatures reaching 33 °C. Further examples are Alikes (22 °C), Kanava (50 °C), Paliochori (50 °C), Mandrakia (54 °C), Adamas Haros (42 °C), Provata, Tria Pigadia and many others elsewhere. The chief salt present in the waters is sodium chloride, but the sulphates of calcium, magnesium, potassium and ammonia are also found, along with carbon dioxide and silicon dioxide. The spa waters are considered to remedy rheumatoid arthritis, as well as gynecological and dermatological illnesses among others. These areas also exhibit elevated land temperatures. Over millions of years, these high surface temperatures have resulted in the formation of rich mineral deposits, which are now extensively

Fig. 5.18 Columns of Andesitic lava in Glaronisia. *Source* http://milos-greece.com

mined. Furthermore, after long term research involving extensive drilling by the Public Electricity Corporation and the Institute of Geological Research, it was found that there is a geothermic field under the island with highly pressurized steam at an approximate depth of 1100 m. This could prove to be a valuable geothermal resource for the generation of electric power in the future.

5.2.4 What to See in Mílos

In Plaka, Mílos' capital, there are quite a few things to see. You can visit the archaeological museum where various artefacts from the excavations on the island are displayed, as well as a copy of Aphrodite of Mílos. There is also a folklore museum, where you can see how the people of Mílos used to live and dress. The Venetian Kastro is also worth a visit. Just above Klima there is a big archaeological excavation site, and here you will see many ancient ruins, as well as the spot where Aphrodite of Mílos was found. At Sarakiniko (Fig. 5.19) the ancient town Fylakopi is being excavated, and the landscape here resembles the moon. The Christian catacombs from the first to third century where about 2000 people were buried are definitely worth a visit (Fig. 5.14). In Adamas, the harbour of Mílos, there is a mining museum which would be of interest to geotourists.

Fig. 5.19 Sarakiniko beach is famous for its white tuff formations sculpted by erosion. *Source* www.greecetravel.com

Fig. 5.20 Map showing the different places of interest of the Mílos Island. *Source* www.kostantakis.gr

Fig. 5.21 Rocks at Kleftiko beach. *Source* www.greecetravel.com

5.2.5 Excursions

Several small boats which sail from Adamas (Fig. 5.20) can take you on daily trips round the island, stopping at a number of beaches to swim and in Kimolos for lunch. From the beach at Kypos you can take a boat trip to the beaches of Geronta, Kleftiko (Fig. 5.21), Geraka and Paliorema. From Pollonia you can visit the neighbouring island of Kimolos by motor boat or on the small ferry boat.

5.2.6 What to Do

Apart from sunbathing and swimming, this is an excellent island for snorkelling because of its many caves. There are water sports at some of the beaches, and fishing is also a nice pastime. The sunsets are among the best in Greece. There are over 70 beaches and bays on Mílos. There are all kinds of beaches: sandy, rocky or pebbly. It is worth mentioning the sandy beach Tsigrado, the Sarakiniko and Kleftiko beaches with the caves and Voudia which has warm upwelling waters. Maps, touristic guides and books with photos of the island can be found everywhere, in bookstores, kiosks, super markets as well as in the souvenir shops.

5.2.7 Conclusions: Volcanic Hazards on Mílos Island

The reactivating of this volcano accompanied by hazardous explosive hydro-volcanic eruptions (the maximum expected event) cannot be excluded. So, the most probable event at Mílos is that of a hydrothermal explosion, thus constituting a considerable natural hazard for the area. Nevertheless, volcanic risk in Mílos island is relatively very low (Vougioukalakis and Fytikas 2005) considering the low probability of occurrence of volcanic events in the near future. The permanent population of the island is 5,000 persons, which during summer time increases to 10,000–12,000 through tourism. Social and economic activity is thus relatively low, although quarrying is intense.

References

Botz, R., Stuben, D., Winckler, G., Bayer, R., Schmitt, M., & Faber, E. (1996). Hydrothermal gases offshore Mílos Island, Greece. *Chemical Geology, 130*, 1161–1173.

Dando, P. R., Hughes, J. A., Leahy, Y., Niven, S. S. J., Taylor, L. J., & Smith, C. (1995). Gas venting rates from submarine hydrothermal areas around the Island of Mílos. Hellenic volcanic arc. *Continental Shelf Research, 15*(8), 913–929.

Decker, R. W. (1990). How often does a Minoan eruption occur? In D. A. Hardy (Ed.), *Proceedings of the 3rd International Congress. Thera and the Aegean world, III* (vol. 3, pp. 444–452). London: The Thera Foundation.

Druitt, T. H. & Francaviglia, V. (1990). An ancient caldera cliff line at Phira, and its significance for the topography and geology of pre-Minoan Santorini. In: D. A. Hardy (Ed.), *Proceedings of the 3rd International Congress. Thera and the Aegean world,* III (vol. 3, pp. 362–369). London: The Thera Foundation.

Druitt, T. H., & Francaviglia, V. (1992). Caldera formation on Santorini and the physiography of the islands in the late Bronze Age. *Bulletin of Volcanology, 54*(6), 484–493.

Eriksen, U., Friedrich,W. L., Buchardt, B., Tauber, H., & Thomasen, M. S. (1990). The Stronghyle Caldera: Geological, palaeontological and stable isotope evidence from radiocarbon dated stromatolites from Santorini. In: D. A. Hardy (Ed.), *Proceedings of the 3rd International Congress. Thera and the Aegean world, III* (vol. 3, pp. 139–150). London: The Thera Foundation.

Fontaine, F Jh, Rabinowicz, M., & Boule'gu, J. (2003). Hydrothermal processes at Mílos Island (Greek Cyclades) and the mechanisms of compaction-induced phreatic eruptions. *Earth and Planetary Science Letters, 210*, 17–33.

Friedrich, W. L. (1994). *Feuer im Meer.* Heidelberg: Spektrum Akademischer Verlag. 256 pages.

Friedrich, W., Kromer, B., Friedrich, M., Heinemeier, J., Pfeiffer, T., & Talamo, S. (2006). Santorini eruption radiocarbon dated to 1627–1600 BC. *Science, 312*, p548.

Fytikas, M. (1977). Geology and geothermics of Mílos Island. Thesis, Thessaloniki University, 228 pages (in Greek with English summary).

Fytikas, M. (1989). Updating of the geological and geothermal research on Mílos Island. *Geothermics, 18*, 485–496.

Fytikas, M., Giuliani, O., Innocenti, F., Marinelli, G., & Mazzuoli, R. (1976). Geochronological data on recent magmatism of the Aegean Sea. *Tectonophysics, 31*, T29–T34.

Fytikas, M., Innocenti, P., Manetti, R., Mazzuoli, R., Peccerillo, A. & Villari, L. (1984). Tertiary to Quaternary evolution of volcanism in the Aegean region. In: J. E. Dixon & A. H. F. Robertson (Eds.), *The geological evolution of the Eastern Mediterranean* (vol. 17, pp. 687–699). London: Geological Society, Special Publication.

Fytikas, M., Innocenti, F., Kolios, N., Manetti, P., Mazzuoli, R., Poli, G., et al. (1986). Volcanology and petrology of volcanic products from the island of Mílos and neighboring islets. *Journal of Volcanology and Geothermal Research, 28*, 297–317.

Heiken, G., & McCoy, F. (1984). Caldera development during the Minoan eruption, Thira, Cyclades, Greece. *Journal of Geophysical Research, 89*, 8441–8462.

Heiken, G., & McCoy, F. (1990). Precursory activity to the Minoan eruption, Thera, Greece. In D. A. Hardy (Ed.), *Thera and the Aegean world, III* (Vol. 2, pp. 79–88). London: The Thera Foundation.

Kyriakopoulos, K., Gaki-Papanastassiou, K., Evelpidou, N., Papanastassiou, D., Vassilopoulos, A. & Komianou, A. (2006). *Natural hazards and anthropogenic parameters affecting Mílos Island landscape, Greece: HAZARDS 2006: 11th International Symposium on Natural and Human Induced Hazards*, Patras, Greece, 22–25 June 2006.

Newhall, C. G., & Self, S. (1982). The volcanic explosivity index (VEI)—An estimate of explosive magnitude for historical volcanism. *Journal of Geophysical Research—Oceans and Atmospheres, 87*(C2), 1231–1238.

Pichler, H., & Friedrich, W. L. (1980). Mechanism of the Minoan eruption of Santorini. In C. Doumas (Ed.), *Thera and the Aegean world, II.* London: The Thera Foundation.

Pyle, D.M. (1990). New estimates for the volume of the Minoan eruption. In: D. A. Hardy (Ed.), *Proceedings of the 3rd*

International Congress. Thera and the Aegean world, III (vol. 2, pp. 113-121). London: The Thera Foundation.

Siebert, L., & Simkin, T. (2002). *Volcanoes of the world: An illustrated catalog of holocene volcanoes and their eruptions*. Washington: Smithsonian Institution, Global Volcanism Program Digital Information Series, GVP-3.

Vougioukalakis, G. E. & Fytikas, M. (2005). Volcanic hazards in the Aegean area, relative risk evaluation, monitoring and present state of the active volcanic centers. In M. Fytikas & G. E. Vougioukalakis (Eds.), *South Aegean volcanic arc: Developments in volcanology* (vol. 7, pp. 161–183). Amsterdam: Elsevier.

Volcano Tourism in Turkey

6

Gülpınar Akbulut

Geographical Location of Turkey

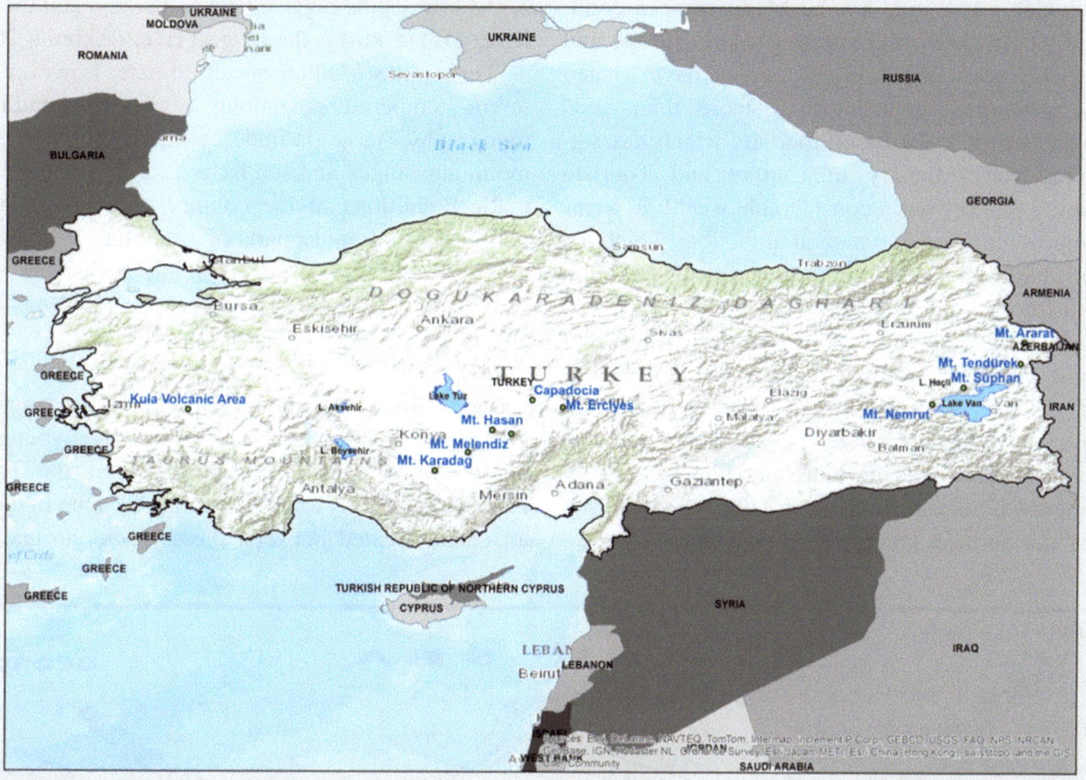

Summit Elevations: e.g. Ararat (5137m), Erciyes (3917m), Hasan (3268m)	
Major attractions: Cappadocia Fairy Chimneys, Lava Flows, Maar Craters and Lakes, Volcanic Hot Springs	
Status of protection: National Parks (Gőreme), UNESCO World Heritage Sites (Gőreme and Rocksites of Cappadocia; Mt. Nemrut Dağ), UNESCO Global Geopark (Kula Volcanic Mountains)	
Last known eruptions: Mt Nemrut (1692), Mt Ararat (1840), Mount Tendürek (1855) **Current status:** Dormant	**Visitor numbers:** Over 7 million visitors per year
Type of volcanoes: Stratovolcanoes, Shield Volcanoes, Lava Domes, Volcanic Fields, Calderas, Cinder Cones	

G. Akbulut (✉)
Cumhuriyet University Sivas, Sivas, Turkey
e-mail: gakbulut@cumhuriyet.edu.tr

P. Erfurt-Cooper (ed.), *Volcanic Tourist Destinations*, Geoheritage, Geoparks and Geotourism,
DOI: 10.1007/978-3-642-16191-9_6, © Springer-Verlag Berlin Heidelberg 2014

6.1 Introduction

Geotourism can be considered as a part of natural area tourism and ecotourism, but is a specialized form of tourism in that the focus of attention is the geosite (Dowling and Newsome 2006). However, one of the most important sectors of geotourism is volcano tourism (Erfurt-Cooper 2010) with many people interested in volcanic environments and therefore it is rapidly being recognised as an exciting new direction for tourism (Wartiti et al. 2007). Turkey is one of the most important tourist attracting countries of the world due to its rich natural and cultural landscapes, which are unique among Mediterranean countries, as well as its location connecting Asia, Europe and Africa. In 2012, over 36.7 million tourists have visited Turkey and generated an income of more than USD 29.4 billion including beach-based tourism, which has significantly improved (Ministry of Culture and Tourism 2011). However, Turkey has a considerable wealth in terms of archaeological, cultural, and natural attractions which are an excellent basis for developing geotourism (Akbulut 2009a). Both foreign and domestic tourists are attracted to the geological and geomorphologic heritage which includes scientific value, geotourism appeal, educational and historical values. In line with worldwide changes, Turkey's tourism policy is transforming by developing alternative tourism sectors such as volcano tourism. This chapter presents the patterns of volcanic tourism in Turkey and outlines the future of this form of tourism for this country.

6.2 Turkey

Turkey is an Eurasian country bounded by the Black Sea in the north, the Mediterranean Sea, Syria, and Iraq in the south, Armenia, Georgia and Azerbaijan (Nahjivan) in the northeast, Greece and Bulgaria in the northwest, the Aegean Sea in the west, Iran in the east, and has a total coastline of 8,333 km including that along the Sea of Marmara (see Map on previous page). Turkey has an area of 814,578 km^2 and a population of over 70 million people. Mt Ağrı (Ararat 5,137 m) is the highest mountain in Turkey; Lake Van, which is the largest soda lake in the world and covering a surface area of 3,712 km^2, is the biggest lake in Turkey, and the Kızılırmak River (1,355 km) is the longest river (Akbulut 2009a).

Turkey has Mediterranean climate; however, it has different temperatures among regions depending on the topography, large latitude, altitude, and sea. Especially mountain ranges and sea have a large influence on the climatic conditions of the country with continental climate dominating the inner parts of Anatolia.

6.3 Geology

Turkey is located in the Alpine-Himalayan orogenic belt. It has a complex geological history, with a young continent-continent collision occurring between the Arabian and Eurasian plates (Karaoğlu et al. 2005). Due to the closure of an ocean located between these blocks around 65 million

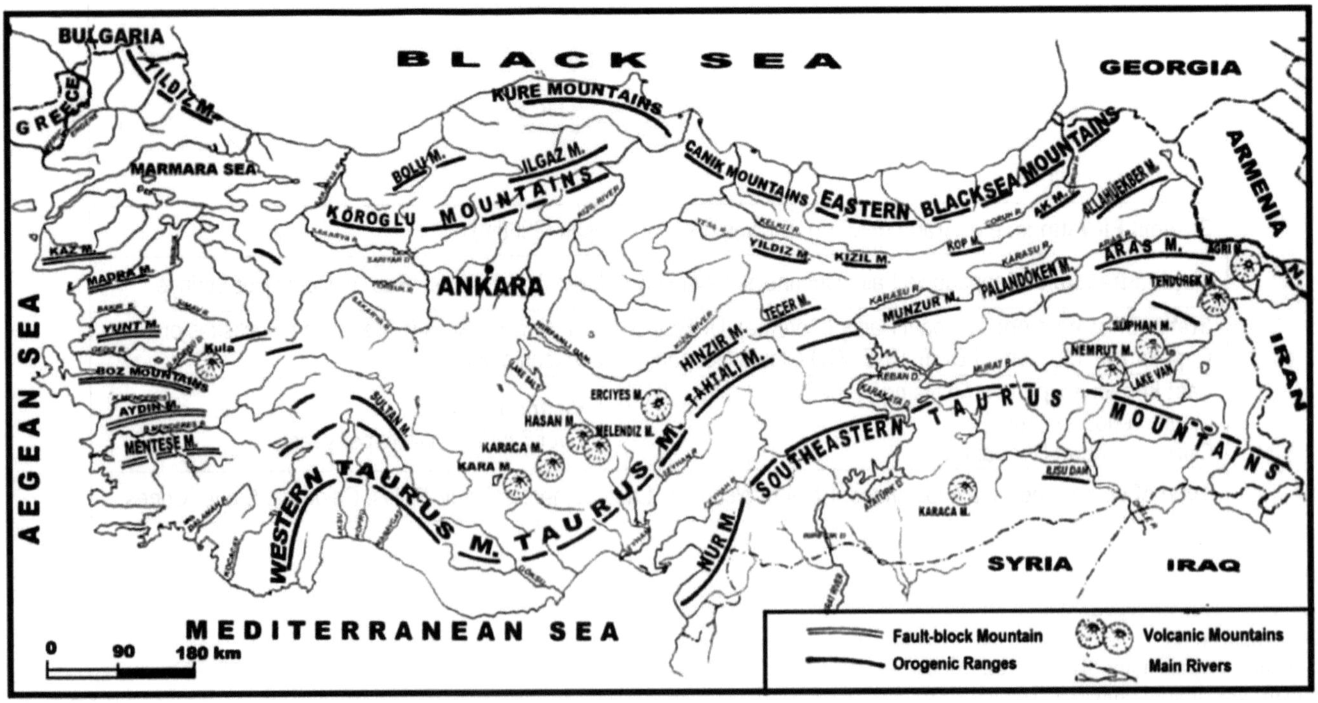

Fig. 6.1 The main mountain ranges of Turkey. *Source* Atalay (2002b)

years ago, the main geological features of Turkey were established. Two main mountain ranges extended on the northern and the southern parts of Anatolia dominated the tectonic and stratigraphic features of the country. These mountains are the Northern Anatolian Mountains and the Taurus Mountains. Besides these, fault-block mountains with uplifted blocks along the normal faults are found in the western part of Anatolia (Atalay 2002a). Volcanic mountains in Turkey exist in various parts of the country. During the end of the Tertiary and Early Quaternary periods eruptions occurred and, as at result of these, volcanic cones were formed both in Eastern Anatolia, and in Inner and South-Eastern Anatolia (Atalay 2002b). Today, Ağrı, Süphan, Erciyes, Hasan, and Karacadağ are among the most important volcanic mountains of Turkey (Fig. 6.1).

6.4 Volcanic Tourism in Turkey and Its Importance

Volcano tourism is an important part of geotourism. People would like to gain more knowledge about volcanoes and thousands of tourists are visiting these areas with different aims. Volcanic mountains are among the most important tourist attractions of Turkey. In particular, Mount Ağrı, Erciyes, Süphan, and Nemrut hold significance for geotourism and winter sports (Table 6.1). Mount Ağrı (Ararat) is located in the Eastern Anatolia Region of Turkey and is a young stratovolcano that formed during the Pliocene. Its 5,137 m summit is the highest peak in the country. Above the height of 4,200 m the mountain mostly consists of rocks covered by an ice sheet (Güner 2000). Mount Ağrı has many natural attractive features such as the Cehennem Valley, Lawrence Cave, The Fountain of the Prophet, Fish Lake, and Ice Cave (Arınç and Kaya 2004). Mount Ağrı is suitable for various sports such as trekking, camping, climbing, mountain biking and slope parachuting (hang gliding) and attracts hundreds of tourists every year who climb the different slopes of the mountain and go camping (Figs. 6.2 and 6.3). The mountain's rich heritage of historical and cultural values is very significant in both national and international dimensions, especially regarding the legend of Noah's Ark, a large wooden ship built by Noah to save his family as well as a male and female of every species from the worldwide flood. Therefore the mountain is sacred (Akbulut 2009a, 2013). Many scientists have come from different countries to find Noah's ark on Mount Ağrı, but have not yet reached a significant result. Other touristic attractions are the meteor depression in Doğubayazıt, the İshak Paşa Palace, and the Bayazıd Mosque (Akbulut 2009b). For instance, according to the Ağrı Cultural and Tourism Ministry data in 2011, the total tourist numbers for Ağrı are 1,023,371 including 176,180 domestic (17 % of total) and 847,191 foreign (83 % of total) visitors.

Mount Nemrut (2935 m) is one of the most interesting volcanic areas in Turkey from aesthetic, scientific, natural, and educational points of view (Fig. 6.4). It is extremely important for geotourism and understanding the development of the volcanic heritage of Turkey (Akbulut 2012). Mount Nemrut is named after King Nimrod who lived in this area around 2100 BC. Mount Nemrut is situated on the western coast of Lake Van in the East Anatolia Region of Turkey. This mountain is a young stratovolcano and continued to be active until 1597 AD (Aydar et al. 2003).

Table 6.1 Important volcanic features in relation to volcanic tourism in Turkey

Volcanic features	Region	Volcanic features	Region
Ağrı (Ararat) (5137 m)	East Anatolia	Karadağ	Inner anatolia
Süphan (4058 m)	East Anatolia	Acıgöl ve Kalecitepe Maars	Inner Anatolia
Erciyes (3917 m)	Inner Anatolia	Lake Aygir and Maar	East Anatolia
Lesser Ararat (3896 m)	East Anatolia	Çora Maar	Inner Anatolia
Tendürek (3584 m)	East Anatolia	Gölcük Maar	Mediterranean Anatolia
Hasan (3268 m)	Inner Anatolia	Kutören Maar	Inner Anatolia
Kısır (3197)	East Anatolia	Meke Maar	Inner Anatolia
Aladağ (3138 m)	East Anatolia	Lake Nar and Maar	Inner Anatolia
Akbaba (3040 m)	East Anatolia	Pazarcık basalts	Inner Anatolia
Nemrut Mountain (2935 m) and Caldera	East Anatolia	Kula volcanoes	Western Anatolia
Melendiz (2963 m)	Inner Anatolia	Mount Spil and pillow lavas	Western Anatolia

Fig. 6.2 A general view of Mt. Ararat. *Photo* Gülpınar Akbulut

Fig. 6.3 A large group of mountaineers are about to reach the peak of Mt. Ararat. *Photo* Gülpınar Akbulut

Among the greatest geological values of Mount Nemrut is the Nemrut Caldera, one of the largest in the world and which was formed by the collapse of the peak of volcano (Fig. 6.5). Fumarole activity from several outlets takes place in the caldera which is approximately 450–500 m deep (Akbulut 2009b) and includes lakes, 10 maars, 12 lava domes, 3 lava flows, mineral springs and trees such as oak and birch communities (Ulusoy et al. 2008). Especially, two

Fig. 6.4 Mount Nemrut is a UNESCO world heritage site based on its outstanding universal value. *Photo* Gülpınar Akbulut

Fig. 6.5 Aerial view (from SW) of the Nemrut caldera and lake Nemrut. *Photo* Abdurrahman Aksoy

Fig. 6.6 North east view of lake Nemrut and Lake Ilıgöl inside the Caldera. *Photo* Bitlis cultural and tourism manager

big lakes within the caldera are important for volcano tourism (Figs. 6.6 and 6.7). One of them is located in the western part of the caldera and measures about 12 km^2. The second lake is a small warm lake with temperature reaching up to 60 °C (Gürbüz 1995) and these characteristics make it very important for the tourism industry with a winter sports center and hotel accommodation.

The surrounding areas of Mount Nemrut offer natural, historical and cultural heritage attractions such as Akdamar, Çarpanak, the Yaka islands, and the Seljuk Cemetery (Figs. 6.8 and 6.9). Akdamar Island has an important historical church, and the Seljuk Cemetery in Ahlat is the largest Turkish-Islamic graveyard in the world with important cultural value for tourism (Akbulut 2012).

Mount Süphan is the second highest volcanic mountain of Turkey with its 4,058 m peak and is located in the north of Lake Van. The summit of the mountain has a small glacier at about 3700 m. Besides, there is a caldera and a lava dome with a diameter of approximately one kilometer (Atalay 1994) and Lake Aygır, an explosion maar on the south eastern slope of Mount Süphan. Lake Aygır is about 1.5 km^2 with its deepest point at 5 m and many tourists go camping in its surroundings in summer (Gürbüz 1995).

Fig. 6.7 A group of people collecting Obsidian at Lake Nemrut. *Photo* Gülpınar Akbulut

Fig. 6.8 Akdamar Island (Lake Van, Anatolia) with a small Armenian church built in the 10th century. The outside walls are decorated with art depicting Biblical stories. *Photo* Gülpınar Akbulut

Mount Tendürek (3584 m) is located between Doğu Beyazıt and Çaldıran in the southwest of Mount Ağrı with a 400–500 m diameter lake in the east crater at an elevation of about 3,000 m (Atalay 1994). Mount Tendürek is known as the most recently active volcano of Turkey and not developed for tourism because of safety concerns. However, volcanic regions in Eastern Anatolia have a significant potential to be developed for volcano tourism. With more than one million visitors per year, tourism is growing in the volcanic belts, although the tourists are usually domestic visitors. For instance, according to the Cultural and Tourism Ministry data in 2010, the total tourist numbers for Van are 736,242 including 653,023 domestic (88.6 % of total) and 83,219 foreign (11.4 % of total) visitors. The total tourist numbers of Bitlis reached 64,669 including 59,529 domestic (92.05 % of total) and 5,140 foreign (7.95 % of total)

Fig. 6.9 Headstones at the Seljuk Cemetery (Ahlat, Lake Van) are beautifully decorated with Arabic and Persian floral and geometric patterns. *Photo* Gülpınar Akbulut

Fig. 6.10 Badlands topography of the Rose Valley in Cappadocia. *Photo* Patricia Erfurt-Cooper

visitors. International tourist numbers are comparatively low owing to security concerns in this region.

Many geological and geomorphologic heritage sites are located in Central Anatolia with Cappadocia as a particularly significant geosite in terms of geotourism. The area of Cappadocia extends over 7,000 km^2 inside the triangle of Kayseri-Aksaray-Niğde Provinces in Central Anatolia (Aydan and Ulusay 2003). The geological history of Cappadocia is largely characterised by the creation of volcanic landforms and deposition of ignimbrites during the Neocene-Quaternary periods and extends as a belt in NE–SW direction with a long axis of about 300 km (Aydan and Ulusay 2003). The area is covered by almost horizontal, several hundred meters thick, acidic volcanic tuffs and lavas from Erciyes, Melendiz, and Hasandağ volcanoes (Erinç 1971; Koçyiğit and Erol 2001; Kopar 2007), from which many special morphological features such as fairy chimneys, badlands, valleys and caves have developed due to the erosive effects of precipitation and the Kızılırmak River and its branches (Figs. 6.10 and 6.11).

Fig. 6.11 Fairy chimneys in Cappadocia. *Photo* Patricia Erfurt-Cooper

Fig. 6.12 Mt Erciyes (3916 m) is the highest volcanic mountain in Cappadocia. *Photo* Gülpınar Akbulut

Mount Erciyes is located near Kayseri city and at a summit height of 3,917 m is the highest peak in Central Anatolia (Fig. 6.12). Mount Erciyes is a young stratovolcano with many small volcanic cones (Atalay 2002a; Gönençgil et al. 2003; Akbulut and Gülüm 2012). The mountain is rich in endemic and relic animal and plant species and well known for winter sports. There are five hotels at Mount Erciyes and according to the Kayseri Tourism Manager's records, 274,322 domestic and 67,899 foreign tourists (total 324,221 tourists) visited this mountain and Kayseri city in 2010.

Mount Hasan is the second highest mountain in Central Anatolia at 3,268 m (Akbulut 2009b) and features lava tunnels which have been used for stone mining (Fig. 6.13). Mount Melendiz is another volcanic mountain (2935 m). Further, there is Lake Nar, a relatively small (about 0.7 km^2) and shallow (about 26 m) crater lake (Fig. 6.14), which is located approximately 36 km southeast of Aksaray (England et al. 2008). Mount Karacadağ is located between Konya and Karapınar and surrounding this volcano are the maar craters Meke Maar and Acıgöl which are still full of

Fig. 6.13 The stratovolcano Hasandağ is the second highest mountain in the Central Anatolia and is currently dormant. To reach the summit takes around 6 h. *Photo* Gülpınar Akbulut

Fig. 6.14 To reach Lake Nar takes a short 30 min drive from Aksaray. *Photo* Gülpınar Akbulut

water. The Meke Maar (Lake Tuzla) is an explosion crater 2 km northeast of Acıgöl in Karapınar with a volcanic cone at the centre. However, this maar has environmental problems owing to the use of its water for agriculture. At Acıgöl there are also hot springs and a crater of approximately 1.5 km in diameter (Atalay 1994). Another volcanic mountain is Mount Karadağ (2271 m) which is located in Karaman in the southwest of Mount Karacadağ.

In addition to these volcanic mountains, there are special morphological features such as 'Fairy Chimneys', which are quite a rare landform (Fig. 6.15). Some of the Fairy Chimneys have a conical appearance, but many are domed. Originally, they developed out of the vertical channels and fissures in the tuff layers caused by rainfall, creating pinnacles. These formations are protected by their harder rock capping (Bowen 1990) and featured in Göreme, Ürgüp (Fig. 6.16) and Avanos, of which Göreme is the only National Park. Fairy chimneys have various colours; some of them are pink due to their formation of rhyolitic tuff (Doğaner 1995). Some fairy chimneys resemble animals (Figs. 6.17 and 6.18), such as seals, camels or humans (Akbulut and Gülüm 2012).

Fig. 6.15 Erosion has created the Fairy chimneys from thick tuff layers capped by harder basalt flows. *Photo* H. Baykal

Fig. 6.16 Ürgüp City in the Central Anatolia region is a popular tourist destination. *Photo* Gülpınar Akbulut

Fig. 6.18 Rock formations reminding of a group of shouting men in Göreme. *Photo* Gülpınar Akbulut

Fig. 6.17 Unique rock formations resembling a seal can be viewed between Göreme and Ürgüp. *Photo* Gülpınar Akbulut

Besides the fairy chimneys, there are many important valleys such as Göreme, Zelve, Kızıl, Güllüdere, Bağlıdere, Görkün creek, Zemi, Pancarlık, and Halac creek which constitute the major attractions for tourism (Emiroğlu 1975; Doğaner 1995; Somuncu and Yiğit 2009). These valleys, in which hundreds of impressive dwellings (Fig. 6.19), catacombs and rock-cut churches were carved, are very typical for Cappadocia (Aydan and Ulusay 2003).

Cappadocia has not only important volcanic features, but also historical structures of significance. 'There are many underground cities and semi-underground settlements, most of which are more than 1500 years old, in the Cappadocia Region' (Aydan and Ulusay 2003). Underground cities in Cappadocia were built as hiding places by early Christians escaping from the persecution of the Roman Empire (Göney

Fig. 6.19 Houses in Cappadocia are carved out of thick volcanic tuff beds. *Photo* Patricia Erfurt-Cooper

Fig. 6.20 Aerial view of the Kula volcanic field in the Manisa Province in western Turkey. *Photo* Gülpınar Akbulut

1988; Doğaner 1995). At the same time, they were used for protection from the cold climatic conditions, as well as for the preservation of food and supplies (Akbulut and Gülüm 2012). Due to the fact that people have lived in the underground cities for long periods without being able to go out, they were enriched with storage rooms for provisions,

Fig. 6.21 The perfect shape of the Sandal volcanic cone, locally known as 'devlit'. *Photo* Gülpınar Akbulut

Fig. 6.22 The Fairy Chimneys can be visited in the Kula Geopark area. *Photo* Gülpınar Akbulut

ventilation chimneys, wineries, churches and abbeys, water wells, toilets and meeting rooms (Doğaner 1995). Today, there are 22 known large-scale underground cities, among those the most importants are Derinkuyu, Kaymaklı and Özkonak. Today, new underground settlements of different sizes and shapes are used to a limited extent compared to past usage for housing, storing foods, mushroom plantation, and hotel accommodation (Aydan and Ulusay 2003).

Volcanic mountains, fairy chimneys, and other features in Cappadocia are very interesting with their aesthetic, ecological, educational, cultural, historic, scientific and recreational values which make this region one of the most famous tourist destinations in Turkey. With more than 5 million people per year tourism is rapidly growing in Cappadocia. For instance, although Nevşehir has a population of 284,025, according to the Nevşehir Tourism Managers' records, 548,275 domestic (32.5 % of total) and 1,138,487 foreign (67.5 % of total) tourists have visited the region in 2009. International tourists include Japanese, Spanish, French, German, Italian, and American, and those visitors are generally well educated. The main tourist activities in the Cappadocia region include hot air ballooning, trekking, cycling, and horseback riding (Erdoğan and Tosun 2009). Besides, there are touristic cave houses, rock-cut structures converted into restaurants, hotels, shops, and pubs in Cappadocia. There is no accommodation problem in the region.

'Katakekaumene' or Kula volcanics in Western Anatolia is one of the important tourist destinations of Turkey (Fig. 6.20). The area between Kula and Karataş settlements and the Demirköprü dam-lake have been formed by extensive tectonics and represent a landscape with volcanic elements such as volcanic cones, explosion craters, lava flows, fairy chimneys and tuff layers with impressions of footprints of prehistoric humans (Figs. 6.21 and 6.22). All these features dominate the landscape and the structural surface (Koçman 2004). Volcanic activity in Katakekaumene continued from the middle Miocene to early Holocene periods. The soil surface has a dark colour for kilometres causing the vegetation to be very scarce on these volcanic formations, which gives the land a decidedly eerie appearance (Akdeniz 2011). There are 68 volcanic cones as well as fairy chimneys in this volcanic area. The volcanic cones are locally known as 'divlit' and 'devlit' basalt lava flows (Koçman 2004).

The most interesting feature in the Kula volcanic area however are the footprints of prehistoric humans (Fig. 6.23), which were discovered during the construction

Fig. 6.23 A View of unique prehistoric footprints in a layer of volcanic tuff in the Kula Volcanic Global Geopark. *Source* http://tr.geoparkula.org and volkan filik

of the Demirköprü Dam (1954–1960). In the first stages of research, 200 footprints were found with 60 of these prints removed and taken to the depository of the Natural History Museum in Ankara. At least one footprint was taken to the Manisa Museum and the Ege University Natural History Museum. Today, there are no less than 12 different size human footprints on this land (Akdeniz 2011). The Kula volcanic area has been officially recognised by the UNESCO as a Global Geopark and is a member of the European Geopark Network (September, 2013) based on the region's scientific and universal values. Moreover, this area has also a cultural heritage including Kula houses, traditional handicrafts, a historical bazaar, and special local cuisine.

When we evaluate the distribution of volcanic areas in Turkey, it is important for the development of tourism to determine the benefits and disadvantages (using a SWOT analysis, see Fig. 6.24) of these natural resources as tourist attractions.

6.5 Conclusion

Tourism in Turkey is mostly based on sun, sea and sand; however, alternative tourism has increased recently. In particular, it has a big potential in terms of volcano tourism due to volcanic mountains and features in Eastern and Central Anatolia. Some of the destinations offer nature based activities such as winter sports, climbing and trekking, with a number of them having specialised as geoparks, such as Mt Ararat, Mount Nemrut and Caldera, Cappadocia and the Kula volcanoes.

Nevertheless, there are some problems regarding volcano tourism in Turkey. One of the major issues facing volcano tourism today is the evaluation of what is significant in geological heritage, land use planning and future use of geotourism resources. Another issue is the lack of an integrated conservation approach and management strategy as these geosites and their resources should be preserved for future generations. To protect the unique scenery, it must have significant natural and cultural values and benefit the local people for their participation in this type of tourism. Geotourism would also raise their awareness about geography, geological sciences, and environmental matters through organised education programs. Materials that can be used for the advertisement of geosites should include books, displays, natural museums, and computer animations. For the development of geotourism, it is important to find financial support and qualified human resources.

Fig. 6.24 The benefits and disadvantages of volcanic environments as tourist attractions. *Source* Gülpınar Akbulut

Strengths	Weaknesses
• The richness of volcanic areas and its features such as volcanic mountains, craters, calderas, maars and volcanic lakes • Different transportation system usage • The richness of natural and cultural heritages • Endemic and relic flora and fauna • Educated tourist profile • Unique scenery • Suitable for different nature based sports (climbing, trekking, biking, skiing and balloon) • Geomorphological structure suitable for education and scientific field trips • Traditional handicraft, foods and wine	• Geotourism investment is insufficient • Accommodation in some regions is insufficient • Financial resources are insufficient • Local people have no knowledge about volcanic areas, geoparks and geotourism • Land use issues • Negative climatic conditions in some regions • Transportation problems in some regions
Opportunities	**Threats**
• Big cities in the vicinity of volcanic mountains and features • New job opportunities and income sources • Geological and cultural structure • The development of transport infrastructure • The usage of Communication and Internet resources • Geotourism potential	• Destruction of natural areas by tourists • Negative effect on cultural and social life of host communities • Environmental problems, especially litter and pollution on tourist routes • The decrease of the green park • Destruction of the endemic and relic flora and fauna

Acknowledgments This work has been supported by Akbulut's other studies. I would like to thank Abdulrahman Aksoy, Hakan Baykal, Ömer Ünsal, Bitlis Province Cultural and Tourism Manager for permission to use the photos and Patricia Erfurt-Cooper for correcting the text and improving the English.

References

Akbulut, G. (2009a). The main geotourism resources of Turkey, geographical diversity. In K. Donert, Y. Arı, M. Attard, G. o'Reilly & D. Schmeinck (Eds.), *Proceedings of the Herodot Conference in Ayvalık, Turkey* (pp. 227–233). May 28–31, 2009.

Akbulut, G. (2009b). *Suggested Geoparks in Turkey: Volcanic Mountains, New Challenges with Geotourism Proceedings of the VIII European Geoparks Conference* (pp. 264–269). Portugal.

Akbulut, G. (2012). Volcanic Features in the around lake van in the eastern anatolia region of Turkey as a suggested geopark. In *The 10th European Geoparks Conference* (p.12–19). Langesund, Norway, Sep 16–18, 2011.

Akbulut, G. & Gülüm, K. (2012). A suggested geopark site: Cappadocia, *The Journal of Balkan Ecology, 15*(4), 415–426.

Akbulut, G. (2013). The geotourism potential of Ardahan-Kars-Ağrı-Iğdır provinces. I. Uluslararası Iğdır Sempozyumu Sosyal Bilimler Bildiri Kitabı (19–21 Nisan 2012) (Ed.: Cengiz Atlı), Iğdır (in Turkish), pp. 251–279.

Akdeniz, E. (2011). Some evidence on the first known residents of katakekaumene (burned lands). *Mediterranean Archaeology and Archaeometry, 11*(1), 69–74.

Arınç, K. & Kaya, F. (2004) Doğubayazıt İlçesi'nin Turizm Potansiyeli, *Güneşin Doğduğu Yer: Doğubayazıt Sempozyumu*, Çekül Vakfı, pp. 433–450.

Atalay, İ. (1994) *Türkiye Coğrafyası*, Ege Üniversitesi Basımevi, İzmir.

Atalay, İ. (2002a). Türkiye'deki Dağlık Alanların Oluşumu, Yapısal ve Ekolojik Özellikleri, Türkiye Dağları, *I. Ulusal Sempozyumu (25–27 Haziran 2002)*, Orman Bakanlığı Yayınları. Vol 183, pp. 12–23.

Atalay, İ. (2002b). Mountain ecosystems of turkey. *7th International Symposium on High Mountain Remote Sensing Cartography, ICA, 2002*, 29–38.

Aydan, Ö., & Ulusay, R. (2003). Geotechnical and geoenvironmental characteristics of man-made underground structures in Cappadocia, Turkey. *Engineering Geology, 69*, 245–272.

Aydar, E., Gourgaud, A., Ulusoy, I., Digonnet, F., Labazuy, P., Sen, E., et al. (2003). Morphological analysis of active mount nemrut stratovolcano, eastern Turkey: evidences and possible impact areas of future eruption. *Journal of Volcanology and Geothermal Research, 123*, 301–312.

Bowen, R. (1990). The future of the past at Göreme in Turkey. *Environmental Geology and Water Sciences, 16*, 35–41.

Doğaner, S. (1995). Peribacalarının Turizm Bakımından Önemi. *Türk Coğrafya Dergisi., 30*, 25–39.

Dowling, R., & Newsome, D. (Eds.). (2006). *Geotourism: Sustainability, impacts and management*. London: Elsevier Butterworth Heinemann.

Emiroğlu, M. (1975). Ihlara Vadisi. *Coğrafya Araştırmaları Dergisi, 7*, 45–108.

England, A., Eastwood, W. J., Roberts, C. N., Turner, R., & Haldon, J. F. (2008). Historical landscape change in cappadocia (central Turkey): A palaeoecological investigation of annually laminated sediments from Nar Lake. *The Holocene, 18*(8), 1229–1245.

Erdoğan, N., & Tosun, C. (2009). Environmental performance of tourism accommodations in the protected areas: Case of Goreme historical national park. *International Journal of Hospitality Management, 28*, 406–414.

Erfurt-Cooper, P. (2010). Introduction to volcano and geothermal tourism. In P. Erfurt-Cooper & M. Cooper (Eds.), *Volcano and Geothermal Tourism—Sustainable Geo-Resources for Leisure and Recreation*. London: Earthscan.

Erinç, S. (1971). *Jeomorfoloji II*, İstanbul Üniversitesi Yayınları No: 1628, İstanbul Üniversitesi Coğrafya Enstitüsü Yayınları No.23, İstanbul.

Gönençgil, B., Kayacılar, C., and Diğerleri. (2003). *Erciyes Dağı'nın Ekosistem Özelliklerinin Belirlenmesi ve Dağ Alanları Yönetimi (DAY) Açısından Değerlendirilmesi*. TÜBİTAK-YDABAG 101Y055 Numaralı Proje.

Göney, E. (1988). *Nevşehir Yöresinde Yeraltı Kentleri*. Turizm Yıllığı, pp 93–104.

Güner, İ. (2000). Touristic Potential of Ağrı Mountain. *Eastern Geographical Review, 4*, 385–387.

Gürbüz, O. (1995). Turizm Coğrafyası Açısından Nemrut Kalderası. *Türk Coğrafya Dergisi, 30*, 255–265.

Karaoğlu, O., Özdemir, Y., Tolluoğlu, A. U., Karabıyıkoğlu, M., Kose, O., & Froger, J. L. (2005). Stratigraphy of the volcanic products around nemrut caldera: Implications for reconstruction of the caldera formation. *Turkish Journal of Earth Sciences, 14*, 123–143.

Koçman, A. (2004). Yanık Ülke'nin Doğal Anıtları: Kula Yöresi Volkanik Oluşumları. *Ege Coğrafya Dergisi, 13*, 5–15. İzmir.

Koçyiğit, A., & Erol, O. (2001). A tectonic escape structure: Erciyes pull-apart basin, Kayseri, Central Anatolia Turkey. *Geodinamica Acta, 14*(1–3), 133–145.

Kopar, İ. (2007). *Hasan Dağı ve Yakın Çevresinin Fiziki Coğrafyası*, Gündüz Yayıncılık, Ankara.

Nevşehir, Ağri, Van, Bitlis and Kayseri Tourism Managers 2009–2011 tourist records.

Somuncu, M. & Yiğit, T. (2009). Göreme Milli Parkı ve Kapodokya Kayalık Sitleri Dünya Mirası Alanı'ndaki Turizm Sürdürülebilirlik Perspektifinden Değerlendirilmesi. *V. Ulusal Coğrafya Sempozyumu 2008* (pp 387–402). Ekim 16–17, 2008.

The Ministry Cultural and Tourism. (2010). Nevsehir, Agrı, Van, Bitlis and Kayseri Tourist statistics (2009–2011).

The Ministry Cultural and Tourism. (2011). Tourist statistic, Ankara.

Ulusoy, İ., Labazuy, P., Aydar, E., Ersoy, O., & Çubukçu, E. (2008). Structure of the Nemrut caldera (Eastern Anatolia, Turkey) and associated hydrothermal fluid circulation. *Journal of Volcanology and Geothermal Research, 174*, 269–283.

Wartiti, M., Malaki, A., Zahraoui, M., Ghannouchi, A. & Gregorio, F. (2007). Geosites inventory of the northwestern tabular middle atlas of morocco. *Environment Geology*, 2–7.

7 Volcanic Heritage of the Carpathian–Pannonian Region in Eastern-Central Europe

Szabolcs Harangi

7.1 Introduction

The Carpathian-Pannonian region in eastern-central Europe appears to be a geologically calm area, where no disastrous earthquakes, no devastating volcanic eruptions occur. However, this is just the present status of this area; the past 20 million years were much more different (Harangi, 2011). This was the period when the Carpathian basin or Pannonian basin as the scientists call it, was formed behind the uplifting orogenic chain of the Carpathians. The reasons why this region is considered as one of the most important natural laboratories of the Earth are the supposed subduction along the foreland of the Carpathian chain with a still hanging deep vertical slab beneath the south-eastern Carpathians, the lateral movement and rotations of microplates, a dramatic thinning of the continental crust and the lithosphere, as well as major subsidence and the consequence formation of deep basins. In addition this complex tectonic progression was accompanied by varied types of volcanic activities, further subsidence and a final tectonic inversion with coeval general uplift. The long-lasting volcanism formed extended ignimbrite plateaus, large composite volcanoes, coalescing lava-dome complexes and different basaltic volcanic landforms from shield volcanoes to deep maars (Lexa et al. 2010; Harangi 2011). The extensive volcanism has gradually calmed down and the volcanic landforms have changed considerably, leaving the eroded remnants of the volcanic edifices. However, this transformation provided a unique benefit, i.e. a spectacular insight into the nature and the structure of the inner parts of the volcanoes.

There are many opportunities to start a volcano discovery tour in this region. One can begin with an amazing walk observing different types of pumiceous pyroclastic flow deposits, i.e. ignimbrites and the fabulous conical fairy chimneys or beehive stones as the local people call them. Then, the visitors can take a path along one of the most destructive volcanic deposits, formed during nuée ardente events. These valley-fill block-and-ash flow deposits are now in an inverted position and build mysterious cliff towers. Next, they could see also the scar remained after the huge sector collapse of an andesitic composite volcano. Let's have a look beneath the volcanoes! It is not so difficult in this region since many volcanoes eroded deeply exhibiting their root zones. A spectacular journey into the inside of a basaltic volcano along a volcano path starts at the gate of the first volcano park in this area.

In the Carpathian-Pannonian region, seemingly there are no active volcanoes, yet it is an open book for visitors, who want to have an exciting insight into the wide range of volcanic phenomena within short distances. Furthermore, this is the area where volcanic heritage meets historic, cultural, gastronomic and winery pleasures, among others. Nevertheless, this area is a challenging destination also for those people who want to visit only active or potentially active volcanoes. The last volcanic eruption occurred here at the dacitic Ciomadul volcano at 30 ka ago. A large subplinian eruption resulted in a deep explosion crater where now the picturesque St. Anna Lake is found. The last basaltic volcanic eruption took place at the northern part of this region within a deeply eroded andesitic volcanic area. This occurred at about 100 ka following approximately 6 Ma quiescence. Basaltic volcanic fields are unique types of volcanoes with long-lasting, but intermittent activities. It is difficult to claim that such a volcanic field is already extinct even after several 100's ka of repose time. The youngest basaltic volcanic field of this region is located close to Ciomadul. The first eruptions in the Perşani area occurred at 1.2–1.3 Ma and continued intermittently until 500–600 ka. A recent geophysical research indicates that

S. Harangi (✉)
MTA-ELTE Volcanology Research Group, Pázmány sétány 1/C, H-1117, Budapest, Hungary
e-mail: szabolcs.harangi@geology.elte.hu

S. Harangi
Department of Petrology and Geochemistry, Eötvös Loránd University, Pázmány sétány 1/C, H-1117, Budapest, Hungary

P. Erfurt-Cooper (ed.), *Volcanic Tourist Destinations*, Geoheritage, Geoparks and Geotourism,
DOI: 10.1007/978-3-642-16191-9_7,

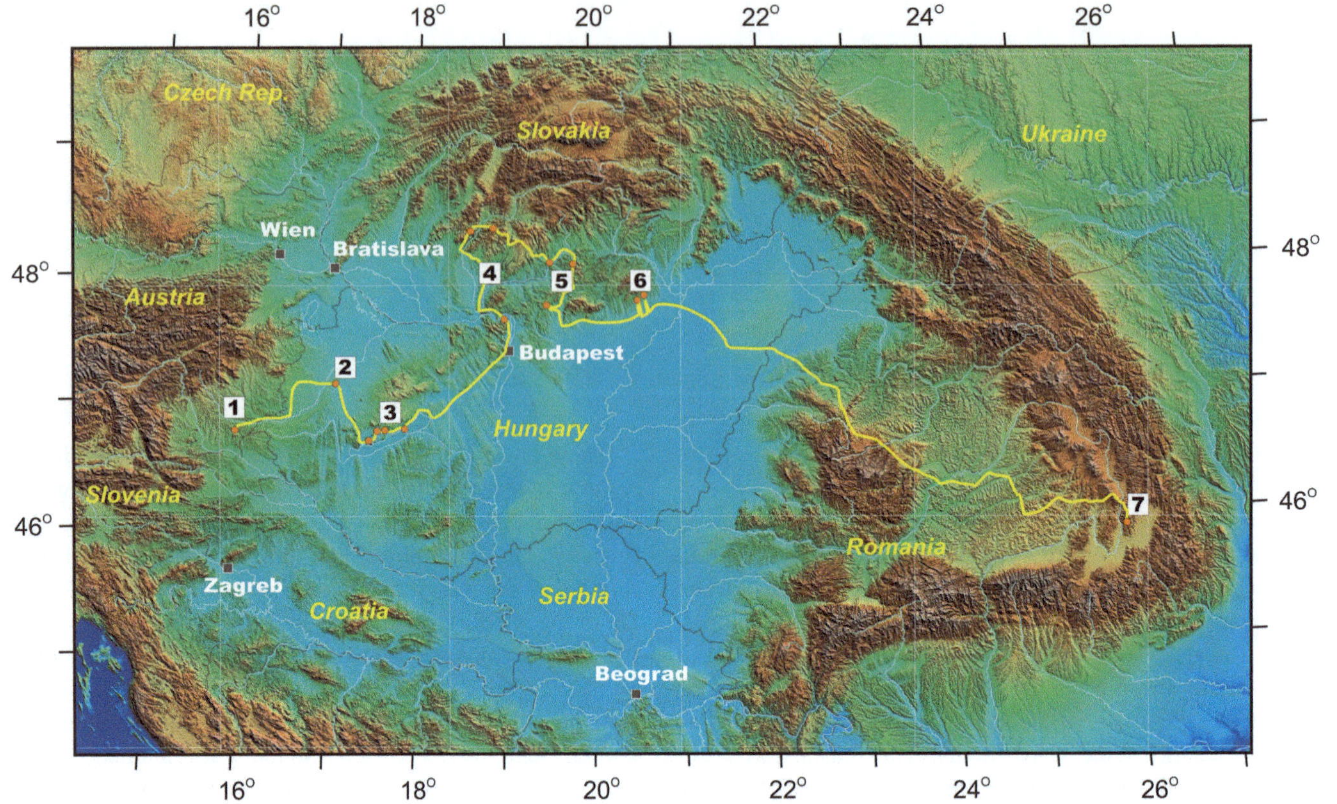

Fig. 7.1 The suggested volcano route in the Carpathian-Pannonian region with selected stops of volcanic spectacles. *1* Styrian volcanic field; *2* Kemenes Volcano Park, *3* Bakony-Balaton Geopark, *4* Visegrád Mts. and the Banska Štiavnica area, *5* Novohrad-Nógrád Geopark, *6* Bükkalja, *7* Ciomadul volcano. *Source of the topographic map* http://geophysics.elte.hu/atlas/geodin_atlas.htm

seismic anomalies could exist both beneath this volcanic field and beneath the Ciomadul volcano, so continuation of the volcanism cannot be excluded.

It is not easy to select only a few places which could be on the must-see list, but this chapter may serve as an introductory review to come and look around this region to discover volcanic heritage and have a rest in this area where hospitality is a tradition. This chapter provides a possible route from west to east with several suggested stops (Fig. 7.1).

7.2 Volcanoes, Chocolate, Wine and Much More: The Steirische Vulkanland

Our volcano tour starts at the Styrian monogenetic basalt volcanic field at the westernmost margin of the Pannonian basin in Austria. Eruptions occurred here from 3.7 to 1.7 Ma building small shield volcanoes, scoria cones, polygenetic basalt volcanoes and maars. Many of them contain high-quality lava rocks and therefore they are intensively quarried. The best known volcano of this area is located at Kapfenstein, where the volcanic deposits contain large amount of mantle-derived ultramafic xenoliths and olivine xenocrysts (Fig. 7.2). There are also abundant rounded quartz pebbles, while juvenile clasts are very rare. This indicates maar-forming phreatomagmatic explosive events occurring probably in a fluvial valley. The energetic explosions generated deep excavations of the subsurface rock beds while the volcanic jets accompanied radially expanding pyroclastic surges, which moved along the fluvial deposits and incorporated large amount of pebbles. A 2 km long geotrail with 11 stops provides a pleasant and easy walk around the eroded remnants of the volcano. The uniquely prepared explanation panels with German text give a brief overview of the main features of the volcanic deposits, the formation of the volcano as well as the further Styrian basaltic volcanoes close to Kapfenstein (Fig. 7.3). Olivine is a kind of trademark of this locality and therefore it is not a surprise that a local winery (Winkler-Hermaden) produces a fantastically concentrated Zweigelt called Olivin (Fig. 7.4).

The beauty of the volcanic landscape such as another remnant of a maar volcano with a castle called Riegersburg on its top (Fig. 7.5) inspired the local people to construct a common philosophy for development and that is the

Fig. 7.2 Ultramafic xenoliths, olivine xenocrysts and abundant lithic clasts indicate a typical maar volcanic deposit in Kapfenstein. *Photo* Szabolcs Harangi

Fig. 7.3 Explanation panels with German text provide an overview of the main features of the volcanic heritage. *Photo* Szabolcs Harangi

Steirisches Vulkanland (www.vulkanland.at/). There are many tourist attractions within this framework, such as the Zotter Chocolate Manufakturer, where the visitor could get information all about chocolate, from the bean to the cult. The guided tasting tour around this special world of chocolate provides valuable information as well as numerous creatively arranged titbit/treat stations. There are many more culinary experiences in this region as well as wide range of hand-made products such as magma-slippers and volcano-hats, among other things.

Fig. 7.4 The famous Zweigelt called Olivin provided by the local Winkler-Hermaden winery. *Photo* Szabolcs Harangi

Fig. 7.5 A beauty of the Steirisches Vulkanland: the Riegersburg Castle is sitting on the eroded remnant of a maar volcano. *Photo* Szabolcs Harangi

7.3 A Journey into the Inside of a Basaltic Volcano: The Kemenes Volcano Park

About 130 km from Kapfenstein, a newly developed volcano park called Kemenes Volcano Park (www.kemenesvulkanpark.hu/) which opened officially in 2013 provides a must-see stop for volcano lovers. The idea for the construction of a volcano park in the middle of a flat plain (called Kisalföld, i.e. Little Hungarian Plain) was raised in the 1990s, when intense volcanological research of the basaltic volcanoes commenced in Hungary. This started with the study of the Ság volcano by the researchers and students of the Department of Petrology and Geochemistry, Eötvös University. The Ság hill was extensively quarried for the high quality basalts (many of the roads in Vienna and in Budapest were made by these cube-shaped basalt

Fig. 7.6 The main crater of the 5 Ma old Ság volcano, with irregular fan-shaped basalt columns and red spatter deposit on the top. *Photo* Szabolcs Harangi

blocks) and as a result, the inner part of the 5.3 Ma old basalt volcano has been exposed (Fig. 7.6). The unique remaining cliff walls convey the great moments of the volcanic history. The birth of the Ság volcano started with phreatomagmatic eruptions in a fluvial-marshy environment and this resulted in the formation of a wide tuff ring. Subsequently the role of mixing with external water-rich sediments decreased significantly and the style of volcanism changed to pure magmatic explosive type (Strombolian to Hawaiian). Scoria cones developed within the tuff ring and later on fire fountains were spewing along a 400 m long fissure. The accompanying lava flows filled slowly the crater of the volcano and formed finally a lava lake without breaching the tuff ring wall. In summary, the volcanic history of Ság hill involves all the principal types of basaltic volcanic activity and left a wide variety of basaltic volcanic products. The more then 50 years long quarrying revealed not only this volcanic heritage, but exposed additional spectacular volcanic phenomena such as impact sags beneath ballistic blocks (Fig. 7.7), peperites, i.e. intimate mixture of basaltic magma fragments and fine-grained volcaniclastic sediments along the contact of a feeding dyke, funnel–shaped vents and peculiar bread-structured basalt spatters, among others.

A volcano path with 12 stops including instructive explanation panels (with Hungarian text, but guides with English and German language could be obtained at the Ság Museum) helps the visitors to understand the main volcanic phenomena and the formation of a basaltic volcano (Fig. 7.8).

Fig. 7.7 Detail of the initial phreatomagmatic volcanic activity—alternation of base surge and fall deposits with ballistic bombs. *Photo* Szabolcs Harangi

Fig. 7.8 Walk on the new Volcano path with explanation panels. *Photo* Szabolcs Harangi

Fig. 7.9 An out-door activity place with plate tectonic puzzle and volcano columns at the Ság Museum. *Photo* Szabolcs Harangi

Each panel provides a question related to the relevant stop and the answer can be found on the following panel. In addition, the cartoons showing Volcano-Fellow, the symbol of the volcano path, in the explanation panels nicely reflect the main volcanic feature at each stop. Immediately at the foot of the volcano is a newly renovated museum with an interactive exhibition, which shows the volcanological, historical, botanical, ornithological and winery aspects as well as the quarrying techniques performed in the Ság hill. Around the Ság Museum a small play-ground provides a good place for the children, where they can test their knowledge on volcanoes by using the rotatable cubes of the volcano columns and where they can use a large wooden-puzzle to set up the plate tectonic picture of the Earth (Fig. 7.9). The outdoor rock-exhibition shows the most common volcanic and plutonic rocks of Hungary with the explanation plates telling of their age, locations and origin. But there is still more to visit here!

The nearby Volcano House (Fig. 7.10) was opened in 2013 and this is the first permanent interactive exhibition devoted to volcanoes in Eastern Central Europe. The visitors

Fig. 7.10 The Volcano House at the foot of the Ság hill provides an instructive insight into the nature of volcanic processes. *Photo* Szabolcs Harangi

Fig. 7.11 One of the most attractive parts of the exhibition in the Volcano House of the Kemenes Volcano Park shows the hidden secret of the extraterrestrial volcanoes. *Photo* Szabolcs Harangi

can test how is a walk on different types of lava flows and can see 3D maquette of the Etna and the volcanic areas of the Carpathian-Pannonian region. The exhibition provides an interesting tour in the world of volcanoes with explanation panels in English and German. One of the main attractions is a visual journey into the extraterrestrial volcanoes (Fig. 7.11). All in all, it is an intense full day program for volcano friends! In addition, there are more tourist destinations around the Kemenes Volcano Park. The famous Spa and Wellness Centre of Sárvár is only 20 km away, whereas Kissomlyó (10 km from Ság) provides a nice quiet place to experience the country life and go on a short walk along another volcano path of the nearby basaltic hill.

7.4 The Land of Calmed-down Volcanoes and Dinosaurs: The Bakony-Balaton Geopark

The heart of the Bakony-Balaton Geopark (www.bakony-balaton-geopark.hu/) area is only an 80 km easy drive (along the Route 84) from the Ság hill. The first unique

Fig. 7.12 The beautiful volcanic landscape of the Tapolca basin in the Bakony-Balaton geopark: eroded remnants of various basaltic volcanoes. *Photo* Szabolcs Harangi

Fig. 7.13 Cross-stratified base surge deposit at the wall of the Barátlakások of the Tihany maar volcano. *Photo* Szabolcs Harangi

place along the road, where the visitor has to stop is the beautiful medieval Sümeg castle sitting on a steep conical Cretaceous reef-limestone. The Geopark has an extent of 3100 km^2 and comprises 171 different geological formations of various ages. One of the main geologic attractions of this area is the spectacular basalt volcanic field formed from 7.9 to 2.6 Ma. More than 50 basaltic volcanoes developed during this time range from shield volcanoes (e.g. Kab-hill, Agár-tető) through complex polygenetic volcanoes (e.g. Badacsony, Szent-György hill) to maar volcanic complexes (e.g. Tihany, Pula; Martin and Németꞏ 2004). The original volcanic edifices however are strongly affected by the climatic fluctuations during the Pleistocene. Wind erosion and frost disintegration, accompanied by watercourses fed by abundant rainfall during the interglacial periods, carried away a significant amount of loose sediments, mostly consisting of phreatomagmatic deposits of the initial tuff rings. As a result of this strong erosional impact, only basalt-capped volcanic hills (butte) have remained, providing the unique landscape such as seen in the Tapolca basin (Fig. 7.12).

The volcanism took place mostly in a marshy environment following the long period, when the entire area was covered by an extensive lake (Pannon Lake). Semi-consolidated, water-saturated sediments were widespread and as hot basaltic magma mixed with them at shallow depth, violent phreatomagmatic explosions occurred. Occasionally, this magma-water interaction was taking place gradually at deeper level as diatremes propagated downwards. There, the type of water-storage system, i.e. the hydrological condition, changed from the upper porous media aquifer to karst fracture-filling aquifer. That means increasing amount of water mixed to the uprising basaltic magma. In this condition, the eruption jets were loaded by abundant fragments of deep-seated basement rocks, such as

Fig. 7.14 From (*left*) The Lavender House in Tihany is the central interpretive site of the Bakony-Balaton geopark. Pushing a button and you can see how an impact sag is formed beneath a ballistic block. A scoria cone at the corner of the Lavender House, where occasional eruptions occur. *Photos* Szabolcs Harangi

the Permian red sandstone and the Silurian schist. This unique change of the explosion locus can be intimately followed in the maar volcanic sequence of the Barátlakások (Monks cells) at the north-eastern margin of the Tihany peninsula (Fig. 7.13). Cells and a cave-church were carved out here in the phreatomagmatic sequence by Greek monks in the tenth to eleventh centuries and as a result this place offers a fine three-dimensional view of the volcanic formations. In this unique outcrop, the cross-layered structure of the base surge deposits and the asymmetrical impact sags present a special opportunity to reconstruct the former vents (eruptions occurred both at the present position of the Külső-tó, i.e., Outer lake and at the present Füredi-öböl, i.e. Füred bay; note that at that time Lake Balaton did not exist, it formed only about ten thousand years ago). The Tihany peninsula provides further beautiful attractions such as the post-volcanic hot-spring deposits, generally classed as geyserites, which are found mostly at the south-eastern and southern parts of the peninsula. A good walking path, called Lóczy-Geyser Nature Trail, guides visitors to these peculiar formations. Information about the zoological, botanical and landscape values along the path can be read in the explanation boards, both in English and Hungarian. The main visitor centre of the Geopark, the Lavender House, was also built in Tihany and was opened in 2011 and offers a highly instructive exhibition about the volcanic phenomena as well as the cultural and biological diversities of the area (Fig. 7.14).

In addition to Tihany, there are many other spectacular places of volcanic heritage in this new Geopark. The beautiful landscape of the Tapolca basin shows the typical erosional remnants of the former basaltic volcanoes (Fig. 7.12). The basalt-capped butte was formed where lava lakes filled the wide craters of initial tuff rings and this hard rock was resistant against the subsequent erosion (Fig. 7.15). The soft pyroclastic rocks eroded and as a result the columnar jointed structure of these lava lake rocks has been exposed. The tall

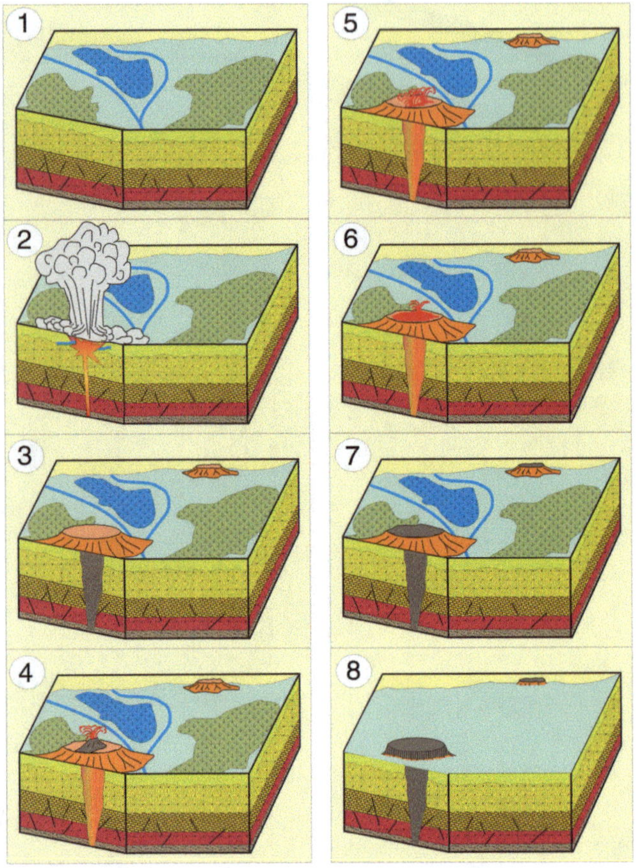

Fig. 7.15 Formation of the typical basalt-capped volcanic hills (*butte*) of the Kemenes and Tapolca basin. *1* Fluvial-marshy environment, *2* and *3* Violent phratomagmatic eruptions due to magma-water interaction forming tuff rings with wide craters, *4* and *5* Strombolian and Hawaiian explosive eruptions within the tuff ring, *6* Formation of lava lake, *7* Termination of the volcanism, *8* Subsequent erosion removed the loose pyroclastic deposits of the volcanoes and only the inner hard basalt lava lake rocks resisted and remained forming the present typical landform. *Source* Harangi (2011)

basalt columns were rounded primarily due to the frost disintegration and they look like giant organ pipes, as local people call them. The most famous ones can be seen at the Badacsony and at the Szent György hill (Fig. 7.16). Good walking trails at both hills make it easy for visitors to see these particular creations of nature and which lead to viewing points, where the panorama of the unique volcanic landscape of the Tapolca basin can be seen.

The nearby Káli basin offers at least two must-see places for volcano-friends. Szentbékkálla is a picturesque small village at the northern margin of the basin. It is famous for the abundance of mantle-derived ultramafic xenoliths found in the phreatomagmatic volcanic products (Fig. 7.17). However, just next to the old church at the northern part of the village a peculiar outcrop is found. In the lower part, a massive basaltic hydrovolcanic flow deposit is exposed with gas segregation pipes and with a wide range of colourful rock fragments. They are mostly lithic clasts, which represent almost the whole lithosphere from the upper mantle part (green-coloured peridotites) through rare granulites from the lower crust to the many rock types of the shallower crust. There are also rounded quartz pebbles indicating that the pyroclastic flow swept a gravely fluvial deposit in a paleo-valley. The steam of the water of that creek penetrated through the pyroclastic flow deposit forming the gas segregation pipes. This massive rock unit is overlain by cross-laminated deposit representing the diluted overbank facies (i.e. pyroclastic surge) of the flow. In addition to this peculiar volcanic formation, there is another geological spectacle west to the village. This is called 'Sea of stones' and contains big boulders and groups of rocky cliffs formed along the shoreline of the Lake Pannon. The sand, pebbly sand and gravel beds were cemented by silica. These blocks have been tilted frequently from their original position with one of them resulted in a specific balanced rock.

The half-cut hill of Hegyestű at the eastern margin of the Káli basin can be seen from far and presents another breath-taking exposure of the tall basalt columns. It is a deeply eroded neck of a 7.9 Ma old basaltic volcano and has remained after the quarrying of the basalts. The abandoned quarry serves now as one of the most important geological exhibition sites of the Geopark. The 336 m peak of the Hegyestű rises more than 200 m high above the shore of Lake Balaton and therefore its top is an excellent viewing point with a 360° panoramic view. A small exhibition shows the basic geology of the Balaton Upland area and the history of quarrying with a picnic area next to it.

Last but not least, from a palaeontological point of view one of the most valuable sites of the Geopark is a dinosaur locality at Iharkút (High Bakony), which was discovered in the early 2000s. Fossil vertebrate remains found here are unique in the world. Based on more than 10,000 isolated bone and tooth fossils, the presence of 30 vertebrate groups was revealed in the Late Cretaceous (85 million years old) sandy-clayey sediments of fluvial–alluvial plain facies.

Fig. 7.16 Variations of columnar jointing: the volcanic neck at Hegyestű and the so-called basalt organs representing the margin of a lava lake rock body at Szent György hill. *Photos* Szabolcs Harangi

Fig. 7.17 Colourful rock fragments in the unique basaltic hydroclastic flow deposit at Szentbékkálla: note the abundance of mantle derived ultramafic xenoliths. *Photo* Szabolcs Harangi

7.5 Andesitic Volcanoes of Northern Hungary and Southern Slovakia

7.5.1 Royal Palace on a Huge Volcanic Avalanche Deposit and Walks in the Roots of the Volcanoes

The formation of the Pannonian basin was accompanied by extensive calc-alkaline andesitic-dacitic volcanic activity, mostly along the inner side of the Carpathians. They are traditionally thought to be related by subduction of an oceanic plate; however, there is a lack of corroborated geological and geophysical evidence for this. Alternatively, they could have been related to the main extensional phase (12–18 Ma) of the evolution of the Pannonian basin, when partial melting of metasomatized lithospheric mantle and mixing of mantle-derived mafic magmas and crustal derived silicic melts resulted in the formation of calc-alkaline intermediate magmas. This volcanism commenced at about 16 Ma and terminated at 10 Ma in the northern part of the region, whereas at the east it lasted from 12 up to 30 ka. The volcanic activity built up groups of composite volcanoes mostly by summit lava dome outpourings and associated pyroclastic flows such as can be seen presently at Merapi or at the Kamchatkan volcanoes. As a result, steep-sided conical volcanoes developed that were subsequently undergoing catastrophic sector collapses (Karátson 2007). The remnant of one of these associated landslide-debris avalanche deposits can be seen superbly just beneath the walls of the Royal palace at Visegrád in the picturesque Danube-bend. The Visegrád Mountains however offer even more points of volcanological interest. The highest point of the mountain is Dobogókő (700 m a.s.l.), which is a place of cultic significance for some Hungarians. Just beneath this ridge are fabulous tall cliffs standing steeply in the thick forest and represent the deposits of devastating nuée ardente block-and-ash flows. Thus, the current highest point once was a deep valley, where such pyroclastic flows could rush down. This also indicates that at that time (at about 15–16 Ma) a volcanic cone developed in this area, the Keserűs volcano. After a sector collapse, further erosion revealed the deeper structure of this volcano. The deep ravines bordered by narrow cliffs of volcaniclastic deposits such as the Rám-gorge offer challenging excursions inside the volcanic heritage (Fig. 7.18). Another unique product of the erosion can be seen in Kő-hill at the eastern margin of the Visegrád hill, where big andesite blocks are sitting amazingly on the top of narrow rock columns (Fig. 7.19).

An additional spectacular place to visit and see what is under the large calc-alkaline andesitic composite volcanoes is in Southern Slovakia. The Banska Štiavnica and Kremnica areas are famous for their ore deposits. Their history is closely linked to the exploitation of its abundant resources such as silver and gold. The first mining settlement was founded by Celts in the 3rd century. In the Middle Ages, this area was the main producer of silver and gold in the Kingdom of Hungary and in 1735, one of the first mining schools in the World was founded in Banska Štiavnica (Fig. 7.20). The Banska Štiavnica Geopark (www.

Fig. 7.18 The Rám-gorge provides challenging excursions within deeply eroded volcaniclastic deposits. *Photo* Szabolcs Harangi

Fig. 7.19 An amazing product of erosion: spectacular 'rock tables' in Kő-hill. Big andesite blocks are sitting on narrow rock columns. *Photo* Szabolcs Harangi

Fig. 7.20 Banska Štiavnica, an old mining town is hidden in the erosional remnant of a calc-alkaline volcanic complex. Within the town, the eroded neck of a 6 Ma old basaltic volcano is found (Kálvária-hill, i.e. the Calvary hill) where volcanic and cultural heritage meets. *Photo* Szabolcs Harangi

geoparkbs.sk), which is presently not an official member of the European Geopark Network, offers an excellent educational outline about volcanism and its related ore generation and mining history of the area. A part of this history is found within the town in the rich Mining Museum; however, there is also an instructive geological path with 19 stops starting in the Červená studňa saddle in the Paradajs Mountain. The explanation boards provide a broad insight into the formation of the Štiavnický stratovolcano. There is a marked green touristic sign leading from the Square of St. Trinity which connects with the sign of the instructive path in Červená studňa. The undemanding route (about 2.5 h long) is also suitable for children and elderly people, free from any orientation problems and it is conceived in the form of circuit around the Paradajs Mountain. Finally, it is worth to mention that the youngest basaltic volcano of the entire Carpathian-Pannonian region is found close to Banska Štiavnica. The Putikov volcano formed at about 100 ka near the present village of Nová Baňa. Presumably tall lava fountains built a spatter cone within the erosional remnant of the andesitic volcanic landscape and finally the basaltic lava travelled several kilometres burying the older terraces of the river Hron. It is noteworthy that this volcanic activity is suggested to have occurred after several million years of quiescence in this area.

7.6 Where Volcanic Heritage meets Historic and Cultural Events: The Novohrad-Nógrád Geopark

The Novohrad-Nógrád Geopark (www.nogradgeopark.eu/) is the first 'across border' geopark situated in northern Hungary and southern Slovakia. One of the main driving forces for the application to become officially recognised as a geopark was the rich volcanic heritage of this area which includes:

- damaging pumiceous ash-flows;
- submarine and subaerial lava flows;
- one of Europe's largest coherent lava plateaus;
- exposed subvolcanic bodies and volcanic vents, maars and diatremes;
- 'petrified' gas bubbles and lava spatters;
- platy and columnar jointed basalts and andesites including a unique 'andesite-slide';
- garnet in the volcanic rocks, and
- fragments from the upper mantle.

All of this within a restricted, small area makes it without doubt an excellent place to gain a unique insight into volcanogenic processes. The volcanic phenomena here are strongly linked to cultural and historical heritage of the region. Beautiful medieval castles (e.g. Fiľakovo, Somoskő, Salgó and Hollókő) are built on top of the former vents of the basalt and andesite volcanoes and contribute to the rich cultural tradition of the area (Fig. 7.21).

While not all the spectacular sites can be discussed here, there are a few must-to-see places that are certainly unique, such as the Nature Reserve of the Ipolytarnóc fossils and the curvilinear columns of basalts and andesites at Somoskő and Bér, respectively. The name of Ipolytarnóc became famous after a petrified tree that is almost 100 m long and has a circumference of eight metres. This makes it probably one of the largest petrified pine trees in the world. The tree was washed out from a volcanic layer deposited following a large devastating eruption at about 17.5 Ma. Hot pumiceous ash layers buried a vast area and destroyed the subtropic vegetation, but on the other hand, remnants of the rich flora as well as many footprints of various animals at a watering place were nicely preserved. Based on this particular preservation, and the intricate research work, which reconstructed this unique place, the Ipolytarnóc Paleontological Site protected area (http://osmaradvanyok.hu/index.php?p=hu_home) received the European Diploma in 1995. The newly reshaped visitor centre close to the village offers an interesting outline of this geological heritage and a movie theatre with world-class 3D animation introduces visitors to the prehistoric past. The associated trail network consists of five introductory study paths starting from the visitor centre with guided tours available. The geological study path in the Borókás creek reveals a treasure of paleontology, including shark teeth in the shallow sea sediments, the palaeosurface of an ancient dry land dotted by thousands of footprints of mysterious creatures and remnants of a petrified forest, all recalling the memories of a prehistoric past. The trail is accessible even in rainy weather, including by wheelchair. The pumiceous ignimbrite is well exposed above the footprint sandstone and contains many charcoal fragments indicating the high temperatures at the time of deposition. The footprint sandstone as well as part of the large petrified pine tree is exposed in covered exhibition halls and in a protective cellar. At the end of the trail, the Tasnadi Kubacska Hall provides a spectacular outline of the richness of the various footprints and the high-tech 3D movie projecting the prehistoric track gives a unique interpretation of a vanished era.

The next site recommended for a visit is located just at the Hungarian-Slovakian border, which is now free to cross. The medieval castle of Somoskő was built on a volcanic neck of a 3 Ma old basaltic volcano with exposed convex curvilinear columns which are likely to represent the outer margin of the neck (Fig. 7.22). Beneath the 'rock-cascade' is a wide 'stone-sea', where the eroded fragments of the steep cliff have accumulated. A short study path (free of charge) with explanation panels guides the visitors explaining both the main geological and the biological features. There is no entrance fee either to visit the ruins of

Fig. 7.21 Castles on volcanoes: Somoskő (**a**) and Filakovo (**b**). *Photo* Szabolcs Harangi

the medieval castle of Somoskő. Another unique curvilinear columnar jointed rock is exposed 60 km further south close to the village of Bér (Fig. 7.22). As far as we know, there are only a few examples in the world where such a good example of a concave curvilinear columnar jointed structure is preserved. It is locally known as the 'andesite-slide' and is exposed in the Nagy-hill, where a short path leads visitors to the site. This structure could have formed either in a volcanic neck or in a lava flow burying a narrow deep valley as the andesitic magma slowly cooled down.

Fig. 7.22 Variation of curvilinear columns: convex (**a**) and concave (**b**) columnar jointed rocks. The basaltic rock-cascade in Somoskő and the andesite-slide in Bér. *Photo* Szabolcs Harangi

7.7 The Land of Devastating Ignimbrite Deposits, Inviting Wine Cellars and the Fabulous Fairy Chimneys: The Bükkalja Area

The Miocene to Quaternary volcanism of the Carpathian-Pannonian region started with explosive eruptions of large volume silicic magmas. This volcanic activity coincided with the initial thinning of the continental crust and lithosphere beneath the area. The caldera-forming eruptions resulted in dominantly thick pumiceous pyroclastic flow deposits, called ignimbrite. In some places the thickness of the ignimbrite unit exceeds 200 m! In such thick ignimbrite sequences welded facies were often formed by sintering of pumices and silicic glass shards at high temperature and compressive loads. As a result, elongated fiamme occurs frequently in volcaniclastic rocks. The southern foreland of the Bükk Mountains (called Bükkalja) provides an excellent insight into the nature of the ignimbrite deposits. It is indeed a volcanological paradise of ignimbrites showing various facies, i.e. non-welded to welded types of such silicic pyroclastic flow deposits. The repetitive violent eruptions for an about 7 million year long period (from 20 to 13 Ma) could have resulted in an ignimbrite plateau. The subsequent erosion dissected this plateau and deep creeks were formed (a good example of this bad-land landscape can be seen at Kazár, close to Somoskő) which then widened, becoming broad valleys and decreasing significantly the extent of the ignimbrite surface. One of the peculiar results of such erosion is the formation of fabulous fairy chimneys at the retreating steep margins of the valleys. These conical geomorphologic features are locally called beehive stones (Fig. 7.23), because of the rectangular 'windows' carved into them up to heights of 10–12 m. There are dozens of such conical cliffs in Bükkalja (72 beehive stones with 473 chambers are currently known in this area) and almost all include the enigmatic carved 'windows'. The origin of these 'windows' is still unclear and highly debated, but the most

Fig. 7.23 The land of devastating ignimbrite deposits: (*left*) A fabulous conical fairy chimney in Bükkalja—typical erosional forms of the un-welded ignimbrites. (*right*) Fairy chimney castle of un-welded ignimbrite at Szomolya, one of the most spectacular places of beehive stones in Bükkalja. *Photo* Szabolcs Harangi

Fig. 7.24 Wine cellars in the ignimbrite. They hide barrels of well-known wines of the Eger area. *Photo* Szabolcs Harangi

widely supported view is that the carved compartments were used for beekeeping. The local name of these fairy chimneys refers to this supposed origin of the 'windows'. Another idea for their mysterious origin suggests that idol statues were placed into them, or they were used as an urn cemetery. The most beautiful group of beehive stones in Bükkalja is found at the northern edge of the village of Szomolya. Explanation boards and direction indicators clearly show the location of the exceptional site right from the centre of the village. Along the road, wine cellars excavated in the ignimbrite can be seen that are another typical feature of this volcanic area (Fig. 7.24).

Fig. 7.25 The cave-dwellings of the Little America at Cserépfalu. *Photo* Szabolcs Harangi

The local wine is widely respected. Viticulture began in the region as early as the eleventh century. The cultivation of wine was not interrupted even during the 91 years of Turkish occupation, when wine was an important source of income for the Turks. The typical wines of the Bükkalja (or Eger) region is called Leányka, Királyleányka, Hárslevelű (Linden-Leaf), Olaszrizling (Italian Riesling), Muskotály (Muscatel), Tramini, Szürkebarát and Chardonnay. The spiciness, fieriness and relatively high acidity of the red wines of Eger is what typifies Bikavér (Bull's Blood) which is why local wine-growers put most of their energy into producing it. The method of preparing this noble wine is based on age-old traditions and was finally set down in the Bikavér Codex. Walking in this ignimbrite field, one can often find wine-cellars carved into the ignimbrite at the outer parts of the villages and towns and one has a good chance to find them open for a quick (or long?) wine-tasting.

In order to have a good outline of the various types of ignimbrite facies (unwelded and welded), probably the best is to take a short route from the northern edge of Bogács (where of course excellent wine cellars are also found) to north to the village of Cserépfalu. Just above the wine-cellars of Bogács (carved into unwelded ignimbrite), good outcrops of fiamme-bearing welded ignimbrite can be found following the tourist path indicated by a red line. Further north, leaving the village of Cserépfalu, a dirt road leads to the Mész-hegy hill, where an explanation board indicates the direction of a forest nature trail where good examples of beehive stones can be found with the so-called famous 'Devilstower'. Just at the other side of the dirt road, another small forested hill, called Túrbucka, offers a good opportunity to have a closer look at shiny, fiamme-bearing welded ignimbrite rocks. Back to Cserépfalu, it is recommended to visit the so-called Little America (eastern side of the village at Berezdi walk; the direction is marked from the village centre). These cave-dwellings (Fig. 7.25) served as home for poor people, who could not immigrate to the American continent during the economic crisis which affected the world during the beginning of the twentieth century. They could reach only this place instead of America and therefore people said that they were able to emigrate only to 'Little America'.

7.8 The Youngest Volcano of the Region: The Ciomadul

It is about 650 km (about 9 h) drive from the Bükkalja region to reach the youngest volcano of the Carpathian-Pannonian region, near the village Tusnádfürdő (or Baile Tusnad, an area in Romania where mostly Hungarian people called Székely are living. Therefore we give both the Hungarian and Romanian name of the geographic places). This volcano is called Ciomadul (or Csomád in Hungarian), a dacitic lava dome complex with two picturesque craters in the central part (Fig. 7.26). The older one (Mohos swamp) is covered by a 10-meter-thick peat layer and contains numerous colourful tiny pools of water. Entrance is only permitted for guided tours and visitors can follow a wooden boardwalk just above the swamp. The younger crater is filled by the 6 m deep St Anna Lake, the best preserved crater lake of Europe. This is an explosion crater formed presumably by a violent subplinian eruption at about 30 ka. The eruption cloud shifted south-eastwards and deposited a

Fig. 7.26 The youngest volcano of the region, the Ciomadul—could it be reactivated in the future? The Ciomadul lava dome complex from north. *Photo* Szabolcs Harangi

Fig. 7.27 The colourful Mohos swamp, the older crater of the Ciomadul formed by phreatomagmatic eruption. *Photo* Szabolcs Harangi

15 cm thick layer of 2–3 cm sized pumice clasts as far away as 20 km (the outcrop is found in an abandoned sand quarry at Kézdivásárhely/Târgu Secuiesc). The proximal deposits of 1–2 m thick pumiceous fallout layers are found mostly around the Mohos swamp (Fig. 7.27).

The volcanic activity of Ciomadul started about 150 ka and occurred intermittently with fairly long quiescence periods. Outpourings of viscous, crystal-rich dacitic magma formed steep-sided lava domes, which coalesced into a larger volcanic edifice. This kind of volcanism was occasionally interrupted by explosive eruptions which formed the two craters. Explosive and gravitational collapse of lava domes resulted in devastating nuée ardentes. Their block-and-ash flow deposits may contain charcoal fragments allowing the determination of their formation age. These outcrops are found mostly at the southern periphery of the volcano along the dirt road from Sepsibükszád (Bixad) to the St. Anna Lake (Fig. 7.28). There are many forest tourist paths around the craters, where one can observe the huge blocks of the lava dome rocks. Note, that bears are living in the forests of this area and therefore certain precautions are needed, although the danger is limited.

The present state of the Ciomadul volcano is unclear regarding whether further eruptions can be expected. In fact, there are many signs suggesting that this volcano could be potentially active, even though the last eruption occurred at about 30 ka. Geophysical research indicates that the magmatic body beneath the volcano could contain some amount of silicic liquid. Furthermore, there are intense gas emissions (mostly CO_2 with some H_2S) around the volcano. One

Fig. 7.28 St Anna Lake—a deep explosion crater formed by a subplinian eruption at 30 ka. *Photo* Szabolcs Harangi

Fig. 7.29 One of the mofettes around Ciomadul, where CO_2 gases are emitted. Their isotope composition suggests magmatic origin. *Photo* Szabolcs Harangi

peculiar site is located a couple of kilometres south of the Ciomadul volcano (Büdös/Potorusu-Bálványos/Balvanyos hills) where abundant mofettes and CO_2-rich springs are found (Fig. 7.29). Many of them are utilised for therapeutic purposes in nice wooden houses. The Smelly/Stinking or Sulphurous cave (Büdös-barlang) is a unique place of such gas emanation. Jókai, the famous Hungarian writer called it as the 'Portico of Hell'. It is located at the south-eastern slope of the Büdös-hill just above the Hotel Bálványos, which was once a sanatorium for tuberculosis patients.

Several relatively steep tourist trails are marked with blue and red spots leading from the hotel to a cave that was formerly the entrance to a sulphur mine. The upper level of the CO_2 gas accumulation is clearly indicated by the yellow colouring of the cave wall and can be unambiguously checked using a pocket-lighter or match-stick. The strong odour which can be smelled even far from the cave is caused by the 0.38 % of sulphurous hydrogen in the emitted gas. There are more such caves around the hill, including the Alum (Timsós) Cave and the Killer Cave, which can be visited following the blue trail. However, one should keep in mind that entering into these caves is highly dangerous! Close to the Killer Cave, there is a depression filled with gas, which is known to cause the death of wild animals (therefore it is called as the Birds' Cemetery). At the end of the trail, there is a small peat bog, called Buffogó-láp. The maximum thickness of the peat is around 4 m. There are smaller and bigger pools of water girded with grass and flossy sphagnum and carbon-dioxide gas emissions can be also recognized here as continuous bubbling. Geochemical investigation of the gas emissions revealed, that it has a magmatic component that also underlines the notion that although this area is seemingly quiet, renewal of volcanism cannot be unambiguously excluded in the future.

In conclusion it can be said that the Carpathian-Pannonian Region is indeed akin to a natural laboratory where the present state of the tectonic environment allows us to gain an insight into the volcanic processes of the past. The sheer abundance of volcanic features and their integration into the historic and cultural traditions offers volcano tourists an extensive geodiversity to explore. The remarkable volcanic heritage has been embraced by the tourism industry and includes not only the dormant volcanic landscapes, but also attractions such as the local historic architecture, the vineyards established on the fertile volcanic soils as well as the natural mineral springs, which are utilised for health and wellness purposes.

References and Further Reading

Bakony-Balaton Geopark (2013). Where to find the Bakony-Balaton Geopark? http://www.bakony-balaton-geopark.hu/home/bakony-balaton-geopark/bevezetes. Accessed 23 July 2013.

Geodynamic Atlas of the Pannonian Basin (2013). http://geophysics.elte.hu/atlas/geodin_atlas.htm. Accessed 30 July 2013.

Geopark Banská Štiavnica (2013). http://www.geoparkbs.sk/geopark-banska-stiavnica.html. Accessed 23 July 2013.

Harangi, Sz. (2011). Vulkánok. A Kárpát-Pannon térség tűzhányói. [Volcanoes. The Volcanoes of the Carpathian-Pannonian region] (p. 440). Geolitera Szeged, ISBN: 978-963-306-110-7. [in Hungarian].

Ipolytarnóc Paleontological Site (2013). Üdvözöljük honlapunkon! http://osmaradvanyok.hu/index.php?p=hu_home. Accessed 23 July 2013.

Karátson, D. (2007). A Börzsönytől a Hargitáig—Vulkanológia, felszínfejlődés, ősföldrajz. [From the Börzsöny to the Hargita—Volcanology, landscape evolution and paleogeography] (p. 463). Budapest: Typotex Kiadó, ISBN 978-963-9664-66-1 [in Hungarian].

Kemenes Vulkánpark (2013). http://www.kemenesvulkanpark.hu/. Accessed 23 July 2013.

Lexa, J., Seghedi, I., Németh, K., Szakács, A., Konečny, V., Pécskay, Z., et al. (2010). Neogene-Quaternary volcanic forms in the Carpathian-Pannonian region: A review. *Central European Journal of Geosciences, 2*, 207–270.

Martin, U. & Németh, K. (2004). Mio/Pliocene phreatomagmatic volcanism in the Western Pannonian Basin. Geologica Hungarica, Series Geologica tomus 26 (p. 191). Budapest, ISBN:963-671-238-7.

Novohrad-Nógrád Geopark (2013). http://www.nogradgeopark.eu/. Accessed 23 July 2013.

Steirisches Vulkanland (2013). Region Steirisches Vulkanland. http://www.vulkanland.at/. Accessed 23 July 2013.

8 Aspects of Volcano Tourism in Italy: The San Venanzo Volcano Park and Museum

Francesco Stoppa

8.1 Introduction

Italian volcanoes directly attract at least one million visitors per year due to their special concourse of natural and archaeological resources. On the top of the list is the Vesuvius with 0.5 million but other notable examples are the Pozzuoli Solfatara and Campi Flegrei (0.3 million), Etna and Stromboli active volcanoes (Fig. 8.1). The latter two pose questions about the safety of tourists due to the casualties that have occurred in the past. Also the volcanic field of Northern Latium and Southern Tuscany, located in an area of large calderas, is suitable for tourism purposes, but presents risks for large volume releases of CO_2 and H_2S. In this regard, Mefite vent in Irpinia (Campania) is the largest localised emission of CO_2 and H_2S worldwide (Chiodini et al. 2010). Others volcanoes, such as Vulture and Pantelleria deserve to have more attention given as sustainable tourist developments. In addition, there are several indoor and outdoor volcanic museums scattered throughout Italy. In the Umbria Region are the Polino and San Venanzo museums.

The San Venanzo volcano has been famous since the 19th century for its kamafugitic (ultra-potassic) lavas and for rare minerals. At San Venanzo, scientific and environmental information is facilitated by an open-air path and a museum, which show both, the volcano and mineralogical aspects of the area, as well as its peculiar flora. Some archaeological resources are also displayed. To facilitate the understanding of the San Venanzo Park's complex aspects, a volcanology museum has been created with laboratories, local and general naturalistic collections. Several models illustrate the nature of volcanic phenomena and their importance for the environment, wildlife and people living on a volcano. In recent times economic interests conflicting with the park have been converted to sustainable activities and increasing numbers of visitors have been recorded. San Venanzo also attracts investors by means of fine arts exhibitions, library and conferences, as well as its volcanic legacy.

8.2 Italian Volcanism

Italy has several volcanic areas of great touristic interest. The country ranks second in Europe, after Iceland, for the highest number of active volcanoes. One-third of the national territory has volcanic-geologic monuments, such as giant calderas, strato volcanoes, and intrusion-related contact mineralisation (Guest et al. 2003; Scarth and Tanguy 2001). These localities allowed the formation of historical collections of volcano-related material such as Monticelli's or Hamilton's collections (Jenkins and Sloan 1996). In addition, there is a rich historical and anthropologic heritage linked with the literature, myths, legends and Saints related to volcanoes in Italy. There are several volcanology museums, but only a limited number of volcanic areas have been accorded special protection that safeguards their specificities, and which takes the fragile environment and its geologic aggressiveness and dynamics into account. In Italy volcanic resources are easily accessible, because these territories are densely populated and at relatively low elevations. Umbria is among the regions that have best enhanced their geologic-volcanic monuments, according to the available resources. Two geo-volcanic museums are situated next to nature trails of volcanic inspiration, Polino and San Venanzo. Specific features are to be found in the monogenic apparatus and range of volcanic forms, but also in the rare rock-types and minerals in these museums.

San Venanzo is a municipality of about 2,300 population in the Province of Terni in Umbria, located about 30 km southwest of Perugia and about 45 km northwest of Terni.

F. Stoppa (✉)
Earth Sciences Department, Gabriele d'Annunzio University, Chieti, Italy
e-mail: fstoppa@unich.it

P. Erfurt-Cooper (ed.), *Volcanic Tourist Destinations*, Geoheritage, Geoparks and Geotourism,
DOI: 10.1007/978-3-642-16191-9_8, © Springer-Verlag Berlin Heidelberg 2014

Fig. 8.1 Locations of Italian Volcanic Parks and monuments

The San Venanzo volcanic park and museum is the largest and the best organised in the country, and is open from March to September every Sunday, other days by appointment only (Orvieto Online 2013). The volcano complex is composed of a maar, a volcanic tuff ring that has produced two lava flows and a small diatreme or pipe (Fig. 8.2).

An intrusion with crystallised minerals of extraordinary size and beauty, also very interesting from the collection and scientific point of view, is on display here (Fig. 8.3). The rock-types include: kamafugite (olivin leucite kalsilite melilitite), carbonatite, melilitolite and also peralkaline phonolite. This volcanic complex has been well known since 1870, but until the 1960s the studies focused mainly on the lava flows of venanzite and floristic endemisms (Stoppa 1996; Stoppa et al. 2004).

A museum is located in an historical building, which offers an educational and experimental path adapted to the available space. A multimedia room where several services are available and three thematic sections have been arranged; accessible also to people with participation restrictions. The internal access stairway has been transformed into a stone display with the rocks and stones arranged from the deepest to the most superficial crustal rock-types, so that visitors can better understand the crust stratigraphy. The space devoted to geology, palaeontology and history of the region is located on the second floor. The mineralogy area is

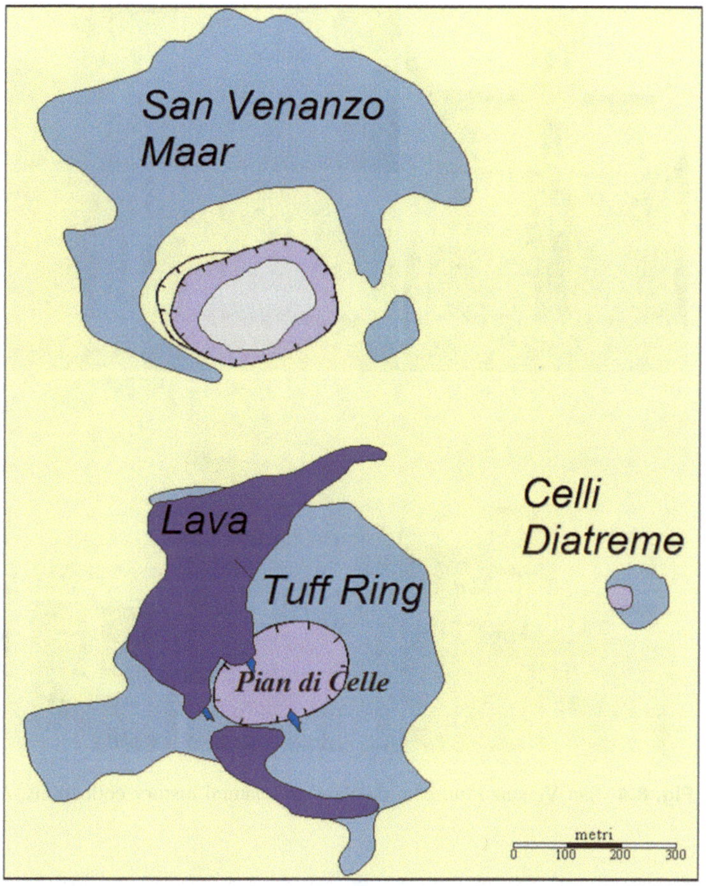

Fig. 8.2 Schematic map depicting the volcanic complex. *After* Stoppa (1994)

Fig. 8.3 Chabazite, San Venanzo, Terni, Umbria, Italia, 0.6 mm coll. *Photo* L. Mattei

Fig. 8.4 San Venanzo museum displays and natural history collections. *Photos* Francesco Stoppa

Fig. 8.5 Gate entrance to the nature trail and regulations for visitors in case of earthquake. *Photos* Francesco Stoppa

followed by a section that is used to showcase volcanology in general, with displays illustrating the volcano distribution, the main volcanic forms and risk. The route ends with a room reserved for the geology and petrology of the volcanoes in San Venanzo (Fig. 8.4).

There is an outdoor trail (Fig. 8.5) equipped with all the facilities and is approximately 1,700 m long (as against an original design of 3,000 m in the project). The trail is perfectly accessible to people with participation restrictions and does not interfere with the waterways and the trails of the local fauna. In fact bridges and walkways have been put in place to avoid any such problems. The path mainly follows the melilitite flow and takes advantage of some artificial cuts, resulting from old extraction activities. Illustrative boards with archaeological, botanic, volcanic, petrologic and mineralogical explanations have been placed along the trail, and amongst the facilities are a parking area, an earthquake safe area, and public conveniences.

8.3 Conclusion

The San Venanzo Volcano Park also shows the commitment of an isolated community that has taken advantage of a co-financed regional project to enhance its natural heritage and safeguard this from any damage or exploitation, while taking advantage of the desire of visitors to see and understand the attractions. At San Venanzo, scientific and environmental information is facilitated by an open-air path and a museum. These show both, volcano and mineral aspects, as well as the flora and fauna. Some archaeological sites are also depicted. To facilitate the understanding of the San Venanzo park complex, a volcanology museum has been created with laboratories, local and general naturalistic collections, and several models which illustrate the nature of volcanic phenomena and their importance for the environment, wildlife and people living on a volcano.

The experience so far acquired from the development and operation of this park can be re-applied by exporting this model to other communities that are willing to establish twinning agreements with San Venanzo. Economic interests conflicting with the park have been converted to sustainable activities and increasing numbers of visitors have been recorded. San Venanzo also attracts investors by means of fine arts exhibitions, library and conferences. An effort for the networking of Italian volcano-related tourism centres is also now ongoing, and is led by San Venanzo museum.

References

Chiodini, G., Granieri, D., Avino, R., Clairo, S., Costa, A., & Minopoli, C. (2010). Non-volcanic CO_2 earth degassing: Case of Mefite d'Ansanto (southern Apennines), Italy. *Geophysical Research Letters, 37*, 1–4.

Guest, J., Cole, P., Ducan, A., & Chestre, D. (2003). *Volcanoes of Southern Italy* (p. 284). Bath: The Geological Society.

Jenkins, I., & Sloan, K. (1996). *Vases and volcanoes: Sir William Hamilton and his collection*. London: British Museum Press.

Orvieto Online (2013). *San Venanzo Volcano park and museum*, Retrieved August 20, 2013 from www.orvietoonline.com/san_venanzo_museo_vulcanologico.html.

Scarth, A., & Tanguy, JC. (2001). *Volcanoes of Europe* (p. 243). England: Terra publications.

Stoppa, F. (1996). The San Venanzo maar and tuff ring, Umbria, Italy: Eruptive behaviour of a carbonatite-melilitite volcano. *Bulletin of Volcanology, 57*, 563–577.

Stoppa, F. (1994). Area vulcanica di San Venanzo. In *Guide Geologiche Regionali, Socità Geologica Italiana* (Vol 15 pp 175–176).

Stoppa, F., Cirilli, S., Rettori, R., Cresta, S., Brozzetti, F., Lavecchia G., et al. (2004). *La poesia del Divenire: Il paesaggio geologico dell'Umbria meridionale* (p. 250). Roma: Bardi Editori.

9 Volcanic Geotourism in France

New Perspectives for Geotourism Development

Nathalie Cayla

9.1 Introduction

France, land of volcanoes! Hard to imagine? Yet metropolitan France and its overseas territories offer traces of various volcanic episodes that have occurred for more than 2 billion years. More or less spectacular and until a few years ago often travelled only by specialists, many volcanic geotopes are now the subject of tourism development for the wider public. France, with 83 million foreign visitors welcomed in 2012, to which must be added the number of French residence tourists, is the world's leading tourism destination. Of course much of the tourist flows are captured by a few large sites that are predominantly urban, but the development of geotourism provides quality attractions in rural areas (Duval and Gauchon 2010). Among these locations, the sites associated with active or extinct volcanoes are very popular. In this Chapter two types of examples will be assessed to show the diversity of experiences related to the protection and development of geotourism in natural areas associated with the discovery of past volcanic activity.

9.2 Pillow Lava Outcrop Along the Geological Transpyrenean Route

France is home to several mountain ranges. The Armorican Massif and the Massif Central, which are witnesses of the Hercynian orogeny and the Alps and the Pyrenees resulting from the Alpine orogeny. In these ranges, the pillow lavas formed by submarine basaltic flows can be observed. At the bottom of the Pyrenees, near the village of Ogeu-les-Bains, a former quarry provides access to an outcrop of pillow lavas. The site was recently prepared as part of a large scale geotourism project, the Transpyrenean Geological Road. Road signs guide the tourists to the outcrop where interpretative panels explain the origin of these lavas (Fig. 9.1). These pillows-lavas result from basaltic flows some 100 million years ago during a period of distension related to the opening of the Atlantic Ocean.

The Geological Transpyrenean Route is an interpretative bilingual (french-spanish) 200 km route, between Riglos (Aragón, Spain) and Oloron (Pyrénées Atlantiques, France). It offers an original reading of an evolving 400 million years landscape combining geology, environment and culture. It is divided into 24 stages, each stop being materialised through interpretative panels and rock samples. This extensive project involved many institutional actors from local associations as well as researchers in the geosciences (Venzal 2012). The road was completed in 2008 and first estimates show that the Geological Transpyrenean Route has already attracted more than 5,000 visitors. While the main economic resource of the Pyrenees is tourism, mainly related to winter sport in the southern area, this project has created new infrastructure and media to showcase a touristic heritage that was not exploited in the area before—the Geology. The conception of a cross-border itinerary shows us that nature does not have borders and that imaginary lines do not have to separate people, projects and resources that share the same environment.

9.3 Seabed Remnants on Mount Chenaillet

In the Southern Alps, a few miles from Briancon near the Italian border, a perfectly preserved ophiolite suite can be observed in the Chenaillet Mountain. This particular type belongs to the series l'herzolites Ophiolite Type (LOT) which corresponds to the slow-spreading Atlantic Ridge type. The rise of Chenaillet (2,634 m) (Fig. 9.2) allows the discovery of rocks, some of which were formed about 30 km deep in the heart of the mantle (Winterflood 2001).

N. Cayla (✉)
Universe and Environment, EDYTEM Laboratory,
Environnements, DYnamiques et TErritoires de la Montagne,
Chambéry, France
e-mail: Nathalie.cayla@univ-savoie.fr

P. Erfurt-Cooper (ed.), *Volcanic Tourist Destinations*, Geoheritage, Geoparks and Geotourism,
DOI: 10.1007/978-3-642-16191-9_9, © Springer-Verlag Berlin Heidelberg 2014

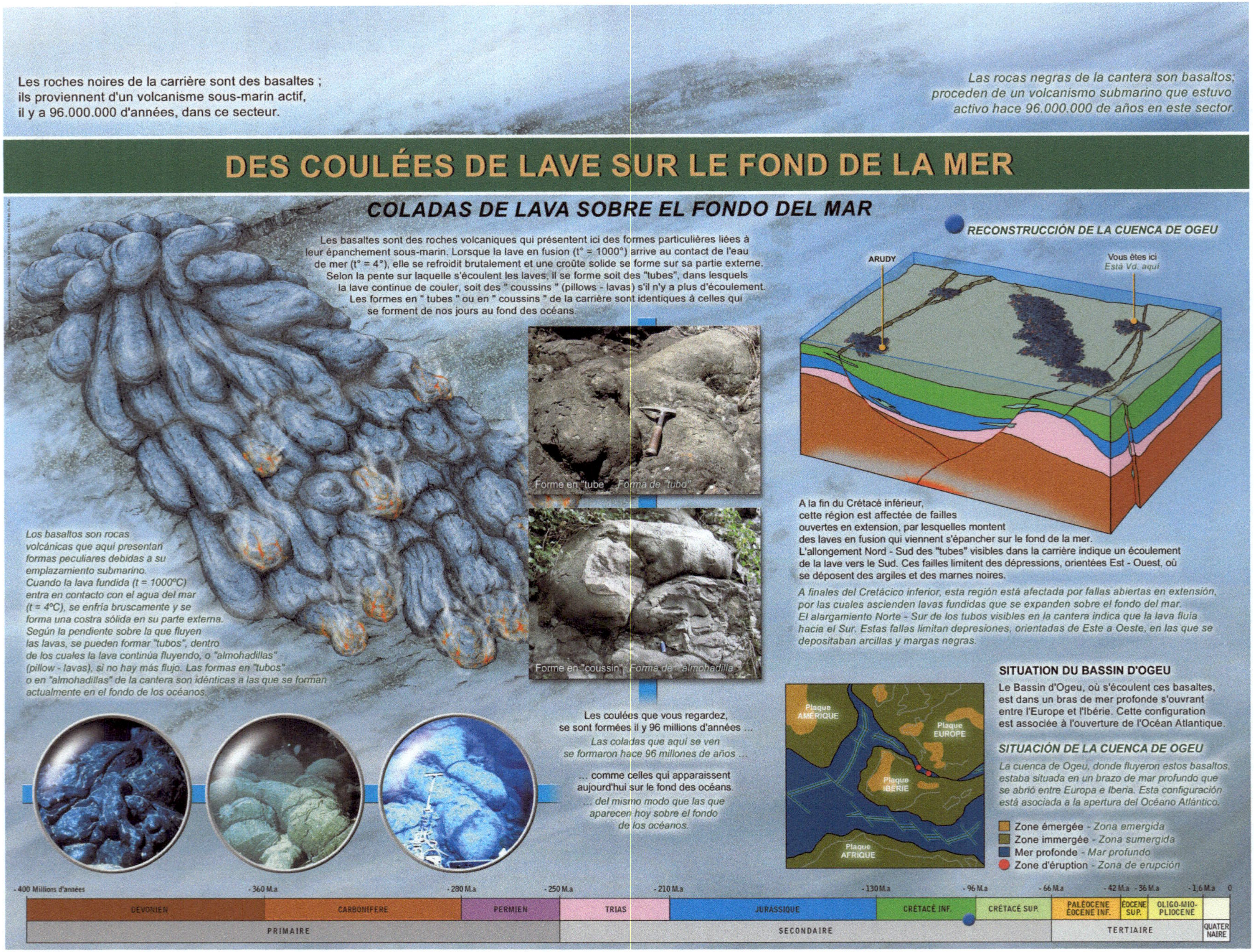

Fig. 9.1 Interpretative panel on the Geological Trans-pyrenean Route. *Source* www.routetranspyreneenne.com

Fig. 9.2 Mount Chenaillet seabed (geological interpretation, geotrail, pillow-lavas details). *Photos* N. Cayla

For example Peridotites (mantle rocks) altered to serpentinite, gabbros that crystallized at the bottom of the magma chamber, and finally the pillow basalts at the top of the series. These pillow lavas are particularly spectacular because of their perfect preservation, leaving it easy to imagine the events which have taken place in the depths of the ocean (Barféty et al. 1995).

Radiometric dating has assigned an age of 165 million years to the ocean floor that formed at the opening of the Ligurian-Piedmont ocean. It took a long time for these volcanic formations to be fully explained. In 1861, Charles Lory (1823–1889), a geologist from Grenoble, led a special excursion of the Geological Society of France into the Massif. The group of French geologists explored the area without understanding at that time the origin of these spectacular outcrops. Jacques Debelmas and Marcel Lemoine in the 1960s attributed the origin of these rocks to the subduction of a portion of oceanic crust (Chalot-Prat 2005). For over 20 years the association 'Centre Briançonnais de Géologie Alpine (CBGA)' has developed numerous projects related to geotopes in the region to train students and teachers in the field of geology, which is part of the curriculum in France.

Meanwhile, training of mountain guides has helped to broaden the knowledge of people who are interested in exploring the geology of Briançon. Gradually the local authorities of the city of Briançon, the ski resort of Montgenevre as well as the Italian valleys of Sestriere and Cesana gathered in the network 'Conférence des élus des Hautes Vallées' and started to develop new touristic offers. Indeed, the tourism economy is the largest economy in the region, with nearly 6 million annual overnight stays in Briançon, which is split evenly between the discovery of the natural environment in summer and the ski season during winter.

An extensive program of border cooperation 'Tourisme dans le Géoparc des Alpes Cottiennes' began in 2007. This project provides extensive tourist facilities with the aim to eventually obtain the label Global Geopark. Interpretation centres and trails teaching earth science as well as rock gardens, are all concrete achievements that will enhance geosite discovery along the entire border between Briancon and Sestrieres.

Other sites in France exhibit much older pillow lavas: the Locquirec beach in Britanny with infra-Cambrian basalts, but also the Belledonne massif near Grenoble with an outcrop of about 400 million years. These two sites are well known by geologists, but they do not present geotouristic infrastructure.

9.4 The Volcanoes of the Massif Central, a Natural Volcanological Museum

The Auvergne has been one of the cradles of volcanism in the world. In 1751 the geologist Jean-Etienne Guettard (1715–1786) travelled through the mountains of the Auvergne (Gohau 1991). He recognised rocks as identical to those originating from Vesuvius in Italy and released a report 'Memorandum of volcanic mountains in France' at the Royal Academy of Sciences, attributing for the first time a volcanic origin to the mountains of the Auvergne. A few years later Nicolas Desmaret (1725–1815) studied the different basalts found commonly in the Auvergne and in particular demonstrated the volcanic origin of this rock. In the early nineteenth century, Poulett Scrope (1797–1876), an English geologist, discovered that the Cantal was an ancient stratovolcano.

The Auvergne volcanoes occupy the central and eastern part of the Massif Central. They are associated with a series of rifts running through the European plate from the Mediterranean to the North Sea. The Auvergne volcanics are a direct consequence of alpine orogeny. Subduction of the European plate beneath the African plate and the collision that followed are the source of a crustal thinning that caused the appearance of a hot spot beneath the Massif Central. Two forms of volcanism have left their mark:

1. Rift volcanism, associated with the Limagne fault and Velay, which started there about 65 million years ago and still continues; and
2. Hot spot volcanism gave rise to a succession of volcanic structures with the oldest of the Cantal stratovolcanoes thought of having formed about 11 million years ago, while the newer volcanoes of the Chaîne des Puys were still active 5,000 years ago (Boivin and Thouret 2014).

9.5 History of the Tourism Development of the Volcanoes of the Massif Central

Tourism around the volcanoes of the Massif Central began with the discovery of remarkable landscapes with the ensuing development of scenic locations for spa guests around Clermont-Ferrand. The creation of the Regional Natural Park of the Auvergne Volcanoes in 1977 showed a desire to utilise geotourism to showcase the discovery of the volcanic phenomena from field observations. These were supplemented by the development and construction of interpretative centres. The current inclusion of the volcanoes and the Limagne fault in the UNESCO World Heritage list is part of a new dynamic process, which recognises the unique value of landscapes for the development of geotourism in the Auvergne. There are currently hundreds of volcanic geosites which are developed in this region with signs and themed trails, twenty museums and interpretative centres to cater for visitors, as well as several disused quarries (Fig. 9.3), which are now open to the public (Aubert et al 2003; Lavina and Monestier 1996).

The Puy de Dôme dominates the chain of mountains with a height of 1,465 m. The ancient volcano is of Pelean type, was formed 11,000 years ago and consists of two nested trachyte domes. Classified as a natural monument in 1933, the Puy de Dôme has been designated as a *Grand Site de France* in 2008 and attracts around 450,000 visitors annually. The development of this site for tourism began in 1878 with the rehabilitation of the mule trail to the summit. In 1906 a railway line 14.7 km long connected the centre of the city of Clermont-Ferrand with the summit of the Puy de Dôme. In the 1930s this train connection was replaced by a toll road for tourists to limit environmental disturbance. Since the summer 2012, the 'Panoramique des Dômes', a rack railway reached the summit of the volcano with minimal environmental impact. This will be accompanied by a museum to better understand the history of this volcano.

South of the chain, the peak of Puy Mary (1,787 m) overlooks the Cantal volcano, one of the largest stratovolcanoes in Western Europe, which covers over 2,700 km^2. Classified as natural monuments in 1985, they have been designated as a Grand Site de France in 1999 and attract approximately 600,000 visitors a year. The history of this stratovolcano is complex and is thought to date back about 10 million years. The access to the summit from the Pas de Peyrol is under heavy visitor pressure, which threatens the protection of this site. In 2007 a comprehensive redevelopment was funded by the General Council of Puy de Dôme. The Cantal volcano and the site of the Pas de Peyrol are central to a €1.4 million redevelopment project, supplemented by four information and interpretation centres, which were installed in the valleys in the communities of Saint-Julien-Mandailles, Dienne, Falgoux and Claux.

9.6 Two Sites That Welcome Visitors to Discover the World of Volcanoes: Vulcania and Lemptégy

Vulcania (Fig. 9.4) opened its doors in 2002. It is a theme park that was created through the initiative of the Regional Council of the Auvergne and is managed by a semi-public

Fig. 9.3 Map of the major Geosites of the Massif Central. *Source* N. Cayla

corporation. The initial investment of 60 million Euros was intended to create a tourist magnet for the Auvergne, and it was anticipated that tourism would spread throughout the whole region. In the first year Vulcania attracted 650,000 visitors; however, attendance had dropped rapidly by 2006 and has not exceeded 210,000 annual visitors since then. This number is far lower than expected by the park management. To reverse this trend, the Regional Council decided to

Fig. 9.4 The Theme Park Vulcania. *Photo* N. Cayla

revitalise the attraction by making it more dramatic, as well as fun and family oriented with a further investment of 12 million Euros. Vulcania II was opened in 2008 and since then the increase in attendance has been significant and regular; from 313,000 visitors in 2008 visitor numbers rose to 341,000 in 2011 and ensured a turnover of 9 million Euros.

Not far away, the volcano Lemptégy (Fig. 9.5), which is managed by a private corporation, celebrated its 20th anniversary in 2012. Lemptégy now ranks as the third most visited geodestination in the Auvergne and welcomes 107,000 annual visitors. This old pozzolan (volcanic ash) quarry was operated for nearly 60 years, exposing the core of the volcano. Initially mining and tourist visits were conducted simultaneously; however, mining operations ceased in 2007 and the site is now used for geotourism only. Apart from visiting the core of the volcano, the combination of spectacular animations and the educational value complete the attraction.

9.7 The Emergence of a Large Scale Volcano Geotourism Destination

The Regional Natural Park of the Auvergne Volcanoes was created in 1977. It includes 153 municipalities located in two departments, the Puy de Dôme and the Cantal, and contains 91,000 residents on an area of approximately 4,000 km^2. In France there are 47 such regional parks. These protected areas are designed to achieve the sustainable development of territories within them. Economic development should be organised in order of respect for heritage, enhancement of local employment, and the maintenance of the residents living in the area. The regional park is committed to a charter which governs its activities for a period of 10 years. Since its inception, the Park has developed geotourism attractions providing all the facilities to discover the volcanoes both in situ and ex situ such as museums and interpretative centres.

Fig. 9.5 The Puy de Lemptégy attracts over 100,000 visitors anuually. Tourists can walk thorugh the quarry, which offers interpretative panels as well as interesting exhibits including volcanic bombs. *Photo* Dr. Dominique Gasquet

9.8 The Chaîne Des Puys and Limagne Fault Application for World Heritage Listing

The General Council of Puy de Dôme, in partnership with the Regional Natural Park of the Volcanoes of the Auvergne, the University of Clermont-Ferrand, and with the support of State services, the Regional Council of the Auvergne and the Clermont Community, has led the project since 2007 to include the Chaîne des Puys and the Limagne Fault on the World Heritage List of UNESCO under criteria VII (areas of exceptional natural beauty and aesthetic importance) and criteria VIII (significant geological processes and geomorphic features). This was based on the spectacular scenery of the Puys and also on three centuries of geoscientific studies conducted in the region. These studies have interpreted the tectono-volcanic landform and answered fundamental geomorphological questions of the emergence of landforms and the structuring of the continents through a succession of sequences of more than 350 million years. The final approval of this application was expected to be released in 2014. **Update by Editor:** However, based on the IUCN Evaluation Report (2014) the Chaîne des Puys and Limagne Fault were not inscribed on the World Heritage list under the proposed natural criteria. Instead, the IUCN (2014, pp 72-73) recommends to consider the nomination of the area as a Geopark to credit the geological, cultural and natural values of the Chaîne des Puys and Limagne Fault.

9.9 Conclusion

Geotourism is a growing practice on French territory based on the new role given to the natural geological heritage (Cayla 2009). Now recognized, and protected by new legislation, such heritage objects are considered as territorial resources on which it is possible to build local development projects. Often initiated originally by local stakeholders such as associations and quarry operators, this type of project is now widely supported or led by local authorities, who are aware of the interest that can arise from such practices (Peyrache-Gadeau 2010).

References

Aubert, M., Borget, J. N., & Bril, H. (2003). *Le volcanisme en Auvergne*. Editions Chamina.

Barféty, J. C., Lemoine, M., de Graciansky, P. C., Tricart, P., & Mercier, D. (1995). *Briançon, Carte géologique de la France à 1/50000*. Paris: BRGM.

Boivin, P., & Thouret, J.-C. (2014). The volcanic chaîne des puys: a unique collection of simple and compound monogenetic edifices. In M. Fort, & M. -F. André (Eds.), *Landscapes and landforms of France* (pp. 81–91). Netherlands: Springer.

Cayla, N. (2009). *Le patrimoine géologique alpin: de la médiation à la valorisation géotouristique*. Thèse de doctorat de Sciences de la Terre, de l'Univers et de l'Environnement, Université de Savoie.

Chalot-Prat, F. (2005). An undeformed ophiolite in the Alps: Field and geochemical evidences for a link between volcanism and shallow plate tectonic processes. In G. R. Foulger, J. H. Natland, D. C. Presnall & D. L. Anderson (Eds.), *Plates Plumes & Paradigms* (pp. 751–780). Geological Society of America. Special Paper 388.

Duval, M., & Gauchon, C. (2010). Tourisme, géosciences et enjeux de territoires. *TEOROS, 29*(2), 3–14.

Gohau, G. (1991). *A history of geology*. New York: Rutgers University Press.

IUCN (2014) IUCN Evaluations of Nominations of Natural and Mixed Properties to the World Heritage List. WHC-14/38.COM/INF.8B2. IUCN Report for the World Heritage Committee, 38th Session Doha, Qatar, 15–25 June 2014

Lavina, P., & Monestier, H. (1996). *La chaîne des Puys, itinéraires de découverte géologique*. Editions AEDIS.

Peyrache-Gadeau, V. (2010). Renouvellement des ressources et diversité des durabilités territoriales. In Ressources, patrimoine, territoires et développement durable. Peter Lang (Ed). *Coll. Ecopolis* (pp. 237–254).

Venzal, C. (2012). Patrimoine géologique et tourisme transfrontalier dans le massif pyrénéen: évaluation et analyse touristiques de la Route Géologique Transpyrénéenne de la vallée d'Aspe (France) et du Haut Aragon (Espagne). Abstracts Colloque international Routes touristiques et itinéraires culturels entre mémoire et développement, Université de Laval.

Winterflood, I. (2001). An excursion in the French–Italian alps: the 'Atlantic'-type ophiolite at mont chenaillet. *Geology Today, 17*(4), 143–148.

Internet Resources Consulted

La route géologique transpyrénéenne: www.routetranspyreneenne.com.

Le Centre Briançonnais de Géologie Alpine: http://cbga.net.

Un sentier géologique à la découverte du Chenaillet: http://francoise.cp.free.fr/geol/.

Le panoramique des Dômes: www.panoramiquedesdomes.fr/.

Le grand site du Puy Mary: http://www.puymary.fr/.

Le site de Vulcania: www.vulcania.com/.

Le site de Lemptégy: www.auvergne-volcan.com/.

Le site de la candidature UNESCO de la chaîne des Puys faille de Limagne: www.chainedespuys-failledelimagne.com/.

Le site du conservatoire des espaces naturels en Auvergne www.cen-auvergne.fr/.

La Garrotxa Volcanic Zone, Spain

10

Josep M. Prats and Emili Bassols

10.1 Introduction

La Garrotxa Volcanic Zone (Spain) is situated in south-west Europe in the north-east corner of the Iberian Peninsula. Its nearest geographical points of reference are the Pyrenean mountain chain to the north and the Mediterranean Sea to the east, while the well-known city of Barcelona lies just 150 km to the south. This is an upland zone with altitudes that range in height from 200 to 1,000 m above sea level. Its main relief features, along with its wide variety of soils and high rainfall (almost 1,000 mm annual average), ensure generous plant cover over all of its mountain slopes. Most valley bottoms, flattened by the effects of local volcanic activity, are today devoted to traditional agricultural activities. All in all, the resulting landscape is one of great harmony and beauty.

Volcanic activity in La Garrotxa Volcanic Zone can be placed within the framework of the intra-continental rift that crosses Western Europe and in the same geodynamic context as the Eiffel (Germany) and Auvergne (France) volcanic zones. Although some evidence pointing at earlier volcanic activity does exist, chronological data place most of the volcanic activity in La Garrotxa at between 350,000 and 10,000 BP. Today, over 40 volcanic structures stand along the valleys of the rivers Fluvià, Brugent and Llémena, and around the city of Olot and village of Santa Pau, and in total cover an area of around 373 km^2 (Fig. 10.1). The volcanoes of this region are monogenetic, and were formed as a result of single eruptions with a mix of effusive, Strombolian and phreatomagmatic phases (Marti et al. 2012). In general, La Garrotxa Volcanic Zone is home to a great variety of pyroclastic deposits, both Strombolian and phreatomagmatic in formation, and numerous lava flows. Today, the volcanoes are dormant rather than extinct and so, other than the occasional rockfall at a couple of sites, there are no serious threat or risk issues for the local population.

In 1982, this volcanic area was declared a Natural Park (IUCN Category V), which today covers around 150 km^2. The Park's governing board works to protect the zone's geological heritage by promoting scientific research and educational activities aimed at different segments of the population. It also manages land use to ensure that regulations are complied with and monitors human activities conducted within the protected area. Thanks in part to the large number of educational activities that take place every year in the Park for schools and adults, local people give their full support to the protection and conservation of the area and its values.

10.2 Tourism in La Garrotxa

The main attractions in La Garrotxa Volcanic Zone are the volcanoes of Croscat, Santa Margarida and Montsacopa. The former is regarded as the youngest in the Iberian Peninsula (its eruption dated at 13,270–13,040 BP) and during the 1970s was quarried intensely. This provoked much concern amongst scientists and local people alike and the resulting protests eventually led to the declaration of the Park in 1982. Today, the former quarry on the north face of the volcano is one of the most visited sites in the Natural Park.

Santa Margarida attracts visitors due to its almost perfectly circular crater that houses a small 16th century church in its midst. On the other side, Montsacopa rises over the city of Olot (pop. 37,000) and a short climb to its summit provides an excellent view over the city and the surrounding mountains. Also of renown are La Fageda d'en Jordà, a beech forest growing on the lava flows emitted by the Croscat volcano, and the 50-m-high basalt cliffs crowned by

J. M. Prats (✉) · E. Bassols
Technicians of the Management Team of the Natural Park, Garrotxa, Spain
e-mail: wjmprats@gencat.cat

E. Bassols
e-mail: webassol@gencat.cat

P. Erfurt-Cooper (ed.), *Volcanic Tourist Destinations*, Geoheritage, Geoparks and Geotourism,
DOI: 10.1007/978-3-642-16191-9_10, © Springer-Verlag Berlin Heidelberg 2014

Fig. 10.1 Olot, city of Volcanoes. The capital of la Garrotxa is surrounded by volcanoes. *Photo* J. M. Prats

the dramatically situated village of Castellfollit de la Roca (Fig. 10.2).

Tourism was not well developed in La Garrotxa before the Natural Park was declared in 1982. The very fact of protecting something is indicative of its inherent value; in the case of La Garrotxa, once the Park was established the number of visitors and tourist activities began to increase steadily and soon their impact became an important element that had to be taken into account in local conservation management. In 1995, the first steps were made to address this problem and the Park gave its backing to the creation of a local tourism board whose task would be to manage visitor and tourist flows (despite the lack of mechanisms for this in the law that created the Park). This association, Turisme Garrotxa (2013), set up in 1996, is an excellent example of how public and private sectors can work together in the promotion of tourism and protection of the geophysical environment upon which this is based.

One fundamental principle accepted by all the partners of this association is **sustainability**. It was agreed that tourism development in the Park would only be possible if there was a firm commitment to this principle. Tourism

Fig. 10.2 Castellfollit de la Roca, a medieval village built on top of a basaltic cliff. *Photo* J. M. Prats

businesses were quick to accept that visitors were attracted mainly by the well-preserved volcanic landscape and the excellent state of conservation of the other natural values within the Park. However, the fact that many of the Park's attractions are situated within a small area causes a concentration of visitors that can be problematic during holiday periods. The Park's management policies seek to promote alternative attractions at sites that are less well-known, but are nonetheless still well worth visiting. Thus, the providing of information plays an essential part in the Park's management strategy. In addition to the three official information centres run by the Park, since 2001 50 private businesses have received specific training and are recognized as Park Information Points. They cooperate with the Park's mission by providing visitors with accurate information and help improve knowledge of the natural and cultural values that the Park is working to preserve.

This spirit of cooperation has become a hallmark of La Garrotxa and fits in well with the spirit of the European Charter for Sustainable Tourism in Protected Areas Program (ECST). Defined by the Europarc Federation (2013) and based on successful pilot programs run in 10 parks, as well as on the work of a steering committee with representatives from tourism and conservationist organizations, the ECST is both a planning tool and a quality label. Its main aims are to define strategies and five-year action plans in which all relevant stakeholders are invited to participate. This participation is organized through a *Permanent Forum of Sustainable Tourism* in which public as well as private tourism bodies and organizations are represented. The Charter is based on the following 10 sustainable principles:

Working in Partnership

Principle 1 To involve all those implicated by tourism in and around the protected area in its development and management. A permanent forum, or equivalent arrangement, should be established between the protected area authority, local municipalities, conservation and community organizations and representatives of the tourism industry. Links with regional and national bodies should be developed and maintained.

Preparing and Implementing a Strategy

Principle 2 To prepare and implement a sustainable tourism strategy and action plan for the protected area. The strategy should be based on careful consultation and be approved and understood by local stakeholders.

Addressing Key Issues

Principle 3 To protect and enhance the area's natural and cultural heritage, for and through tourism, and to protect it from excessive tourism development;

Principle 4 To provide all visitors with a high quality experience in all aspects of their visit;

Principle 5 To communicate effectively to visitors about the special qualities of the area;

Principle 6 To encourage specific tourism products that guarantee the discovery and understanding of the area;

Principle 7 To increase knowledge of the protected area and of sustainability issues amongst all those involved in tourism;

Principle 8 To ensure that tourism supports and does not reduce the quality of life of local residents;

Principle 9 To increase benefits from tourism to the local economy;

Principle 10 To monitor and influence visitor flows to reduce negative impacts.

Individual tourism businesses can also be awarded the Charter label if on a voluntary basis they commit themselves to implementing a three-year action plan designed to improve their environmental performance and to strengthen their relationship with the Park authorities via collaboration with ongoing programmes. The businesses awarded the label receive in exchange for their efforts certain rewards in the form of activities such as training programmes, marketing and technical support. Most of the awarded businesses appreciate the European scope of the label and the fact of being identified as an environmentally friendly enterprise. As of 2011, 22 tourism businesses in La Garrotxa had been awarded the Charter.

10.3 Tourist Activities

A recent study of the impact of tourism on La Garrotxa has revealed the increasing importance of this sector to the local economy. Since 2001, the total number of businesses offering accommodation (hotels, bed & breakfast, rural accommodation and campsites) rose from 93 to 211, with a corresponding increase in beds from 5,015 to 6,464. The number of restaurants rose from 117 in 2001 to 249 in 2010, while in the same period the number of companies working in complementary activities increased from 12 to 19. Visitor numbers have also increased, from 296,192 in 2001 to 355,735 in 2010, of whom 81.2 % were domestic and 18.7 % from abroad. Of the Spanish visitors, 42 % were

Fig. 10.3 Croscat Volcano: The youngest of the volcanoes in the Iberian Peninsula and the reason for protection is to stop its destruction. *Photo* J. M. Prats

from Catalonia, above all from Barcelona, which is just 90 min away from La Garrotxa. Of the reasons given for visiting La Garrotxa, 70 % of the visitors stated that 'to enjoy nature' was the main reason for their visit, followed by 'to relax' (31.7 %).

It is worth highlighting two important elements of the tourism infrastructure in La Garrotxa. The first is the *Itinerànnia* network of hiking trails, which stretches for 2,700 km and provides access to the whole of La Garrotxa. All the trails are well signposted and include many age-old paths that connect even the smallest hamlets in the region. This network enables visitors to discover every corner of La Garrotxa, from volcanoes and dramatic cliff-top trails to small Romanesque churches, typical farmhouses and picturesque mountain villages. The second such infrastructural element is a cycle path starting in Olot, the capital of La Garrotxa, which runs for 60 km as far as the city of Girona. This path follows the course of the railway line that once linked these two cities and in Girona connects with another such trail that continues on to the shores of the Mediterranean. Thus, it is possible to cycle from Olot to the sea along 100 km of quiet, safe and not too demanding cycle paths.

10.4 Facilities

There are four information centres in the park. The main centre is in El Casal dels Volcans in the city of Olot itself; this building is surrounded by the gardens belonging to the Casal and houses the small Museum of the Volcanoes. The second centre, Can Serra, stands at the entrance to the beech forest of La Fageda d'en Jordà and is the most visited of all. A third centre is located at the entrance to the volcano Croscat (Fig. 10.3), and the last, Sant Feliu de Pallerols, occupies a former railway station and lies next to the Olot-Girona cycle path.

As well, there are a number of tourism offices, for example, in Olot, Besalú and Sant Joan les Fonts, which are run by the local town councils and work in collaboration with the Park visitor centres. As mentioned above, over 50 private businesses reinforce and complement this network of information points and provide visitors with information regarding the full range of visitor possibilities offered by the Park.

10.5 Environmental Education

Due to its enormous educational potential, over 60,000 schoolchildren visit La Garrotxa Volcanic Zone annually. Most come from Catalonia and all local children visit the Park at least once during their school careers. To satisfy this demand, several private companies in addition to the official service provided by the Park are available to guide visitors. Schools can choose between a number of companies that have all received official training and are accredited by the Park, and which use the Park's pedagogical material (Fig. 10.4).

Fig. 10.4 La Fageda d'en Jorda, the beech forest growing on a lava flow, one of the most visited places in la Garrotxa. *Photo* J. M. Prats

10.6 Brochures

The main published material about the region is the annual magazine *Tot Garrotxa* (All Garrotxa), which contains all the relevant information about tourist attractions and services, and the activities offered by local tourism providers. In addition, the Park has published specialized information about its volcanoes (Guide to Volcanism) in English and a number of brochures in different languages.

References

Europarc Federation (2013) European Charter for Sustainable Tourism. www.europarc.org/what-we-do/european-charter-for.

Martí, J. et al. (2012). *Volcanoes—A field guide to La Garrotxa Volcanic Zone*. Garrotxa: Parc Natural de la Zona Volcànica de la.

Parcs de Catalunya (2013) Zona Volcànica de la Garrotxa. www.gencat.cat/parcs.

Turisme Garrotxa (2013) www.turismegarrotxa.com.

The Giants Causeway and Fingals Cave

11

Linked Volcanic Tourist Attractions in Ireland and Scotland

Malcolm J. M. Cooper

11.1 Introduction

Linked volcanism has become a world recognised tourism resource of today in two particular locations. The attractions that are now the *Giant's Causeway* in Northern Ireland and *Fingal's Cave* in Scotland were formed by the same lava flows about 60 million years ago. The Giant's Causeway (*Clochán an Aifir*) is an area of many thousands of interlocked basalt columns located in County Antrim on the northeast coast of Northern Ireland (Fig. 11.1). It was declared a World Heritage Site by UNESCO in 1986 and a National Nature Reserve in 1987 by the United Kingdom's Department of the Environment for Northern Ireland (UNESCO 2009). Most of the columns are hexagonal, as is normal with this form of lava when it cools, although there are some columns with four, five, seven or eight sides (Aydin and Degraff 1988). The tallest columns have been measured at about 12 m in length, and the solidified lava bed is up to 28 m thick in places (UNESCO 2009). The Giant's Causeway is today owned and managed by the UK National Trust and is one of the most popular tourist attractions in Northern Ireland (Northern Ireland Tourist Board 2008).

The second example, *Fingal's Cave (An Uaimh Bhinn*, or "melodious cave"), is a natural sea level cave on the uninhabited island of Staffa (Fig. 11.2), in the Inner Hebrides of Scotland, and has been designated a National Nature Reserve (National Trust for Scotland 2012). As already noted, it is formed from jointed basalt columns similar in structure to the Giant's Causeway and to those of nearby Ulva Island, all part of the same ancient lava flow (Bell and Jolley 1997). Its size and configuration amplify the sounds produced by waves and are said to give it the atmosphere of a natural cathedral (Cooper 1979; Bray 1996).

While little is known of the early human history of Staffa Island, the then uninhabited island was brought to the attention of the public by the naturalist Sir Joseph Banks in 1772 (Cooper 1979), becoming known as Fingal's Cave after the hero of an Irish legend Fionn mac Cumhaill, who is said to have constructed a causeway from Ireland to Scotland out of basalt columns (Keay and Keay 1994).

11.2 Geology and Discovery

Approximately 60 million years ago (UNESCO 2009) the areas that ultimately became County Antrim in Northern Island and the Inner Hebrides of Scotland were the location of intense volcanic activity, when highly fluid basalt lava forced its way through chalk beds to form an extensive lava plateau (the Thulean Plateau; Geoffroy et al. 1996). As the basalt lava cooled rapidly it set in pillar-like structures, a process known as columnar jointing (Jagla and Rojo 2002), which produced the distinctive landforms we see today. This lava flow included the Giants Causeway, Fingals Cave and the Ulva/Mull Island formations (Geoffroy et al. 1996).

The discovery of the Giant's Causeway was first announced in 1693 through the medium of a paper to the UK Royal Society by Sir Richard Bulkeley, a fellow of Trinity College, Dublin. The site received international attention when a Dublin-based artist, Susanna Drury, made paintings of it in 1739 (Fig. 11.3); these paintings won an award from the Royal Dublin Society in 1740, were engraved in 1743 (Arnold 2002), and were subsequently incorporated in a French publication, *The Encyclopédie*. In the caption to the published plates a noted French geologist of the time, Nicolas Desmarest suggested for the first time in print that such structures were volcanic in origin (Encyclopédie, ou Dictionnaire raisonné des sciences, des arts et des métier, 1751–1777; Arnold 2002).

M. J. M. Cooper (✉)
Ritsumeikan Asia Pacific University, Beppu, Japan
e-mail: cooperm@apu.ac.jp

P. Erfurt-Cooper (ed.), *Volcanic Tourist Destinations*, Geoheritage, Geoparks and Geotourism,
DOI: 10.1007/978-3-642-16191-9_11, © Springer-Verlag Berlin Heidelberg 2014

Fig. 11.1 The Giants Causeway. *Photo* Petr Brož

Fig. 11.2 **a** Fingals Cave on the Island of Staffa was a well known volcanic tourist attraction in the eighteenth century. *Artist* John Cleveley. **b** Fingals Cave as tourists see it today. *Photo* Hartmut Josi Bennöhr 2004

Fingal's Cave is located on the island of Staffa, part of the Inner Hebrides group of islands in Scotland (Keay and Keay 1994). Viking settlers gave it this name in the early middle-ages as its basalt columns reminded them of the vertically placed tree-logs that formed the frame of their houses. Its land area is 33 hectares and at the highest point it is 42 metres above sea level. The Island is volcanic in origin, consisting of a basement of tuff covered by two layers of basalt. Slow cooling of the second layer of basalt is what produced the predominantly hexagonal columns which form the faces and walls of the principal caves of the Island (Aydin and Degraff 1988). As noted above, similar formations of basalt columns are found on the island of Ulva and at Ardmeanach on the adjacent Isle of Mull.

11.3 Tourism

Both these sites have a long association with tourism. The Giant's Causeway site first became popular with tourists during the nineteenth century, particularly after the opening of a public tramway that improved access to it

Fig. 11.3 Like Fingal's Cave the basalt columns of the Giants Causeway were already a tourist attraction nearly 250 years ago. *Artist* Susanna Drury (1768)

Fig. 11.4 Basalt columns inside Fingals Cave. *Photo* Karl Gruber (2004)

Fig. 11.5 **a** Basalt on the little island beside the Isle of Staffa, Scotland. *Photo* Hartmut Josi Bennöhr 2004. **b** Landing stage on Staffa Island, a small volcanic island in the Southern Hebrides. *Photo* Sarah Charlesworth (2013)

(Northern Ireland Tourist Board 2008). Fingal's Cave on the other hand was attracting tourists a little earlier, in the late eighteenth century, after a visit in 1772 by the naturalist Sir Joseph Banks (of the exploration of the Southern Pacific Ocean with the navigator Captain James Cook fame) brought it to public notice. He was responsible for the name Fingal's Cave being given to the island's main sea cavern (Fig. 11.4; National Trust for Scotland 2012). This visit was followed by that of many other prominent people throughout the next two centuries, including the English Queen Victoria, the writer Jules Verne, the poets William Wordsworth, John Keats, and Alfred Tennyson, the artist J. M. W. Turner, who painted the cave in 1832, and the German composer Felix Mendelssohn who brought fame to the then uninhabited island of Staffa through his *Hebrides Overture* (Opus 26) (Cooper 1979; Bray 1996). The Island is now in the care of the National Trust for Scotland.

Both of these locations are a haven for sea birds and the rock formations host a number of rare and unusual plants that may interest visitors, but it is primarily the basalt columns that attract the tourists (Watson 2000). Fingal's Cave has a large arched entrance and is accessible by sea or by land. Several tour companies include the cave in sightseeing cruises from April to September, and it is possible to land on the island and walk to the cave overland, where a row of columns just above high-water level allows exploration on foot (Fig. 11.5). There is no visitor centre.

In County Antrim, the Giants Causeway site was also without a permanent visitors' centre between 2000 and 2012 after a fire destroyed the building (Northern Ireland Tourist Board 2008). Nevertheless, some 750,000 people visited the site each year prior to 2012 (Trip Advisor 2012). A new visitor centre was opened during July 2012, built with funding from the UK National Trust, the Northern Ireland Tourist Board, the UK Heritage Lottery Fund and from public donations. However there have been adverse comments on its visitor pricing, design, services and its placement directly across the causeway walk descent made since its re-opening (Trip Advisor 2012).

References

Arnold, B. (2002). *Irish art: A concise history*. New York: Thames & Hudson.

Aydin, A., & Degraff, J. M. (1988). Evolution of polygonal fracture patterns in lava flows. *Science*, *29*(239, 4839), 471–476.

Bell, B. R., & Jolley, D. W. (1997). Application of palynological data to the chronology of the Palaeogene lava fields of the British Province: Implications for magmatic stratigraphy. *Journal of the Geological Society, 154*(4), 701–708.

Bray, E. (1996). *The discovery of the Hebrides: Voyages to the Western Isles 1745–1883*. Edinburgh: Birlinn.

Charlesworth, S. (2013). *Staffa island landing stage*. Creative Commons. Retrieved March 2, 2013, from www.southernhebrides.com/staffa.html

Cooper, D. (1979). *Road to the Isles: Travellers in the Hebrides 1770–1914*. London: Routledge & Kegan Paul.

Drury, S. (1768). *A view of the Giant's Causeway: East prospect*. Engraving. Retrieved March 2, 2013, from http://en.wikipedia.org/wiki/File:Drury_-_View_of_the_Giant%27s_Causeway.jpg

Encyclopédie, ou Dictionnaire raisonné des sciences, des arts et des métier (1751–1777) *Encyclopédie, ou Dictionnaire raisonné des sciences, des arts et des métier*. Paris: Chez Briasson. Retrieved March 2, 2013, from Lindahall.org

Geoffroy, L., Bergerat, F., & Angelier, J. (1996). Brittle tectonism in relation to the Palaeogene evolution of the Thulean/NE Atlantic domain: A study in Ulster. *Geological Journal, 31*(3), 259–269.

Gruber, K. (2004). *Staffa, fingal's cave—Höhleneingang*. Retrieved March 2, 2013, from http://en.wikipedia.org/wiki/File:Staffa04.jpg

Jagla, E. A., & Rojo, A. G. (2002). Sequential fragmentation: The origin of columnar quasihexagonal patterns. *Physical Review E, 65*(2), 026203.

Keay, J., & Keay, J. (1994). *Collins encyclopedia of Scotland*. London: HarperCollins.

National Trust for Scotland (2012) *Fingal's cave*. Retrieved March 2, 2013, from www.nts.org.uk/Property/Staffa-National-Nature-Reserve/

Northern Ireland Tourist Board (2008) *Giant's Causeway remains Northern Ireland's top attraction*. Retrieved March 2, 2013, from http://archive.is/57yl

Trip Advisor (2012) *Giant's Causeway visitor centre reviews*. Retrieved March 2, 2013, from www.tripadvisor.co.uk/Attraction_Review-g209948-d2281662-Reviews-Giant_s_Causeway_Visitor_Centre-Bushmills_County_Antrim_Northern_Ireland.html

UNESCO (2009) *Giant's Causeway and causeway coast*. UNESCO World Heritage Centre. Retrieved December 19, 2012, from http://whc.unesco.org/en/list/369

Watson, P. S. (2000). *The Giant's Causeway*. O'Brien: Printing Press.

Part III
The Pacific Ring of Fire

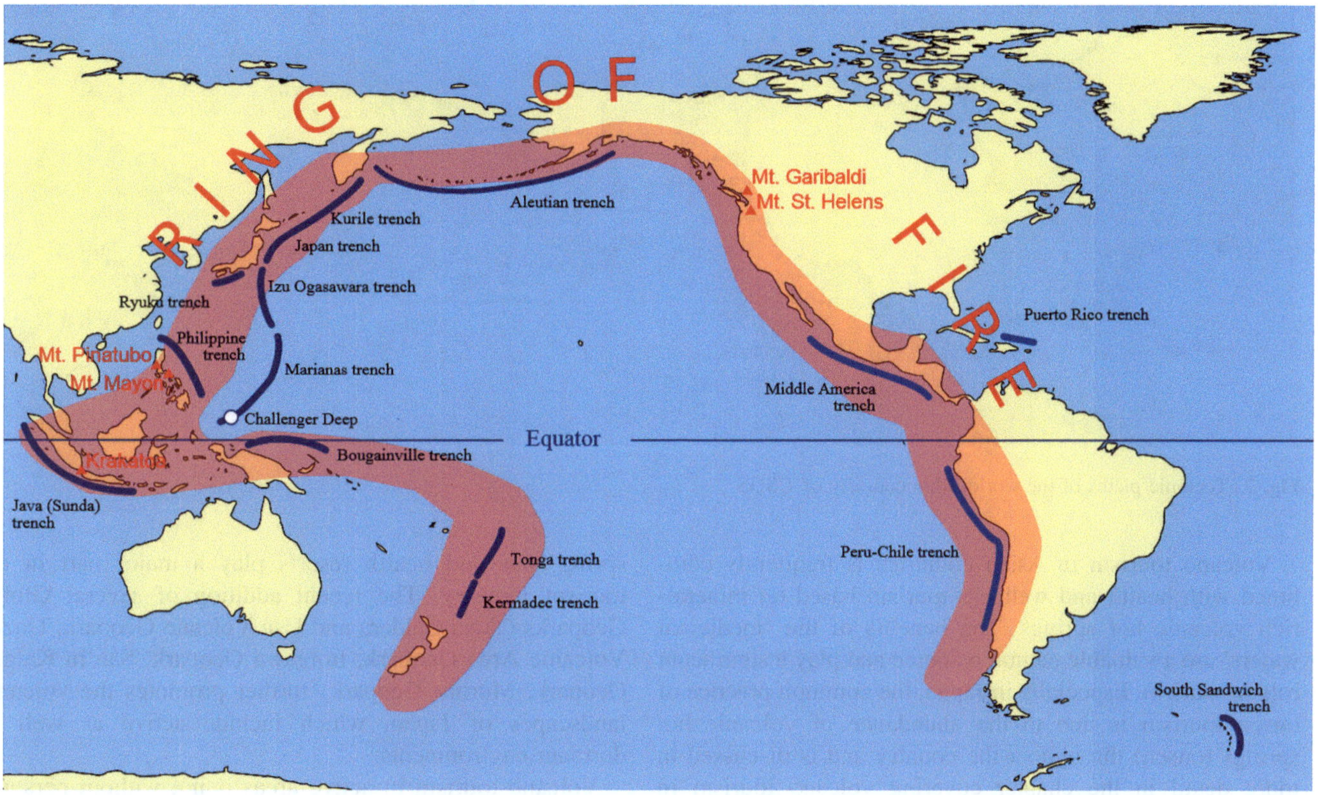

Fig. 1 The Pacific Ring of Fire. *Source* World Databank

The Pacific 'Ring of Fire' (RoF) is an interlinked system of volcanic arcs (Fig. 1), that overlies crustal plate boundaries (Fig. 2). Movement of these plates either through spreading or by subduction is the cause of earthquakes and volcanic activity. From Chile on South America's west coast a long chain of volcanoes follows the Pacific Basin. Through Central America, Mexico and North America the RoF follows the west coast of the United States, where the Cascade Arc alone stretches over a distance of 1,250 km from northern California to the south of British Columbia and includes well-known volcanoes such as Mount St. Helens, Crater Lake, Mt Shasta, Mt Rainier and Lassen Peak among others (Lockwoood and Hazlett 2010). On the northern edge of the RoF, Alaska's outlying Aleutian Island Arc (27 active volcanoes) connects to the Russian Kamchatka Peninsula (29 active volcanoes) and the Kuril Islands (between 40 and 50 active volcanoes, figures vary). Heading further south Japan (108 active volcanoes) and the Philippines (23 active volcanoes listed by the Philippine Institute of Volcanology and Seismology (PHIVOLCS) are part of the western section of the RoF, together with Indonesia (127 active volcanoes, figures vary), Papua New Guinea (67 active volcanoes) and New Zealand including the Kermadic Islands (20 active volcanoes, figures vary). Located in between are many volcanic islands, either active or currently dormant.

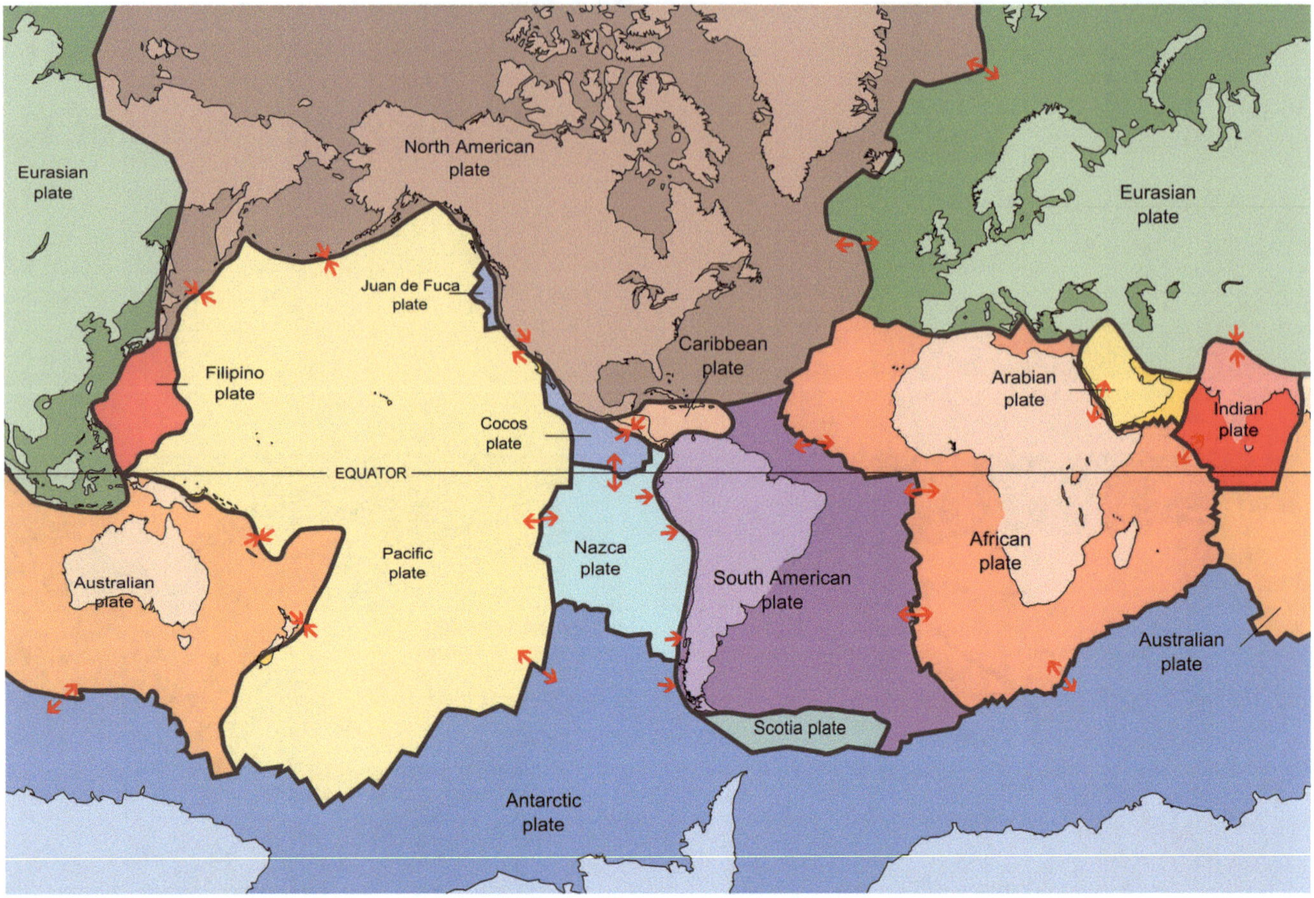

Fig. 2 Tectonic plates of the world. Map courtesy of USGS

Volcano tourism in Asian countries is frequently combined with health and wellness tourism based on mineral-rich volcanic hot springs. The benefits of the 'medicinal waters' are a valuable natural resource and play a significant role in tourism. Especially in Japan, the common practice of onsen tourism is due to the abundance of volcanic hot springs (onsen) throughout the country and is discussed in more detail in the chapter covering volcano tourism in Japan. The majority of Japanese geoparks, national parks, and other protected sites have access to volcanic hot springs and generally cater for their visitors' demand for authentic natural hot spring water. Hokkaido's national parks Akan, Shikotsu-Toya and Daisetsusan would be unimaginable without hot spring resorts, where visitors can combine exploring the volcanic landscape with healthy relaxation in hot springs followed by special regional cuisine. On the island of Honshu it is not any different; the volcanic national parks and geoparks are surrounded by hot spring resorts, which attract millions of visitors annually, who spend their time hiking the volcanic mountains and enjoy their onsen in the company of family, friends and colleagues. Mt Fuji, a volcano that is expected to erupt again in the future, is surrounded by onsen towns, where hot spring spas and health resorts play a major part in the tourism industry. The recent addition of several Global Geoparks (Toya Caldera and Usu Volcano Geopark, Unzen Volcanic Area Geopark, Itoigawa Geopark, San'in Kaigan Geopark, Muroto Geopark) further promotes the volcanic landscapes of Japan, which include active as well as dormant environments.

Volcano tourism in active areas is not without personal risk with the occasional tragic example as evidence. In 1991 on the Indonesian island of Sulawesi a tourist from Switzerland hiking on Mt Lokon (1,750 m) was killed in an eruption (Associated Press 2011; Earthweek 2011; Jakarta Post1 2011). In 2011 Mt Soputan (Sulawesi) erupted and caused the closure of the airport in Sulawesi's capital Manado (Jakarta Post2 2011). Also in 2011 Mt Lokon's activity caused the evacuation of more than 5,000 people on the island of Sulawesi. On the island of Java, Mt Merapi is one of Indonesia's most active volcanoes which erupted in 2010, resulting in an official death toll of 353 people with nearly 400,000 local residents evacuated from the affected area. Despite the potential danger, Merapi continues to be frequented by tourists (activity permitting) with the developing trend towards 'disaster tourism' or 'dark tourism', showcasing villages, rivers and valleys buried in volcanic debris (Lies 2010). Other volcanic destinations in Indonesia such as the new Global Geopark featuring the Batur Caldera seem to be comparatively safe.

The following Chaps. 12–19 are interpretations of volcanic geosites in countries which are located along the Pacific Ring of Fire with its high tectonic activity that extends along the Pacific Ocean basin. One exemption is the Galápagos archipelago, which is located over a mantle plume-fed hot-spot inside the Nazca Plate about 1,000 km offshore Ecuador. Technically this locates the Galápagos Islands, which are a UNESCO World Heritage area and known for their unique natural environment, outside the RoF. However, for in depth clarification of plate tectonics and volcanism, Lockwood and Hazlett's (2010) book 'Volcanoes—Global Perspectives', the 'Encyclopedia of Volcanoes' by Sigurdsson et al. (2000, 2014) as well as Schmincke's (2006) title 'Volcanism' are highly recommended.

In Chap. 12 Harry Keys and Karen Williams share details of the most recent eruption events which involved important volcanic geosites of the Tongariro National Park, New Zealand. Taiwan's volcanic geoheritage is introduced by Angus Robinson and Jiun-Chuan Lin in Chap. 13. The volcanic islands of the Galápagos Archipelago off the Ecuadorian coast are analysed for their geotourism potential by Ross Dowling in a special case study in Chap. 14. The well-known Mount St. Helens on the west coast of North America is another important destination located in the Ring of Fire with its past and present reviewed in Chap. 15 by Chris Newhall, Peter Frenzen and Carolyn Driedger. In Chap. 16 a typical volcano tour to the Mount Batur caldera on the Indonesian island Bali is discussed by Patricia Erfurt-Cooper. David Newsome and Richard Field take a closer look in Chap. 17 at tourism close to the infamous Krakatau volcano located in the Sunda Strait between the Indonesian islands of Sumatra and Java. The following two chapters explore volcano tourism in Japanese National Parks and other protected sites (Chap. 18 by Malcolm Cooper) and the problems for tourism of Papua New Guinea's Rabaul (Chap. 19 by Bruce Prideaux) based on active volcanism.

References

Associated Press (2011). Indonesian Volcano Erupts. Thousands of Residents Evacuated from Slopes of Mount Lokon in Sulawesi Province. Online Document: http://www.guardian.co.uk/world/2011/jul/15/indonesian-volcano-erupts. Accessed 14 March 2013.

Earthweek (2011). Thousands Flee Exploding Indonesian Volcano. Online: http://www.earthweek.com/2011/ew110722/ew110722d.html. Accessed 14 March 2013.

Jakarta Post1 (2011). Evacuated After Mt. Lokon Eruptions. Online: http://www.thejakartapost.com/news/2011/07/17/4889-evacuated-after-mt-lokon-eruptions.html 889(4) .Accessed 14 March 2013.

Jakarta Post (2011). Sam Ratulangi Airport Closed After Volcanic Eruption. Online: http://www.thejakartapost.com/news/2011/07/03/sam-ratulangi-airport-closed-after-volcanic-eruption.html. Accessed 14 March 2013.

Lies, E. and Oblo, D. (2010). After Eruption, Indonesia Turns to Disaster Tourism. Online: http://www.reuters.com/article/2010/12/16/uk-indonesia-tourism-disaster-idUSLNE6BF02620101216. Accessed 25 July 2011.

Lockwood, J. P., and Hazlett, R. W., (2010). Volcanoes: Global Perspectives. Chicester, UK: Wiley-Blackwell.

12 Volcanoes of the Tongariro National Park, New Zealand

Harry Keys and Karen Williams

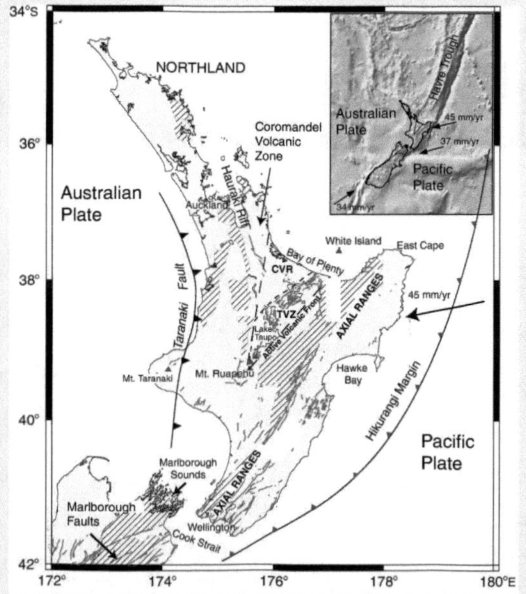

Map 1 courtesy of Tim Stern (Victoria University of Wellington) and Geological Society of London

Geographical location including map co-ordinates: Tongariro National Park, central North Island, New Zealand A 175 24" 08" E; 39 10" 43" N B - 175 25" 49" E; 39 24" 44" N C - 175 33" 29" E; 39 04" 45" N D - 175 41" 43" E; 38 58" 40" N E - 175 46" 45" E; 39 04" 05" N F - 175 49" 04" E; 39 08" 24" N G - 175 44" 26" E; 39 17" 54" N
Type of volcanoes: Composite stratovolcanoes
Last known eruption: 21 November 2012
Major attractions: Tongariro Alpine Crossing and Northern Circuit, Ruapehu Crater Lake, Whakapapa and Turoa ski areas, Round-the-Mountain and other tracks
Status of protection: National Park, World Heritage Area
Visitor numbers: Approximately 1 million per year

12.1 Geographical Location

The volcanoes of Tongariro National Park (TNP) are located in the central North Island of New Zealand. Three active volcanoes, Ruapehu (2797 m), Ngauruhoe (2287 m) and Tongariro (1967 m) (Figs. 12.1a–c and 12.2) and several smaller extinct cones comprise the Tongariro Volcanic Centre most of which lies within the national park. The park is 300 km south of Auckland and 300 km north of Wellington with state highways and the main railway giving easy access. Ohakune and Turangi are the two largest towns at the boundaries of the park while Whakapapa Village lies inside the park on the lower slopes of Mt Ruapehu.The Department of Conservation (2013) has detailed maps of the park, including mapping tools for tracks, and information on huts, visitor centres and places of interest.

H. Keys (✉)
Department of Conservation and Fire and Ice Consultants, Turangi, New Zealand
e-mail: hkeys@doc.govt.nz

K. Williams
Department of Conservation and Fire and Ice Consultants, Taupo, New Zealand

12.2 Geophysical Aspects and Geological History

The Tongariro Volcanic Centre is 50 km long and 30 km wide with a base level of 350–1000 m above sea level. It lies at the southern end of 20–60 km wide Taupo Volcanic Zone (TVZ) which extends 240 km from Ruapehu in the south to the Okataina Volcanic Centre in the north and offshore to White Island volcano and beyond (Map 1). Ruapehu is the highest and with a volume of 150 km^3 the largest volcano (Hackett and Houghton 1989) and the most active over the last 35 years (Fig. 12.3).

P. Erfurt-Cooper (ed.), *Volcanic Tourist Destinations*, Geoheritage, Geoparks and Geotourism,
DOI: 10.1007/978-3-642-16191-9_12, © Springer-Verlag Berlin Heidelberg 2014

(a)

(b)

(c)

Fig. 12.1 **a** Winter aerial view from 3000 m/10,000 feet of the three active volcanoes of Tongariro National Park with fault lines prominent. View is from the northwest with Tongariro in the foreground including Te Maari area on *bottom left* looking towards Ruapehu in the distance. *Photo* Paul Dawson. **b** Summer aerial view from the northeast of the three active volcanoes of Tongariro National Park—five months before the eruption from the Te Maari crater area in the foreground. *Photo* Karen Williams. **c** View from near the summit of Mt Tongariro looking south past Ngauruhoe to Ruapehu with the old volcanic remnant of Hauhungatahi on the *right*. *Photo* Harry Keys

The active volcanoes in New Zealand result from the tectonic setting in which the subducting slab of the Pacific Plate is pushing under the Indian-Australian Plate in an obliquely convergent boundary zone with an eastern expression at the Hikurangi margin and trough just east of the North Island (Map 1). Volcanism in northern New Zealand began about 22 million years ago, consistent with the inception of subduction along the Hikurangi margin

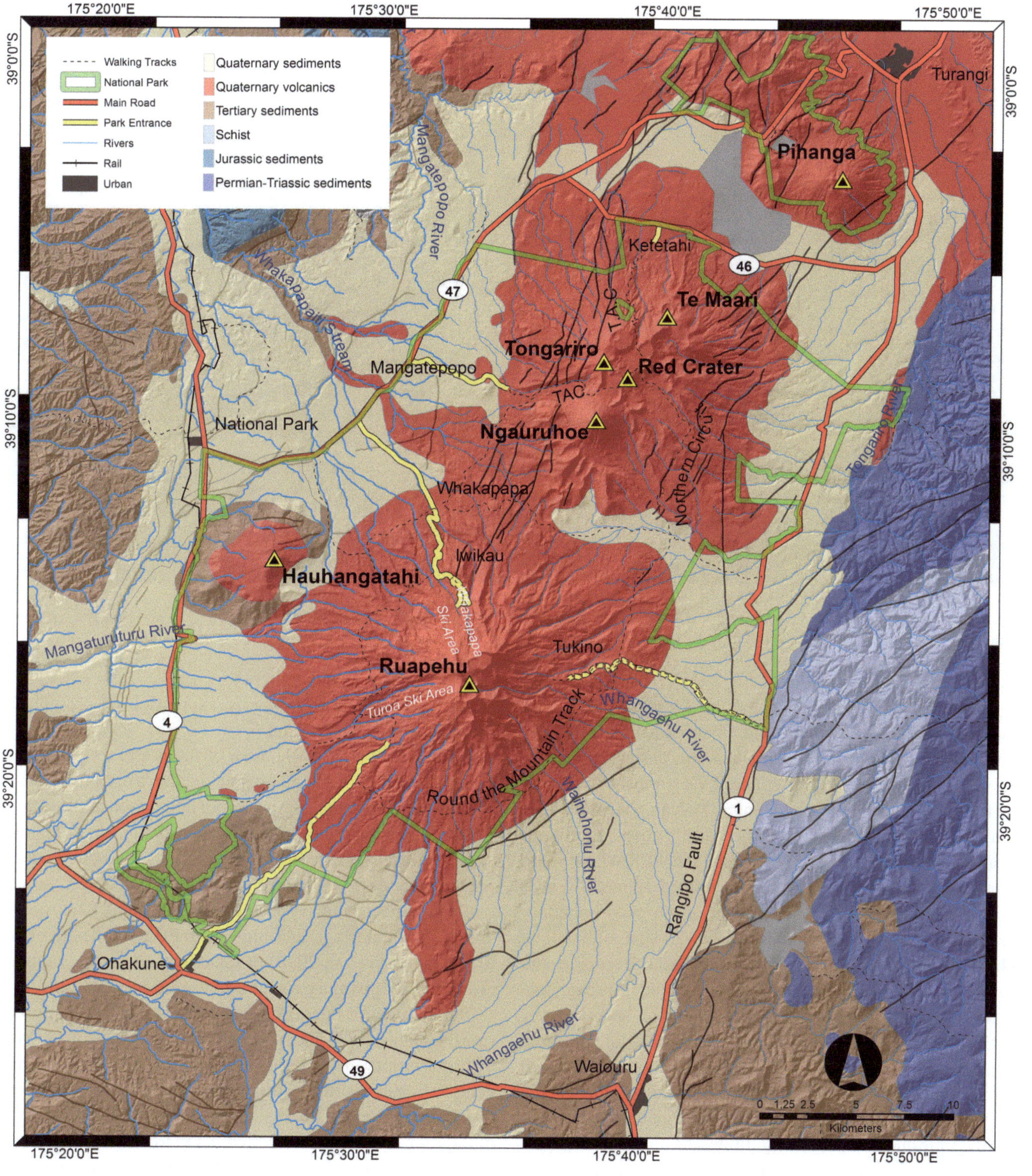

Fig. 12.2 Tongariro National Park with volcanoes and major faults parallel the regional tectonic trend, showing access roads and major tracks including Tongariro Alpine Crossing (TAC). The volcanoes are located at the southern end of the Taupo Volcanic Zone surrounded on three sides by marine sedimentary rocks 4–30 million years old or older *greywacke* and schist >150 million years (Map 1). Map prepared by Jon Procter (Massey University) using regional geology and faults courtesy of GNS Science

(Lamb 2011), and has moved south due to the tectonic motion. The result is a trail of extinct volcanoes that are progressively younger to the east and southeast where the active volcanic front is currently located. Over the last five million years this front has had an apparent southeast migration rate of about 23 mm per year (Stern 2009).

Fig. 12.3 Ruapehu Volcano (2979 m) rises 2000 m above State Highway 1 across the ring plain. Eruptions in 1995 through 1997 provided popular viewing for travellers. *Photo* Harry Keys

Tongariro Volcanic Centre and the TVZ are the current southern part of a 2500 km long chain of active terrestrial and underwater volcanoes and island arcs along the edge of the subducting Pacific Plate. The TVZ forms the eastern portion of the wedge-shaped Central Volcanic Region. This is a thin, spreading segment of the Earth's crust, referred to as a back-arc (Map 1). The region and the TVZ within it are associated with clockwise rotation, hinged apparently near Ruapehu, and southwards movement of the plate boundary over the last four million years (Stern 2009; Lamb 2011). The TVZ became active about two million years ago and has an anomalously high heat flow and widespread geothermal activity. It has been the source of spectacularly explosive rhyolitic eruptions with a pile of volcanogenic material up to 2 km deep in places and pyroclastic material spread out either side of the zone. Typical of the Pacific Ring of Fire however, Tongariro volcanoes are predominantly andesite, with some dacite, basalt and rhyolite.

West, northwest and east of the three main volcanoes in TNP the boundary of the TVZ in this region is marked by prominent active faults (Figs. 12.1a and 12.2). These are controlled by the tectonic pattern but downthrown towards the volcanoes as subsidence has developed in the central axis of the TVZ. The southern termination of the TVZ is 20 km south of Ruapehu's summit at a series of lakes near Ohakune. These lakes occupy extinct craters 30,000 years old that are the most southerly expression of present day volcanism in New Zealand.

Volcanism began in the area of the present TNP about a million years ago. Hauhungatahi in the west is the oldest volcanic feature representing part of a lineation in the Central Volcanic Region (Map 1) that is older than and sub-parallel to the current active volcanic front. It is an eroded andesite volcano dated at ~900,000 years distinctive from the currently low-potassium volcanoes of the active front nearby due to its greater age and high-magnesium andesite chemistry (Cameron et al. 2010). It mantles the top of a sequence of marine sedimentary rock up to 7 million years old.

Ruapehu and Tongariro-Ngauruhoe are active composite stratovolcanoes consisting of alternating layers of volcanic material erupted over the last 300,000 years from multiple vents aligned north-northeast to south-southwest corresponding to the present active volcanic and tectonic trends in the region. Ngauruhoe which is regarded as the youngest cone of Tongariro is now thought to have developed from about 7,000 years ago (Moebis 2011) over some of the oldest Tongariro lavas (Fig. 12.4). In contrast Kakaramea-Tihia and Pihanga comprise a smaller volume, lower and well-vegetated andesitic volcanic massif in the north of the park. They erupted between 230,000 and 20,000 years ago along a line almost at right angles to the present geological trends.

Fig. 12.4 The young Ngauruhoe Volcano (2287 m) developed over an older volcano massif including Tongariro to the north (*right*). Glaciers radiating outwards eroded valleys like this Waihohonu Valley to the southeast. *Photo* Jimmy Johnson

During the last ice age Mt Ruapehu and the Tongariro massif carried glaciers extending down to as low as 1100 m elevation (McArthur and Shepherd 1990). These glaciers and those of the previous two glacial maxima altered the shape of the existing volcanoes by excavating deep valleys (Fig. 12.4) damming and interacting with lava flows and other volcanic material, and depositing material including prominent moraines.

Ruapehu provides a very good example of how a volcano develops over time. It is dissected in places and surrounded by dateable volcanic deposits so that its geological history is well known (Hackett and Houghton 1989; Donoghue et al. 1997; Gamble et al. 2003; Pardo et al. 2012; Price et al. 2012). The oldest dated materials consist of 300,000 year old andesite pebbles deposited on a marine bench now uplifted in the Wanganui district (Neall et al. 1999). These are the remnants of an andesite lava flow which was emplaced following eruptions from an early volcano thought to be in the vicinity of Ruapehu.

Four major phases of growth are recognised with the constructional periods separated by stages of erosion and lower level volcanic activity. On the volcano itself the oldest exposed lava sequences (the *Te Herenga* Formation) are part of a basic andesite cone forming the northern flank of Ruapehu from about 250,000 to 180,000 years ago. Two further major periods of cone growth occurred from 160,000 to 115,000 and 55,000 to 20,000 years ago. After that, lavas younger than 15,000 years old (the *Whakapapa* Formation) were erupted mantling slopes in the northwest, south and elsewhere (Fig. 12.5). Voluminous lahars were generated over much of the life of the volcano, most obviously during the second and third constructional periods and coincident with the last two glacial cycles. Lahar deposits totalling more than 12 km^3 (Neall et al. 1999) have inundated the nearby hill country, creating an extensive ring plain around much of the volcano.

From 27,000 to about 10,000 years ago a sequence of large eruptions (Pardo et al. 2012) produced extensive and revealing layers of tephra around the volcano (Fig. 12.6). These were deposited by explosive eruptions involving up to one cubic kilometre of magma and producing eruption columns possibly up to 40 km high. Most of the youngest lava flows on Ruapehu (Fig. 12.5) were produced during this very active period. The last major eruption in this sequence about 12,000 years ago produced huge pyroclastic density currents down the eastern and western flanks, well past the current State Highway 1. Also around this time a major structural collapse of the northwest flank occurred

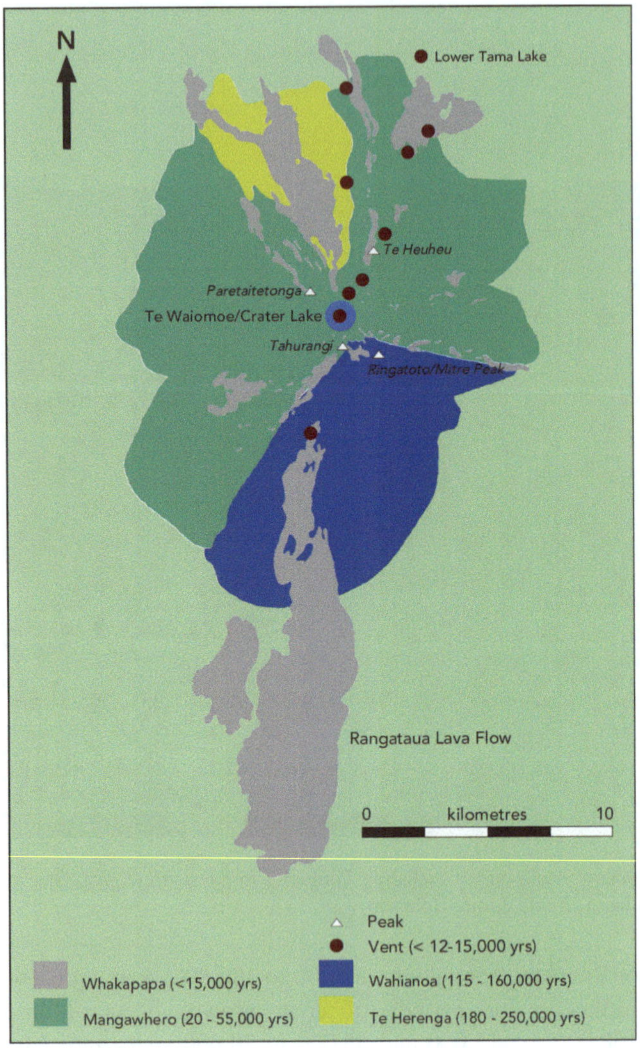

Fig. 12.5 Map of Ruapehu showing extent and recently calculated age ranges of exposed cone-forming lava sequences and young vents, from Hackett and Houghton (1989), Gamble et al. (2003) and Price et al. (2012). Young lava flows from a forthcoming 1:50,000 geological map of TNP are shown courtesy of GNS Science. The Rangataua lava flow sequence is the largest in the national park

and a debris avalanche produced the distinctive mounds through which Highway 48 passes en route to Whakapapa Village (Palmer and Neall 1989).

Over the last 10,000 years activity at Ruapehu summit has been generally smaller and less explosive. A sequence of craters were active in the central and northern summit plateau area between about 9,000 and 3,000 years ago (Palmer 1991) producing lavas through the area now occupied by Whakapapa ski area. They are now full of up to 130 m of glacial ice (Holdsworth 1985). Sometime between 4,500 and 3,500 years ago, collapse of an old hydrothermally altered rim around a southern vent produced a large debris avalanche that swept down the Whangaehu and east across the Rangipo 'desert' destroying a widespread beech forest on the southeastern ring plain (Donoghue and Neall 2001). Since deposition of the Taupo Pumice around the volcano's flanks about 1,800 years ago, a series of 19 widespread thin tephra deposits have been locally preserved, and several are younger than 700 years (Donoghue et al. 1997). It appears the modern Crater Lake came into existence around that time. This youngest crater rim is composed of lavas, agglomerates, scoria, breccia, tuff, tephras and sedimentary material (Hales 2000; Hodgson et al. 2007). Eruptions and collapses of this youngest rim have resulted in a large number of lahars with a wide range of sizes. Some were very large with estimated maximum discharge rates of up to 6,000 m^3 per second, which is much greater than the maximum discharge estimated for the largest Ruapehu lahar (1861) in historical time (Neall et al. 1999).

The Tongariro volcanic complex has also grown since about 300,000 years ago. The oldest dated rocks in place are about 270,000 years old near Tama Lakes with subsequent intervals of rapid cone growth for example at 210,000–200,000, 130,000–65,000, and 27,000 years to the present day (Hobden et al. 1996; Fig. 12.7). The most recent major eruption sequence occurred about 11,000 years ago at the end or just after the last major sequence on Ruapehu (Donoghue et al. 1995; Nairn et al. 1998; Pardo et al. 2012) producing mainly andesite tephras up to 1.25 m thick along the route of the current State Highway 1 (Desert Road).

There has been a long debate about whether volcanism at Tongariro and Ruapehu has occurred or occurs at similar times, possibly driven by tectonic forces which might also drive large rhyolite eruptions. There are gross similarities between the initial and some other periods of major activity or growth of these volcanoes, and some approximate coincidence with the major eruption of Taupo 'super-volcano' 26,000 years ago. But there are also major differences. While different rock dating techniques may not produce entirely compatible results, there appear to be a lack of lavas on Tongariro around 134,000 and between about 65,000 and 25,000 years old, and a lack on Ruapehu around 60,000-100,000 years (Hobden et al. 1996; Gamble et al. 2003). This may be due to an absence of data from several key areas as a result of the loss of parts of the geological record during widespread erosion during the last glaciations or due to inundation by more recent volcanic material. But it is also clear that the two volcanoes have magmatic systems that are chemically different and entirely independent for much of geological time (Gamble et al. 2003) and in the modern day (Moebis 2011). Geological interpretations suggest regional extension and rotation rates across the volcanic region have been consistent over the last four million years, ten thousand years and recent decades (Stern 2009; Lamb 2011). During the period from about 25,000 years ago (or earlier) to 14,000 years ago the major Rangipo Fault southeast of Ruapehu (Fig. 12.3) was moving an order of magnitude faster than now (Villamor et al. 2007). This roughly coincides with

Fig. 12.6 Layers of tephra deposited from a series of large eruptions of Ruapehu and even larger eruptions from caldera volcanoes to the north. There are several spectacular tephra sequences like this beside the Tukino Road and many others beside state highways around the volcanoes. *Photo* Harry Keys

the last period of major volcanism in TNP (Pardo et al. 2012; Nairn et al. 1998). But otherwise there is no direct evidence yet for periods of more rapid tectonic extension or other plate kinematics that might have been associated with higher heat flow, generation of magma or dike intrusion geographically extensive enough to drive simultaneous growth and eruptions of these volcanoes.

12.3 Geological Heritage

Tongariro National Park is inscribed as World Heritage under UNESCO's Natural criteria (vii) and (viii), and Associative Cultural criterion (vi). Criterion (vii) involves "superlative natural phenomena or areas of exceptional natural beauty and aesthetic importance" while (viii) involves "outstanding examples representing major stages of Earth's history, including the record of life, significant on-going geological processes in the development of landforms, or significant geomorphic or physiographic features".

Criterion (vi) is "directly or tangibly associated with events or living traditions, with ideas, or with beliefs, with artistic and literary works of outstanding universal significance". The park itself grew from a transaction between the paramount chief of Ngati Tuwharetoa and the colonial government involving the summit areas of the volcanoes to the current extent of protected land. This has become important to New Zealand in the relationships between Maori and other New Zealanders and in a three-way bond with the volcanoes.

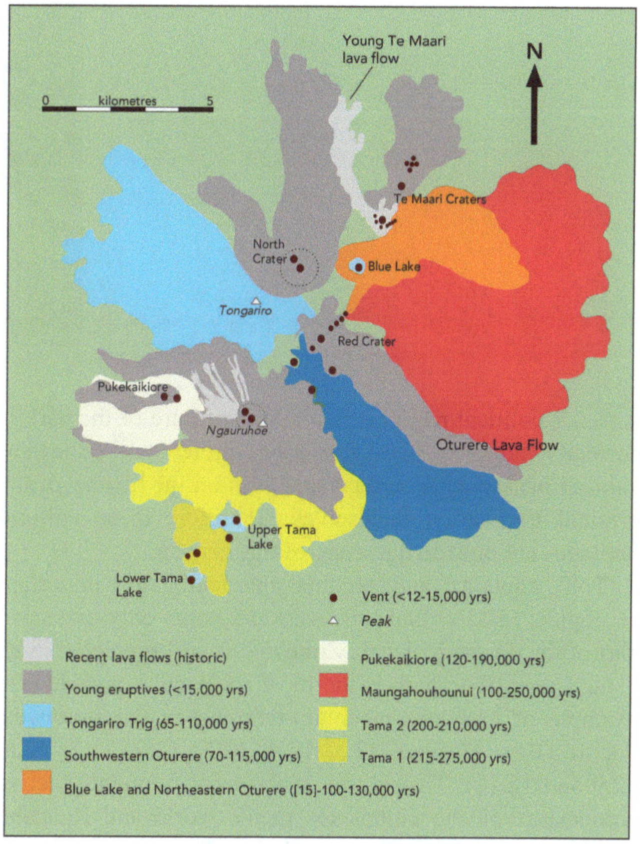

Fig. 12.7 Map of Tongariro-Ngauruhoe adapted from Hobden et al. (1996, 2002), Price et al. (2003) and Nairn (unpublished) showing ages of exposed cone-forming sequences. Young eruptives are from a forthcoming 1:50,000 geological map of TNP courtesy of GNS Science

Fig. 12.8 View from Tukino Road encompassing 1000 m of relief from the summit of Ringatoto Peak (2591 m, *upper right*) to the Whangaehu River in the foreground. This scene includes evidence of recent deepening of the gorge and sediment deposition by the March 2007 lahar, the sequence of 134,000 year old lavas and other material in the middle distance and various younger intrusions, lava flows and deposits. The summit of the glacial horn, Girdlestone Peak, can be seen above the saddle. *Photo* Frank Katavitch

A major political negotiation about the future of the park is currently underway. *Iwi* (Maori tribes) prefer to describe their cultural backgrounds and perspectives on the history of the original transaction themselves. Therefore these cultural heritages are outside the scope of this chapter.

TNP contains and records internationally important geological features and processes and is the only extensive part of the Taupo Volcanic Zone that is still protected in its natural state. The park is the most spectacular and diverse example of andesitic mountain-building in the southwest Pacific (Figs. 12.1a–c, 12.3, 12.4, 12.5, 12.6, 12.7, 12.8, 12.9 and 12.10). This is one of the most frequently active composite volcano complexes in the world and Ruapehu provides a model of volcano evolution. The current volcanic landscape reflects a dynamic balance between cone-building by episodic volcanism and subsequent destruction by glacial, laharic and other erosion processes. Major topographic relief has been produced including the upper Whangaehu Valley which provides a 1,000 m cross-section through the "Wahianoa volcano" and its 300 m thick sequence of lava flows and pyroclastic units around 134,000 years old (Gamble et al. 2003) (Fig. 12.8). A wide variety of volcanic and glacial features, including the world-famous Crater Lake and lesser known features of volcano-glacier interaction, are present.

Reworked volcanic debris from lahars, rivers and landslides, plus airfall and pyroclastic flow deposits, form extensive and classic ring plains surrounding the volcanoes (see Figs. 12.2, 12.3 and 12.6). The park also contains extensive tephra deposits from one of the world's most powerful volcanic eruptions of the last 5,000 years – the Taupo 'super-volcano'. The ring plain contains interesting carbonised examples of forest trees felled by the Taupo blast and lying in a radial direction from it (Fig. 12.9).

Fig. 12.9 Charred and burnt logs of mountain toatoa (*Phyllocladus alpinus*) at Rangipo "desert" lying where they were felled by the hot blast that travelled 80 km from the Taupo eruption in 232 AD. *Photo* Karen Williams

Ruapehu's Crater Lake, prominent in the summit area between the highest peaks, is unique due to its high frequency of eruption, size, warmth, glacial setting and accessibility (Fig. 12.10). Normally it is full to overflowing and 200,000 m^2 in extent, up to 170 m deep with a volume of 7–10 million m^3 of water with a pH level of 1. It is also a *urupa* (burial ground) where chiefs of the *Ngati Rangi iwi* (tribe) were interred in historic time.

Crater Lake is one of only two volcanoes internationally regarded as classic examples of the interaction of magma and lake water. It has become an important volcano "laboratory" for understanding volcanic processes. A notable characteristic of the magmatic-hydrothermal environment of the volcano is the thermal cycling of the lake (excluding the main eruption episodes). Since regular measurements began in the 1960s its temperatures have varied irregularly with time between 9 and 60 °C, at periods ranging between 2 and 4 and 16 months (Fig. 12.11). Correlation between lake temperature and CO_2 discharge from the volcano shows that the heating cycles exhibited in the lake are largely controlled by variations in the rate of magmatic degassing located within about 1000 m beneath the lake (Christenson et al. 2010). Fluid instabilities in the vent system may also contribute (Vandemeulebrouck et al. 2005). Other factors such as the presence of elemental sulphur and other mineral accumulations reducing gas and energy flux up into the lake appear to be secondary to this (Christenson et al. 2010, Fig. 12.12) but do help create anomalous periods with temperatures persistently below 24 °C for 6–12 months (Fig. 12.11). Daily solar heating and cooling cycles of 1–3 °C and approximately monthly cycles of 2–4 °C possibly due to earth tides, are also evident during stable weather conditions.

Phreatomagmatic eruptions, a characteristic of Ruapehu, occur when a significant amount of magmatic fluid interacts with water in the vent or the lake itself with accompanying explosive expulsion of Crater Lake water. In the largest single explosions (or short blast sequences) documented (10 March 1895, 24 April 1975) about 20 % of the lake was expelled not counting the ejected water that flowed back into the lake. Expulsion occurs as jets of tephra (Surtseyan jets) including blocks moving on ballistic trajectories and water (Figs. 12.13 and 12.14). Smaller events referred to as phreatic and hydrothermal eruptions occur with smaller amounts or overpressures of magmatic gas, upwellings of steam or brine from the vent system, or density-driven overturnings in the lake itself. The least energetic of these expel small amounts of water relatively passively over the lake outlet with the smallest measured volume (13 July 2009) equivalent to 0.2 % of the lake.

Lahars, potentially destructive fluid mixtures of volcanic debris and water, are another characteristic of Ruapehu volcano. They are produced by interactions involving volcanic (especially magmatic) processes, lake water and surrounding terrain. It has become clear that most lahars on Ruapehu are formed by explosions through the lake and typically form as dirty snow slurries when erupted water mobilises seasonal snow (Houghton et al. 1987, Cronin et al. 1996). Lahars are also formed by hot volcanic ejecta interacting with snow, tephra deposits mobilised by rain or meltwater, and various kinds of collapse of the rim or breakout of Crater Lake or slopes of the volcano (Manville et al. 2000; Hodgson et al. 2007; Keys and Green 2008). In the last 66 years there have been 17 lahar episodes documented, at intervals of 3 to >10 years, with one to more than 30 lahars per episode. Crater Lake drains into the Whangaehu River so this is the main lahar path but other valleys are occupied depending on eruption magnitude and direction as well as lahar formation process (Fig. 12.15).

The potentially devastating nature of lahars and our experiences with them are well demonstrated by recent history regarding Crater Lake before and after the lahar

Fig. 12.10 The warm Crater Lake (Te Waiomoe) occupies the active crater of Ruapehu and is currently about 100 m deep. In 18 March 2007 the outlet and normal overflow were re-established following a breakout though weak tephra deposits from the 1995-1996 eruptions that had dammed the lake (see also Fig. 12.16). *Top photo* Paul Bradshaw. The *second photo* (b) shows upwelling of dirty sediment-laden water. *Photo* Ben Goddard

hazard was understood. Breakout of Crater Lake in December 1953 resulted in a severe lahar which severely damaged a railway bridge across the Whangaehu River leading shortly afterwards to the death of 151 train passengers in the so-called 'Tangiwai Disaster'. This was a significant event in New Zealand's history. The most recent breakout occurred on 18 March 2007 when a mantle of tephra produced in the 1995–1996 eruption episode, collapsed (Fig. 12.16 a–c) as predicted after the lake refilled, and produced a lahar with the largest discharge rate since 1861 (Keys and Green 2008; Williams and Keys 2008; Massey et al. 2010; Fig. 12.17). Active management based on planning that took a long term view of risks, natural processes, and cultural and world heritage values into account, ensured no-one was injured and damage to infrastructure down the Whangaehu Valley was negligible. There is now little foreseeable volcanic risk to infrastructure below 1600 m down this valley or beside the river. Research on the lahar (Manville and Cronin 2007) will have long term benefits for the management of lahar hazards.

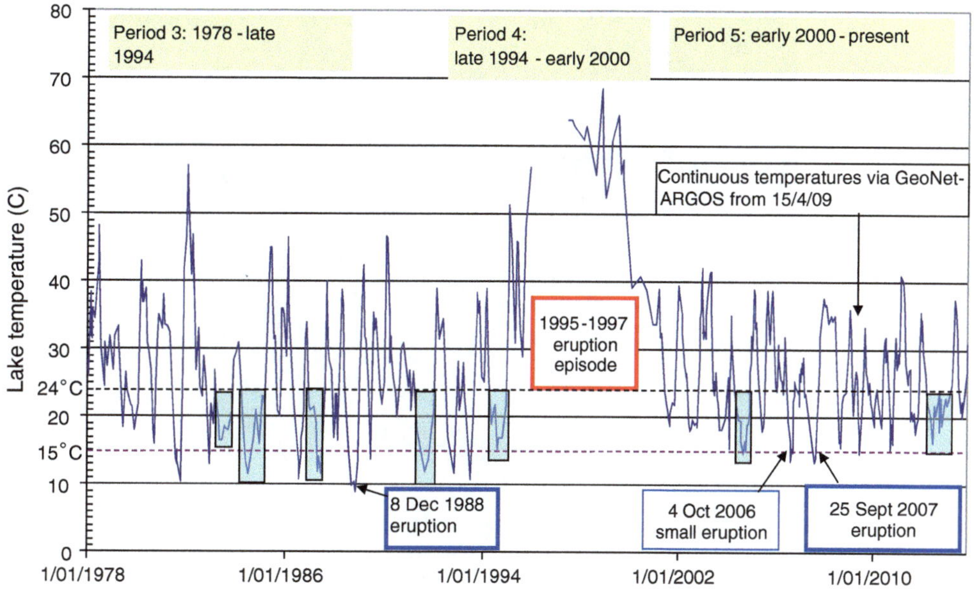

Fig. 12.11 Crater Lake water temperatures measured between 1978 and November 2013 eruptions (data collected by the Department of Scientific and Industrial Research, GNS Science and Department of Conservation). The record can be divided into periods of similar or differing thermal regimes. Anomalous long cool periods (highlighted boxes) are not yet fully understood although there is an increased probability of eruptions during the coolest periods

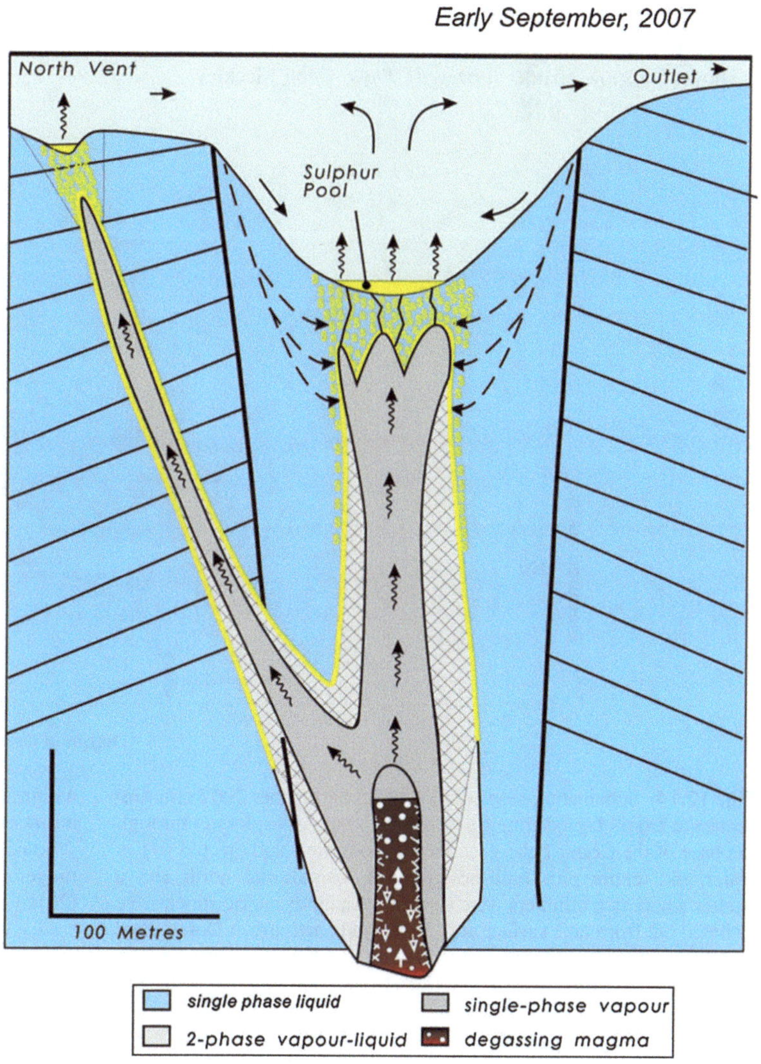

Fig. 12.12 Cross section beneath Crater Lake showing degassing magma high in the vent system, the two vents and the hypothesised "shroud" of elemental sulphur and other minerals (*yellow*). These create a potential for over-pressuring by reducing permeability where it accumulates around the margins of the vapour region (Christenson et al. 2010)

Fig. 12.13 Phreatomagmatic eruption in October 1997 with 400 m tall jets of water and tephra and ballistic bombs trailing steam and almost swallowing a snow boarder (*arrowed*). *Photo* Geoff Mackley

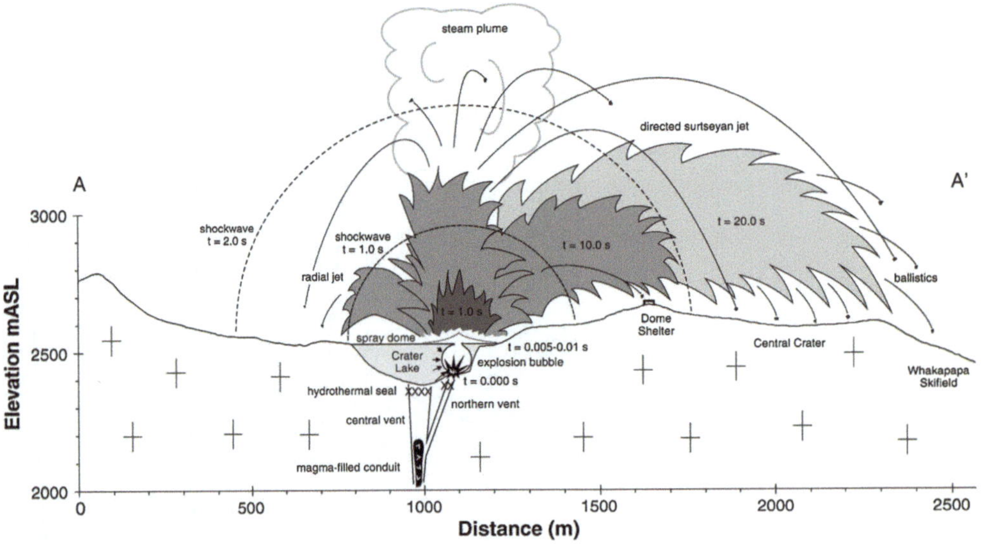

Fig. 12.14 Schematic reconstruction of 25 September 2007 eruption sequence from elapsed time t = 0 to t = 20 s. An explosion through the base of the Crater Lake generated a shockwave and ejected jets of water and tephra plus ballistics directed towards the north and a shelter where two climbers were sleeping. An atmospheric shockwave radiated out from lake surface at c. 1200 km/h, arriving at Dome Shed within 2 s of the explosion, blowing the door open. Within 10 s of the initial explosion Dome Shed had been engulfed by spray and ballistics seriously injuring one climber, while the jets reached their maximum range of c. 2 km in around 20 s. Courtesy of Kilgour, Manville et al. (2010)

Fig. 12.15 Dirty snow and debris mark the path of the lahar from the 25 September 2007 eruption out of Crater Lake down the Whangaehu Glacier. *Photo* Harry Keys

Fig. 12.16 **a** (*before* breakout) **b** and **c** (*after* breakout) showing the 60 m wide and 7.5 m high breach in the tephra dam created on 18 March 2007 and which released 1.3 million cubic metres of Crater Lake water forming the lahar with the largest flow rate since 1861. *Photos* Harry Keys (**a**, **c**), Herb Christopher (**b**)

12.4 Educational Aspects

Numerous school and university groups from New Zealand and overseas visit the park each year to examine aspects of the volcanoes, geology, biology and cultural and social aspects including volcanic risk management. The field guide "Volcanoes of the South Wind" (Williams 2013, now also called "A volcanic field guide to TNP") is used as a text book at many schools. LEARNZ (www.learnz.org.nz), a programme of free virtual field trips has featured TNP on many occasions. The Ruapehu 2012 trip had two sets of background pages available online plus a set of 10 pages fully translated into *te reo* Maori.

12.5 Recent History of Activity

Ruapehu is generally the most active volcano in New Zealand. Over the last 150 years, there have been at least 20 eruptions that produced lahars (Gregg 1960; Otway et al.

Fig. 12.17 Aerial view of the 18 March 2007 lahar at Tangiwai just after the flood peak, with the rail bridge in the foreground and the 1953 memorial area and highway bridge in the middle background. Six years later visible evidence of the lahar passage includes the concrete floor of a toilet block that was swept away, twisted cables of a safety barrier, sediments, and the newly exposed remains of the rear bogie of the locomotive involved in the 1953 disaster. *Photo* Harry Keys

Fig. 12.18 Ruapehu erupting in June 1996. The lobate deposit on the right is from a tephra jet and pyroclastic density current cloud observed the day before. *Photo* Harry Keys

1995; Keys and Green 2010). Many small and even medium-sized eruptive events have not been documented (B Scott, pers. comm.) so statistics are biased towards larger events.

During the two largest and longest recent eruption episodes (1945 and 1995–1996) the lake water was completely boiled/evaporated or ejected out of the crater onto surrounding glaciers, and/or propelled across the outlet and into the Whangaehu catchment. The water displacement was accompanied (1945) or followed (1995) by lava influx up into the crater and then by dry ash eruptions intermittently for five months (1945) and 11 months (1995–1996, 0.07 km^3 of tephra, Fig. 12.18). Water accumulates in the summit crater when these eruptions are over, most rapidly during the summer melt season (Keys and Green 2008). Phreatic eruptions,

Fig. 12.19 Swiss tourist on the slopes of Ngauruhoe at the interpretation and hazard sign featuring the pyroclastic current deposits from the 1975 eruption. *Photo* Karen Williams

periods of increased heat flow producing dense steam plumes, and some hazardous secondary (rain-induced) lahars followed these eruption episodes for three to seven years.

Most eruptions of Ruapehu in the last 100 years have been short-lived one-off events. The most recent of these in September 2007 (Figs. 12.13, 12.14, and 12.15), October 2006 and December 1988 all occurred near minima in the lake temperature cycle (Fig. 12.11). Some earlier events (April 1968, April 1971, November 1977) occurred near maxima in the cycle. Others also occurred during this 1965 to 1976 period when average heat flow and hence average temperature were high, but near the minima of the temperature cycle at the time (June 1969, April 1975). The 1975 event was the largest of these, and the only one documented as being followed by a second small event (Nairn et al. 1979).

Ngauruhoe used to be regarded as New Zealand's most active volcano but since its last major eruption in 1974–1975 has experienced its longest period of quiescence since the earliest written records in 1839. At that time one large deep vent formed the crater. Subsequent eruptions created a sequence of sub-craters. Eruptions in 1870, 1949 and 1954 produced the only modern lava flows in New Zealand, close to where the Tongariro Alpine Crossing traverses up the lower north-western slopes of the volcano. The 1954–1955 eruption built up a large cone of scoria and lava spatter which buried the earlier sub-craters and the north-western third of the old crater rim. Explosive ash eruptions in January and March 1974 and particularly February 1975 produced columns of ash up to 13 km high, small rocks landed up to at least 2 km away, and numerous avalanches of hot debris (pyroclastic density currents, "block-and-ash flows") travelled down the north-western slopes (Fig. 12.19) to near the floor of the Mangatepopo Valley. Huge blocks of lava in the old crater valley at the base of the 1954–1955 cone and several metres of red welded lava ("spatter") forming the rim of this cone are obvious remnants of this last eruption. Since then temperatures and gas emission have declined to the point where only faint plumes of steam are sometimes seen from fumaroles on the old north-eastern rim and upper north-north-western slopes. Earthquake activity 1–2 km below the northwest slopes commenced intermittently from 2005, probably due to geothermal fluid activity or phase changes in a cooling magma body (Jolly et al. 2012).

Prior to 2012, the last confirmed eruptions from the upper Te Maari crater on the northeast flank of Tongariro were in 1896–1897. Red Crater is thought to have last erupted in the 1920s and 1930s and Te Maari may have had minor activity in this time-frame as well. (B Scott, pers. comm.) with earlier eruptions documented from there and the Te Maari craters between 1839 and 1892. No Maori oral history has come to light, possibly including tephras thought to come from eruptions of Red Crater in the last 320 years (Moebis, 2011) and a large well-preserved lava flow that descended from upper Te Maari into forest about 500 years ago. Lavas that look very young have also flowed from Red Crater and some are younger than 1800 years. Recent magma from Te Maari was rhyolitic and quite different from other Tongariro and Ngauruhoe magmas (Moebis 2011; S Cronin, pers. comm.). This may relate to the age of these magmas or the long duration of activity of the Te Maari craters which at 14,000 years are among the longest-lived of Tongariro's post glacial vents.

Tongariro erupted on 6 August 2012 and again on 21 November from the Upper Te Maari area. The eruption was the first confirmed non-summit, multiple vent eruption seen in living memory and had the most significant effect on forest in the park since the previous eruption of Te Maari more than a century ago. Three weeks of seismic unrest preceded the initial eruption. A week after this seismicity began there were reports from people who had smelt gas

Fig. 12.20 The upper Te Maari eruption area with new vents, fissures and the large chasm formed during the 6 August eruption. *Photo* Harry Keys

Fig. 12.21 Mountain toatoa forest beside the TAC and near it, burned red-brown by the hot acid surges (pyroclastic density currents) during the 6 August eruption. *Photo* Karen Williams

around the volcanoes, and, by 23 July, GNS sampling confirmed there were enhanced levels of volcanic gas being emitted.

The first and largest eruption in the sequence to date occurred at 11:52 pm on 6 August without any immediate precursors (Jolly and Cronin, "The Tongariro Special Issue", 2014). Reconstruction of events suggests that the eruption was initiated by volcanic gas released from old, but not yet cooled magma and which pressurised (and fluidised) old hydrothermally weathered and weakened rock material until it failed. The resulting landslide "uncorked" the volcano leading to a series of explosions producing a debris flow sequence and several new craters and fissures including a prominent "chasm" about 600 m long and 30 m deep, and reaming out the upper Te Maari Crater (Fig. 12.20). Hydrothermal material and blocks of old lava were ejected up to at least 1 km in all directions. Some of these rocks weighed up to 2 tonnes and were ejected on ballistic trajectories at speeds possibly up to 600–700 km/h as far as 2.3 km in the main blast directions. Many landed around the TAC track and some hit Ketetahi Hut creating numerous craters, holes, spray and rock "shrapnel" deposits and other impacts.

Ground-hugging pyroclastic density currents (surges or blasts) of hot material, gas and steam from multiple vents travelled east, west, northwest and then north extending over 6.4 km^2, inundating 3.7 km of the TAC and causing quite extensive damage to mountain toatoa forest and other trees and shrubs out to 2.7 km from the vents (Fig. 12.21). Deposit morphology suggests maximum velocities up to 500 km/h and average velocities of 40–100 km/h (Gert Lube & Eric Breard, pers. comm.). Investigations of plant dieback suggest temperatures of 50 to <100 °C (Efford et al. submitted) in the nine places where the current crossed the TAC track. Anyone who was on the track might have been in the current for up to 1–5 min. There would have been multiple fatalities if the eruption had occurred in the summer during fine weather.

The last phase of this first eruption was an ash and steam plume that rose 7–9 km high lit up by a dramatic show of lightning with some incandescent rocks and depositing ash up to several centimetres thick around the vents. Most of these phenomena were observed by residents along State Highway 46. Surrounding pine forests and farmland experienced thin but significant ash fall from the plume as it drifted east-northeast as far as the East Coast of New Zealand. Some fluorine and other toxins were present in leachate from this ash.

The early landslide was followed by a debris flow in at least three pulses and together formed a million cubic metre deposits of rocks and mud extending 2 km distant down a valley to the northwest and damaging native forest and shrubs. The upper part of the deposit choked a gorge area and dammed a tributary stream. This developed into a significant lake with a peak volume of about 30,000 m^3, and whose potential breakout threatened up to 1.8 km of the TAC 4–5 km downstream. The dam broke on 13/14 October, during poor weather following heavy rain, causing a hyperconcentrated lahar. This severely damaged 200 m of the track and overflowed a culvert at State Highway 46 depositing sediment and woody debris over the road.

The 21 November eruption occurred from the upper Te Maari crater at 1:25 pm on the Wednesday afternoon in fine clear weather a month after the TAC was fully reopened. Again there were no immediate useful precursors. About 30–70 people were in the 2 km summit hazard zone as close as 1.3 km from the vent with clear views of the eruption that was also recorded by the GeoNet webcam. Tephra jets emerged through dense steam clouds formed moments before. The jets were followed by a vertically ascending plume column and a series of pyroclastic density currents fed partially by column collapse. These surges were observed travelling to the northwest, north then north northwest at average velocities of 50 km/h. The visible component of the currents lifted off the ground and dissipated after 500 m although a Massey University sensor at 900 m detected air warmer than ambient afterwards. No ballistics were observed and no impact craters have since been reported beyond about 200 m at most. The plume rose 3–5 km depositing 15 mm of fine tephra 600 m from the vent and dustings were reported up to 25 km to the north.

The TAC was not impacted during the November eruption. All the significant volcanic hazards were contained well within the 1 km *Rahui* exclusion area (Fig. 12.30) which was established in mid October for eruption risk management purposes by DOC and Ngati Hikairo. However, images of the eruption and young people's reactions were dramatically presented on television and in other media including the internet.

As of late 2013, all tracks on Tongariro, including the TAC are fully open. The crater area is still very active and risks from further eruptions or secondary lahars are not zero. Activity may now be best regarded as declining degassing of a perturbed hydrothermal system (vent temperatures hotter than 400 °C) and occasional weak seismicity (Bruce Christenson and Art Jolly, pers. comm.). The probability of further eruptive activity including more pyroclastic density currents has decreased but not yet reached background levels and a resumption of activity in the next few months or years can not be ruled out. Plumes of steam evaporating to airborne residues of aerosols and particulates are often still prominent in fine weather.

In historic time the TNP volcanoes have seldom erupted simultaneously. Occasionally eruptions do occur at the same time coincidently because they are so active. 1895 was the last time the three volcanoes erupted at the same time.

12.6 Hazards and Risk Factors

Volcanic activity of the current style and size, especially unpredictable eruptions, poses relatively high risks to recreational users on Ruapehu and Tongariro and historically significant risks to travel networks around it. Risks would be even higher and more widespread if activity similar to that occurring 10,000 years resumed (Pardo et al. 2012). Today, erupted rocks and jets of mud and water, and pyroclastic density currents within TNP (Figs. 12.14, 12.15, 12.18, 12.19, and 12.21) in the Summit Hazard Zones (e.g. Fig. 12.27) represent New Zealand's highest volcanic risks to people (groups or individuals). People are also threatened by lahars of mud and rocks or dirty snow slurries flowing

Table 12.1 Near-misses from eruptions in Tongariro National Park 1945–2012 compiled from personal records

Eruption	Hazard	Immediate risk	Outcome
Ruapehu: early July 1945 after a short non-explosive period during the 1945 episode	Explosive blast, flying rocks in crater basin	Two people camping in crater basin	One person seriously injured
Ruapehu: 8 May 1971 during 8 month period with occasional eruptions	Phreatomagmatic eruption, cascading water and rocks in crater basin	Two surveyors on Pyramid Peak (Peretini)	Almost washed off peak, rocks narrowly missed them
Ngauruhoe: February 1975 during the 1974/75 episode	Block-and-ash (pyroclastic density currents or "ground-hugging") flows; rock (lapilli) fallout	School party 1 km away near South Crater; hikers in Mangatepopo Valley	Ranger took excellent photos of the flows. Hikers sheltered from fallout
Ruapehu: 23 Sept 1995, a week after main eruptive period started	Phreatomagmatic eruption, falling volcanic blocks in crater basin, lahars in Whakapapaiti and Mangaturuturu valleys	Climbers outside summit craters but just inside recommended and posted area. Skiers in lahar path	Rocks just missed climbers. T- bars closed for day an hour before lahars so people had moved out of valleys
Ruapehu: 12 Oct 1997 during a recognised period of elevated risk following 1995–1996 eruption	Phreatomagmatic eruption, flying blocks	Snow boarder on crater rim well inside recommended and signed no-go area (Fig. 12.13)	Rocks fell short
Ruapehu: 25 Sept 2007, eruption without useful precursors (Fig. 12.15)	Phreatomagmatic eruption, tephra jet and entrained rocks, lahar	Two climbers camping inside Dome equipment shed. Groomer driver at Far West T-bar	One person in shed seriously injured, rescued in nick of time, lost leg. Driver's rapid evasive action saved himself and gave alarm
Te Maari (Tongariro): 6 August 2012 following 3 weeks of unrest	Tephra jets plus pyroclastic density currents and rocks to 2.7 km, ash on highways	Ketetahi Hut, 3.7 km of TAC, alpine vegetation, highways	Eruption occurred at night, no-one on the TAC or in hut at the time
Te Maari (Tongariro): 21 November 2012 (Fig. 12.32)	Tephra jets plus density currents to 500 m	4 scientists, pilot and helicopter landed 200 m from crater (planned pickup)	Flew out of area 4 min before eruption

down valleys (Fig. 12.15) at measured speeds of up to 90 km/h. About half of these events are not predictable at present but records show that their probability at Ruapehu is slightly higher than normal at high and low points of the Crater Lake temperature cycle (Fig. 12.11). There have been numerous near-misses and some serious injuries on the mountain from lahars and other eruption products (Table 12.1), although present-day monitoring techniques reduce the risk considerably (see below).

Larger volcanic explosions on Ruapehu (e.g. 1975) produce larger lahars, and more valleys are affected. Lahars may enter catchments in Whakapapa ski area during small eruptions directed to the north and in medium or larger eruptions (with an annual occurrence likelihood of at least 6 %); while large or very large eruptions may produce lahars in Turoa ski area (annual likelihood of about 2 %) (Keys and Green 2010). On four known occasions lahars have entered Whakapapa ski area within two to five minutes after an eruption and all but one of these occurred during periods of apparently normal background activity.

Volcanic activity at Te Maari craters in 2012 significantly increased the risks to recreational users on Mt Tongariro. Serious near-misses occurred during the initial eruption and previously elsewhere near the TAC (Table 12.1). A resumption of more frequent activity from Tongariro, including Red Crater or Ngauruhoe could disrupt use of the TAC and economic activity based on it. Previously gas was the only volcanic hazard prevailing but it posed little real threat due to the rarity of windless conditions.

Normal mountain hazards have led to at least 20 fatalities on Ruapehu in the last 30 years. General hazards on all the volcanoes include snow avalanches, icy slopes plus vertical holes formed by waterfalls in the seasonal snow pack (Fig. 12.22), and cold, wet and windy weather, hypothermia, rock fall, falls and snow avalanches. A lack of mountain experience including route-finding problems and inadequate clothing and equipment is a common factor in many injuries and near-misses. The gullies on North Crater traversed by the TAC have the potential for causing a multiple fatality event in a snow avalanche.

Fig. 12.22 A large steep-sided 10 m deep waterfall hole in spring on Ruapehu with members of ski patrol and Ruapehu Alpine Rescue Organisation training in rescue techniques. *Photo* Darryl Jones

Fig. 12.23 The new Waihohonu Hut on the Tongariro Northern Circuit (TNC) Great Walk track. *Photo* Karen Williams

12.7 Major Tourist Attractions, Activities and Numbers of Tourists

The major attractions and activities in TNP all involve recreation in some form. There is an extensive range of short walks or longer tracks serviced by huts (Fig. 12.23) and commercial accommodation at Whakapapa Village and towns around the park. A 2010 survey of the interests of hut users showed that geological and volcanological aspects of TNP are one of the main reasons why people, other than skiers and snowboarders, visit the park (Frimmel 2010).

Tracks on Tongariro-Ngauruhoe (TAC and TNC, Fig. 12.2) are visited by some 80,000 people per year, mainly in the October-April period. Most people start these

Fig. 12.24 Emerald Lakes and Red Crater on the Tongariro Alpine Crossing. *Photo* Sarah Thompson

Fig. 12.25 Spring skiing off-piste on the Mangatoetoenui Glacier on the east side of Ruapehu. *Photo* Karen Williams

treks at the Mangatepopo road end which provides the highest access to this massif. The TAC traverses 20 km of dramatic volcanic and alpine terrain including the base of Mt Ngauruhoe (Fig. 12.19), Red and Emerald craters (Fig. 12.24) and many other young volcanic features. The recent eruption of Te Maari has created many new features that can be seen from the track. These include steaming craters, a drained lake, burnt forest and numerous impact craters formed by flying rocks ejected in the first eruption. Volcanic activity will sometimes mean that parts of the TAC track have to be closed.

Whakapapa and Turoa ski areas provide some of the best and most varied skiing and snowboarding terrain in New Zealand (July–October or November) and the highest serviced runs in the country. They have a total of up to 450,000 visitor days during the ski season. There is extensive off-piste skiing (Fig. 12.25) and good opportunities for ski touring. Avalanche control is undertaken in the ski areas but

Fig. 12.26 Climbers on an overnight trip at Ruapehu carrying a bucket and biodegradable bags for collecting human waste. *Photo* Harry Keys

beyond ski area boundaries there is no active control and people are expected to manage their own risks.

Ruapehu's Crater Lake (Fig. 12.10) is a unique and beautiful location. It can be easily viewed from several points on the rim of Crater Basin which is accessed from the top of ski lifts most easily from December to April and, if one is well equipped, during winter as well. Views across Crater Lake Basin and experiencing the Tongariro Alpine Crossing can be special and enriching. Together with the Ruapehu Round-the-Mountain, Tongariro Northern Circuit and other park tracks these places enable good trekking and walking options and other opportunities for appreciation of nature. Numerous other activities involving the outdoors are also possible in the surrounding areas. Visitors to these places are encouraged to think about and respect the park's cultural and world heritage values, including undertaking appropriate management of waste (Fig. 12.26). This includes not standing on the summits of the main peaks or entering Crater Basin.

12.8 Guides, Tours and Rangers Available

A search of the internet will quickly help people find guides, tours and transport options for these places if needed. Ruapehu and the TAC can be visited safely without a guide. People need to be well-equipped for cold mountain weather, have basic map-reading skills and an awareness of the changeability of local weather and snow conditions. Routes up Ruapehu are unmarked. Ruapehu Alpine Lifts (www.mtruapehu.com) runs tours of the ski areas and beyond during the winter season and to the Crater Lake in summer. The Department of Conservation (2013) provides valuable information about maps, routes, recommended equipment and clothing, weather forecasts, guides and other useful information (a PDF brochure "Walks in and around Tongariro National Park" is downloadable). Visitor Centre staff are available to provide information and answer questions on a daily basis, year-round. Rangers may be available with adequate advance notice to assist with school parties and other groups. Information on Tongariro can also be obtained from www.officialtongariro.co.nz and www.tongarirocrossing.org.nz.

12.9 Visitor Information Centres, Interpretive Centres and Volcano Museums

The Visitor Centre in TNP in Whakapapa Village is located on Highway 48 off Highway 47. It has an excellent 3D scale model of the park, interpretation of the volcanoes, flora and fauna and other aspects of the park, plus books, audio-visuals and an internet-based connection to the GeoNet natural hazard monitoring system of New Zealand. Smaller information or visitor centres are located in Ohakune, Turangi, Taumarunui and Taupo. A dedicated Volcanic Activity Centre for visitors is located at Wairakei 8 km north of Taupo. Taupo Museum, 60 km north of the park, has a small amount of volcanic and geothermal interpretation. The larger Te Papa Museum in Wellington, and the Auckland and Rotorua museums also have excellent volcanic displays.

12.9.1 Signage and Information Brochures

The park tracks mentioned previously and many others are well signposted. Brochures and maps are widely available with information on tracks, other features of the area and safety advice.

12.9.2 Hazard Maps

Five separate hazard maps, designed for public use, are available covering the summit hazard zones of the active volcanoes of Ruapehu, Ngauruhoe and Tongariro and the two main ski areas Whakapapa and Turoa. Figure 12.27 refers to the Summit Hazard Zone of Mt Ruapehu extending to about 2 km from Crater Lake. It shows all three ski areas, indicates hazards, offers advice and contains thumbprints of the Whakapapa, and Turoa hazard maps. The ski area maps are posted in numerous places in buildings across the ski areas and on the website www.mtruapehu.com/winter/Volcanic-Hazards/. They can also be found on Google Images. The maps have been distributed to the ski areas, local businesses providing transport and accommodation for visitors, club lodges at the ski areas and others.

Corresponding volcanic hazard maps for the Ngauruhoe-Tongariro massif can also be found on http://info.geonet.org.nz/display/volc/Tongariro. Two interpretation panels addressing hazards on the Tongariro Alpine Crossing include the map noting the main volcanic dangers of block-and-ash flows (pyroclastic density currents) and rock fall on the main route (and easiest access) up Mt Ngauruhoe.

12.10 Risk Management

Ruapehu has one of the most intensive volcanic risk management systems in the world and new additions have been made in 2012 and 2013. Better application to Tongariro and Ngauruhoe has been an important advance. There are several components to volcanic risk management in TNP (Keys and Green 2010):

1. Infrastructure location and design
2. Volcano monitoring and other research
3. Volcanic Alert and Crater Lake Warning levels
4. Decision-making (e.g. which facilities should be closed temporarily)
5. The Eruption Detection Systems (EDS and TEDS) for Ruapehu and Tongariro, and linked responses
6. The East Ruapehu Lahar Alarm Warning System (ERLAWS) and the Whakapapa Village Lahar Alarm Warning System (VLAWS) and linked responses
7. Electronic light signs at the main TAC car parks and on the TAC itself
8. Increased public awareness and ongoing research to improve response.

Vulnerable infrastructure such as roads, rail and power lines, ski lifts, and buildings should be located at sites away from or less likely to be at risk from lahars and other hazards, or designed in ways to withstand these hazards. The damaged Ketetahi Hut will be removed from the Active Volcanic Hazard Zone around Te Maari and a better designed shelter will be built on the site for day use only. On the ring plain most bridges across lahar paths are now designed appropriately (Fig. 12.28). However, some bridges on State Highway 1 east of Ruapehu would not withstand lahar events as large as those 450 and 850 years ago. The northern part of the TAC within 1 km of the Ketetahi car park has been affected by recent secondary lahars which has required rerouting and repairs to the track.

Monitoring of Crater Lake, Ruapehu and the other volcanoes and resulting advisories are well developed. These continue to expand in scope and intensity. GeoNet is a modern geological hazard monitoring system owned by the NZ Earthquake Commission and run by the GNS Science. Skilled staff detect and respond to volcanic activity relying on a network of geophysical instruments and automated software applications. New Zealand, like most countries with active volcanoes, uses a system of six Volcanic Alert Levels (http://info.geonet.org.nz/display/volc/). GNS Science through GeoNet is responsible for setting these Alert

Fig. 12.27 Poster focusing on Summit Hazard Zone of Mt Ruapehu (GNS Science, Department of Conservation and Ruapehu Alpine Lifts) accessible at http://info.geonet.org.nz/display/volc/Ruapehu

Levels, which give an indication of volcano status, as mandated in the National Civil Defence Emergency Management Plan. The system provides primary support for management decisions about public warnings and facility operations but does not completely drive them. Alert levels reflect the past and present situation and changes do not always occur fast enough for the needs of land managers so good communication between specialists and managers

Fig. 12.28 The road bridge at Tangiwai in the process of being raised 2 m and strengthened prior to the 2007 breakout lahar from Ruapehu. *Photo* Harry Keys

Fig. 12.29 When an eruption is detected EDS speakers on Whakapapa ski area broadcast sirens and pre-recorded messages warning people to move out of valleys in case a lahar is approaching. *Photo* Harry Keys

remains important. Many eruptions have happened at Alert Level 1 on Ruapehu (its normal alert level) as did the 2012 eruptions of Te Maari. Information on the alert status of Ruapehu, Tongariro and other volcanoes is available at the web site www.geonet.org.nz. The record of near-misses (Table 12.1) shows that most of these occurred during periods in which elevated risk was or would be recognisable with present-day monitoring techniques. Therefore this

Fig. 12.30 Signs beside the Tongariro Alpine Crossing installed to help reduce volcanic risk to visitors. *Left* Rahui (exclusion area) sign 300 m from the boundary of the Rahui. The steaming craters were formed during the explosions of 6 August 2012. *Right* Electronic red light sign at the northern entry to the Active Volcanic Hazard Zone 3 km from the Te Maari craters. *Photos* Harry Keys

monitoring by GeoNet/GNS Science and information availability has now reduced present-day risk to individuals.

The three early warning systems at Ruapehu (ERLAWS, EDS and VLAWS) have different objectives but management responses are being steadily integrated. ERLAWS primary objective is to provide warnings of major lahars to agencies managing State Highway 1 and electricity infrastructure nearby, enabling multi-agency response on the upper Whangaehu River. The EDS is designed to warn Whakapapa ski area patrons and staff of a potential lahar via sirens (Fig. 12.29) and broadcasted pre-recorded announcements. Staff are also warned via their radios, within 30 s after a potential lahar-generating eruption. Ski area staff response, a critical factor, has been improved by independent social science research and better training in the last decade (Leonard et al. 2008). VLAWS is designed to give residents and visitors in Whakapapa Village at least 15 min warning of a lahar coming down the stream that passes close to part of the village.

Ensuring effective response of the public to warnings at Ruapehu is difficult and complicated by the transient nature of ski area visitors and their limited familiarity with hazards. Management and research are ongoing to raise awareness and includes the hazard posters and the DVD "Staying Safe" (www.mtruapehu.com/winter/Volcanic-Hazards/). Siren audibility has been tested and improved and changes made to improve comprehension and awareness of the warning message including the addition of more useful content. During tests, a minority continue to demonstrate only moderate awareness of the correct actions to take and fail to respond effectively. At least 10 people have remained "at risk" during many publicised and blind tests of the EDS from 2005. Residual risk will never be reduced to zero but work continues.

To reach this level of mitigation, lessons from previous eruptions and lahars since 1953 and 1975 have been, and are still being applied, to improving management of lahar and other risks especially to ski areas on Ruapehu (Keys and Green 2010). The current Eruption Detection System is the result of regular reviews of its performance combined with ongoing analysis of eruption characteristics and lahar magnitude-frequency relationships. These factors combined with technological advances means the system has been completely upgraded since it was installed in 1983. Inter-agency synergies and agency-readiness are strengthened and robust as a result of the experience post 1995–1997 eruptions of Ruapehu. There is still room for improvement in the response of ski area patrons.

Volcanic risk management became more active for the Tongariro-Ngauruhoe massif in July 2012, with limited success initially (see Table 12.1). Awareness of volcanic hazards was not high before the 2012 eruption (Coomber and Leonard 2005). Risk management zones have been created since the 6 August 2012 eruption and the inner zone at least (Fig. 12.30a) will remain in place for some years. A new eruption detection system for the TAC and the Tongariro Northern Circuit is now in place. Electronic light signs that can be changed remotely if volcanic unrest increases have been installed in four places on the TAC. They provide advice of the level of volcanic risk in near real-time (Fig. 12.30b) supported by information on the webpage www.doc.govt.nz/volcanicrisk. The recent experience shows that it is difficult for managers, guides and other concessionaires to mitigate risks to people who are trekking very close to the active volcanoes vents. Efforts have been stepped up to increase public awareness that there is always some volcanic risk, and to inform people of eruption hazards and that they visit the TAC at their own risk.

Fig. 12.31 Turoa carpark and ski area on a beautiful day with fresh snow. *Photo* Karen Williams

Fig. 12.32 The second eruption of upper Te Maari on 21 November 2012 produced pyroclastic density currents clearly visible from the TAC. These were contained hazards well inside the Rahui volcanic risk management zone. *Photo* Neil Fausett

12.11 Economic Importance

Tongariro National Park and rivers flowing from it have major economic importance locally and regionally. They provide many businesses with income and many employment opportunities, especially in communities around the park. Ruapehu Alpine Lifts (Fig. 12.31) operates two ski areas on Mt Ruapehu. It is the largest concessionaire operating on public conservation land in New Zealand and the largest employer in the central North Island. In total, concession-based tourism in TNP contributed $30 million of direct turnover in 2004–2005 and the net economic impact of tourism concessions was four times the direct impacts of the concessions themselves (Wouters 2011). Tongariro concessions generated 450 full-time equivalent jobs, each of which created another 0.3 jobs. The Tongariro and Whanganui rivers which rise on the volcanoes are nationally important for the generation of hydroelectricity.

The Ruapehu eruption episode of 1995–1997, the Tongariro eruptions in 2012 and planning for the 2007 dam break lahar provided clear evidence of this economic importance. The greatest contribution to the cost of the 1995–1997 eruption, estimated in excess of NZ$130 million, was the impact on the alpine recreation and tourism industry in the central North Island. This industry was essentially nonexistent in 1945 when the mountain erupted in a similar fashion (Johnston et al. 2000). The 1995 season was terminated prematurely by ash fall, while the 1996 season was severely compromised by ash falls into July and also prematurely terminated by ash. Numerous local businesses were adversely affected and the Turoa ski area company and some others did not recover. Other significant impacts were felt by the aviation and electricity-generation sectors, including damage of about $10 million to turbine blades in a local hydro-generation station. The complete or partial closure of the TAC between 7 August and 19 October 2012 and again from 22 November to 8 May 2013 resulted in a 50 % reduction of staff and a 40 % reduction in turnover for Tongariro Expeditions, one of the largest concessionaires involved (Jarrod Thomas, pers. comm.).

Government agencies spent NZ$7 million preparing for the 2007 lahar including $4 million to raise and strengthen the highway bridge at Tangiwai (Fig. 12.28). In contrast, Taig (2002) calculated the event could have costed in excess of $40 million if things had gone wrong. Fortunately they did not (Keys and Green 2008).

Acknowlegments We thank Gill Jolly, Steve Sherburn, Bruce Christenson, Nico Fournier, Graham Leonard, Art Jolly, Karen Britten, Agnes Mazot, Mark Chadwick, Brad Scott and other scientists from GNS Science for their great support over the years, particularly during periods of volcanic activity, enhanced unrest or uncertainty. Thanks also to Shane Cronin, Jon Procter, Gert Lube, Eric Breard and others from Massey University, and Hugh Cowan from EQC. We also thank Paul Green, Blake McDavitt and many others from the Department of Conservation, Dave Mazey, Andy Hoyle, Mark Woods, Steve McGill, Steve Manunui and Mark Blair from Ruapehu Alpine Lifts, Keith Wood from Ngati Rangi and Ernslaw1, and Doug Tucker, Veti Bala, Phil Goulter, Gareth Grey and others from Genesis Energy. Sir Tumu Te Heuheu, Te Kanawa Pitiroi, Te Ngaehu and Hinemoa Wanikau from Ngati Tuwharetoa and Che Wilson and Hanna Rainforth from Ngati Rangi have also engaged with us constructively over the years. We are grateful to the NZ Army, NZ Transport Agency, Transpower, NZ Forest Managers, Kiwi Rail, Communication Networks Management Ltd, Cybernetics Ltd and Peter Otway for their support. Tim Stern from Victoria University and friends and colleagues allowed us to use the acknowledged maps, diagrams and photographs.

References

Cameron, E., Gamble, J. A., Price, R. C., Smith, S., McIntosh, W. C., & Gardner, M. (2010). The petrology, geochronology and geochemistry of Hauhungatahi volcano, SW Taupo Volcanic Zone. *Journal of Volcanology and Geothermal Research, 190*, 179–191.

Coomer, M. A., & Leonard, G. S. (2005). Tongariro Crossing Hazard Awareness Survey: Public perceptions of the volcanic hazard danger, Institute of Geological and Nuclear Sciences Science Report 2005/37, 28 p.

Christenson, B. W., Reyes, A. G., Young, R., Moebis, A., Sherburn, S., Cole-Baker, J., et al. (2010). Cyclic processes and factors leading to phreatic eruption events: Insights from the 25 September 2007 eruption through Ruapehu Crater Lake, New Zealand. *Journal of Volcanology and Geothermal Research, 191*, 15–32.

Cronin, S. J., Neall, V. E., Lecointre, J. A., & Palmer, A. S. (1996). Unusual ''snow slurry'' lahars from Ruapehu volcano, New Zealand, September 1995. *Geology, 24*, 1107–1110.

Department of Conservation (2013). Central North Island – Ruapehu and Tongariro area/district. www.doc.govt.nz/parks-and-recreation/tracks-and-walks/central-north-island/ruapehu/ and .../national-park/Tongariro. Accessed 16 November 2013.

Donoghue, S. L., Neall, V. E., Palmer, A. S., & Stewart, R. B. (1997). The volcanic history of Ruapehu during the past 2 millenia based on the record of Tufa Trig tephras. *Bulletin of Volcanology, 59*, 136–146.

Donoghue, S. L., & Neall, V. E. (2001). Late Quaternary constructional history of the south-eastern Ruapehu ringplain, New Zealand. *New Zealand Journal of Geology and Geophysics, 44*, 439–466.

Efford, J. T., Bylsma, R. J., Clarkson, B. D., Pittari, A., Mauriohooho, K., & Moon, V. G. (submitted). Vegetation dieback as a proxy for temperature within a wet pyroclastic density current: a novel experiment and observations from the 6th August 2012 Tongariro eruption. In Jolly, A. D. & Cronin S. J. (Eds.). The Tongariro Special Issue. *Journal of Volcanology and Geothermal Research.*

Frimmel, S., (2010). Tongariro Hut Visitor Preference Survey. Unpublished summary report for Department of Conservation, DOCDM 904545, 4 p.

Gamble, J. A., Price, R. C., Smith, I. E. M., McIntosh, W. C., & Dunbar, N. W. (2003). $^{40}Ar/^{39}Ar$ geochronology of magmatic activity, magma flux and hazards at Ruapehu volcano, Taupo Volcanic Zone, New Zealand. *Journal of Volcanology and Geothermal Research, 120*, 271–287.

Gregg, D. R. (1960). The geology of Tongariro Subdivision. *New Zealand Geological Survey Bulletin, 40*, 152.

Hackett, W. R., & Houghton, B. F. (1989). A facies model for a Quaternary andesitic composite volcano: Ruapehu, New Zealand. *Bulletin of Volcanology, 51*, 51–68.

Hales, T. C. (2000). The geology of the summit area, Mt Ruapehu. Unpublished BSc thesis, Canterbury University, Christchurch, 123 p.

Hobden, B. J., Houghton, B. F., Lanphere, M. A., & Nairn, I. A. (1996). Growth of the Tongariro volcanic complex: New evidence from K-Ar age determinations. *New Zealand Journal of Geology and Geophysics, 39*, 151–154.

Hodgson, K. A., LeCointre, J. A., & Neall, V. E. (2007). Onetapu Formation: the last 2000 yr of laharic activity at Ruapehu volcano, New Zealand. *New Zealand Journal of Geology and Geophysics, 50*, 81–99.

Holdsworth, R. (1985). *Radio-echo sounding in McMurdo Sound, Antarctica and Mt Ruapehu, North Island*. NZ Antarctic Record: New Zealand. 6.

Houghton, B. F., Latter, J. H., & Hackett, W. R. (1987). Volcanic hazard assessment ofrr Ruapehu composite volcano, TVZ New Zealand. *Bulletin of Volcanolgy, 49*, 737–751.

Johnston, D. M., Houghton, B. F., Neall, V. E., Ronan, K. R., & Paton, D. (2000). Impacts of the 1945 and 1995–1996 Ruapehu eruptions, New Zealand: an example of increasing societal vulnerability. *Geological Society of America Bulletin, 112*, 720–726.

Jolly, A. D., Neuberg, J., Jousset, P., & Sherburn, S. (2012). A new source process for evolving repetitious earthquakes at Ngauruhoe volcano, New Zealand. *Journal of Volcanology and Geothermal Research, 215–216*, 26–39.

Jolly, A. D., & Cronin, S. J. (Eds.) (2014). The Tongariro Special Issue. *Journal of Volcanology and Geothermal Research*. In preparation.

Keys, H. J. R., & Green, P. M. (2008). Ruapehu Lahar New Zealand 18 March 2007: Lessons for Hazard Assessment and Risk Mitigation 1995–2007. *Journal of Disaster Research, 3*, 284–296.

Keys, H. J. R., & Green, P. M. (2010). Mitigation of volcanic risks at Mt Ruapehu, New Zealand. In J.-P. Malet, T. Glade, & N. Casagli (Eds.), *Proceedings of the Mountain Risks international conference, Firenze, Italy, 24-26 November 2010* (pp. 485–490). Strasbourg, France: CERG.

Kilgour, G., Manville, V., Della Pasqua, F., Graettinger, A., Hodgson, K. A., & Jolly, G. E. (2010). The 25 September 2007 eruption of Mt Ruapehu, New Zealand: Directed ballistics, surtseyan jets, and ice-slurry lahars. *Journal of Volcanology and Geothermal Research, 191*, 1–14.

Lamb, S. (2011). Cenozoic tectonic evolution of the New Zealand plate-boundary zone: A paleomagnetic perspective. *Tectonophysics,*. doi:10.1016/j.tecto.2011.06.005.

Leonard, G. S., Johnston, D. M., Patton, D., Christianson, A., Becker, J., & Keys, H. J. R. (2008). Developing effective warning systems: Ongoing research at Ruapehu volcano, New Zealand. *Journal of Volcanology and Geothermal Research, 172*, 199–215.

McArthur, J. L., & Shepherd, M. J. (1990). Late Quaternary glaciation of Mt Ruapehu, North Island, New Zealand. *Journal of the Royal Society of New Zealand, 20*, 287–296.

Manville, V., Hodgson, K. A., Houghton, B. F., Keys, H. J. R., & White, J. D. L. (2000). Tephra, snow and water: complex sedimentary responses at an active, snow-capped stratovolcano, Ruapehu, New Zealand. *Bulletin of Volcanology, 62*, 278–293.

Manville, V., & Cronin, S. J. (2007). Breakout lahar from New Zealand's Crater Lake. *EOS, 88*(43), 441–442.

Massey, C. I., Manville, V., Hancox, G. H., Keys, H. J. R., Lawrence, C., & McSaveney, M. (2010). Out-burst flood (lahar) triggered by retrogressive landsliding, 18 March 2007 at Mt Ruapehu, New Zealand—a successful early warning. *Landslides, 7*, 303–315.

Moebis, A., Cronin, S. J., Neall, V. E., & Smith, I. E. (2011). Unravelling a complex volcanic history from fine-grained, intricate Holocene ash sequences at the Tongariro Volcanic Centre, New Zealand. *Quaternary International, 246*, 352–363.

Nairn, I. A., Wood, C. P., & Hewson, C. A. Y. (1979). Phreatic eruptions of Ruapehu, April 1975. *New Zealand Journal of Geology and Geophysics, 22*, 155–173.

Nairn, I. A., Kobayashi, T., & Nakagawa, M. (1998). The 10 ka multiple vent pyroclastic eruption sequence at Tongariro Volcanic Centre, Taupo Volcanic Zone, New Zealand: Part 1. Eruptive processes during regional extension. *Journal of Volcanology and Geothermal Research, 86*, 19–44.

Neall, V. E., Houghton, B. F., Cronin, S. J., Donoghue, S. L., Hodgson, K. A., Johnston, D. M., et al. (1999). *Volcanic hazards at Ruapehu Volcano Volcanic Hazard Information Series 8, 31p*. Wellington: Ministry for Emergency Management.

Otway, P. M., Hodgson, K. A., & Nairn, I. A. (1995). Whakapapa skifields lahar study. *Institute of Geological & Nuclear Sciences Science Report, 95*(39), 90.

Palmer, B. A. (1991). Holocene lahar deposits in the Whakapapa catchment, Northwestern ring plain, Ruapehu volcano (North Island, New Zealand). *New Zealand Journal of Geology and Geophysics, 34*, 177–190.

Palmer, B. A., & Neall, V. E. (1989). The Murimotu Formation, 9500 year old deposits of a debris avalanche and associated lahars, Mt Ruapehu, North Island, New Zealand. *New Zealand Journal of Geology and Geophysics, 32*, 477–486.

Pardo, N., Cronin, S. J., Palmer, A. S., Procter, J., & Smith, I. (2012). Andesitic Plinian eruptions at Mt.Ruapehu: quantifying the uppermost limits of eruptive parameters. *Bulletin of Volcanology, 74*, 1161–1185.

Price, R. C., Hobden, B. J., Smith, R. T., Briggs, R. M., Gleadow, A. J. W., & Kunz, P. (2003). *Volcanology of the Tongariro Crossing: a virtual field trip on CD-ROM*. Hamilton: University of Waikato.

Price, R. C., Gamble, J. A., Smith, I. E. M., Maas, R., Waight, T., Stewart, R. B., et al. (2012). The anatomy of an andesite volcano: a time stratigraphic study of andesite petrogenesis and crustal evolution at Ruapehu volcano, New Zealand. *Journal of Petrology, 53*, 2139–2189.

Stern, T. A. (2009). Reconciling short- and long-term measures of extension in continental back arcs: heat flux, crustal structure and rotations within central North Island, New Zealand. In Ring, U. & Wernicke, B. (Eds.), Extending a Continent: Architecture, Rheology and Heat Budget. *Geological Society, London, Special Publications, 321*, 73–87.

Taig, T. (2002). *Ruapehu lahar residual risk assessment* (80p). Wellington, New Zealand: Ministry of Civil Defence and Emergency Management.

Villamor, P., Van Dissen, R. J., Alloway, B. V., Palmer, A. S., & Litchfield, N. J. (2007). The Rangipo Fault, Taupo rift, New Zealand: an example of temporal slip-rate and single-event displacement variability in a volcanic environment. *Geological Society of America Bulletin, 119*(5/6), 529–547. doi:10.1130/B26000.1.

Vandemeulebrouck, J., Stemmelen, D., Hurst, A. W., & Grangoen, J. (2005). Analogue modelling of instabilities in crater lake hydrothermal systems. *Journal of Geophysical Research, 110*. doi:10.1029/2003JB002794.

Williams, K. L. (2013). *Volcanoes of the South Wind: A volcanic guide to Tongariro National Park* (5th ed., 280p). Auckland: Random House NZ.

Williams, K. L., & Keys, H. J. R. (2008). *Ruapehu Erupts* (64p) Auckland: Random House NZ.

Wouters, M. (2011). *Socio-economic effects of concession-based tourism in New Zealand's national parks Science for Conservation 309* (90p) . Wellington, New Zealand: Department of Conservation.

13 Taiwan Geotourism: A Volcanic Experience

Angus M. Robinson and Jiun-Chuan Lin

13.1 Introduction

The beautiful island of Formosa, now known as Taiwan, has been described as 'standing at the crossroads of nature and development'. With a population of just over 23 million, not dissimilar in size to that of Australia, Taiwan comprises an area about one half of the area of the Australian island state of Tasmania. Most visitors to Taiwan are familiar with the industrial landscape that extends down the western half of the island, but have not been exposed to the beauty of its central mountain ranges and coastal environs of Eastern Taiwan.

In a geographical sense, Taiwan's outstanding natural heritage epitomises the definition of geotourism—its environment, culture, aesthetics, heritage, and the well-being of its residents (National Geographic 2005). But from a geological tourism perspective which focuses on geology and landscapes, Taiwan offers a diversity of unique experiences, which includes both dormant volcanism and outstanding volcanic landforms.

As Lonely Planet has described Taiwan, 'Mountain peaks puncture a sea of clouds, slick black volcanic rock wraps the coastlines, and waterfalls shroud themselves in mist. Taiwan is a computer-generated Chinese water colour.'

Geographical location of Taiwan
Type of volcano[es]: Strato-volcanoes, fumaroles, steam vents, undersea fumaroles, crater lakes, boiling mud pools, flood lava flows with distinctive columnar jointing
Last known activity: Volcanic activity ranges from early Miocene, through Pleistocene to present
Major attractions: Yangmingshan National Park, Turtle (Guishan) Island, and Penghu Marine Geopark
Status of protection if any: National parks, nature reserves, wildlife refuges, wildlife habitats, forest reserves, ecological parks, national geoparks
Visitor numbers: 4.7 million inbound tourists (2012) engaged in sightseeing principally.

Fig. 13.1 Principal volcanic areas of Taiwan

A. M. Robinson (✉)
Geotourism Sub Committee, Geological Society of Australia, Sydney, NSW, Australia
e-mail: angus@leisuresolutions.com.au

J.-C. Lin
Department of Geography, National Taiwan University, Taipei, Taiwan
e-mail: jclin@ntu.edu.tw

P. Erfurt-Cooper (ed.), *Volcanic Tourist Destinations*, Geoheritage, Geoparks and Geotourism,
DOI: 10.1007/978-3-642-16191-9_13,

13.2 Taiwan: The Place

Lying on the Tropic of Cancer, Taiwan is located 150 km off the southeast coast of mainland China, between cool-temperate Japan to the north, sub-tropical South China to the west, and the tropical Philippines and Indo-Malayan islands to the south. The location, combined with a range of high mountains, with more than 200 peaks over 3,000 m in height, supports a diverse flora of over 4,000 vascular plant species and a spectrum of six forest types. This range of environments in turn supports a rich fauna, flora, and biological diversity. In addition to this big island, there are over 120 outlying small islands falling into a number of groups, each group contains one or several inhabited main islands (or temporary fishing stations or army outposts) and other smaller isles. It is reported that only some 25 of these islands are inhabited. The diversity of these small-island groups emerges from their differences with respect to geological origin, size, remoteness, geographic types, and human culture (Source: Huei-Min Tsai, Taiwan Islands Net).

Of these island groups, three have volcanic origins (refer also Fig. 13.1).

Mid-oceanic archipelagos type: Penghu Archipelago comprising 64 small islands and islets; composed of basalt with a coral reef coastline.

Remote oceanic islands type: Green Island (Lutao), volcanic with coral reef and Orchid Island (Lanyu),volcanic with a sub-tropical rain forest.

Near-shore or connected islands type: Turtle Island (Guishan), active volcanic with accompanying undersea thermal springs, and the five islands off north-eastern Taiwan, Pengchiahsu (Pengchiayu), Mienhuahsu (Mienhauyu), Huapinghsu (Huapingyu), Chilungtao (Chilungyu) and Kueishantao, all volcanic islands formed of either andesite or dacite.

13.3 Geology of Taiwan

Taiwan is a modern mobile belt in the classical sense, and an orogeny formed by the collision of an island arc with the Asian continental margin. The main island of Taiwan comprises in part a metamorphic basement covered by Cenozoic deposits to a thickness of more than 10 km. The north-south axis of deposition has shifted progressively westward with successive orogenic episodes. A considerable part of the Tertiary sediments of Taiwan has been subjected to different grades of induration or metamorphism. Large igneous intrusions are rare, but there are several important volcanic fields in northern Taiwan, eastern Taiwan, and the Taiwan Strait. Taiwan is located on a convergent plate boundary between the Asiatic continental plate and the Philippine Sea oceanic plate (Source: Central Geological Survey of the Ministry of Economic Affairs).

The major rock formations of Taiwan form long narrow belts roughly parallel to the long axis of the island. These rock belts extend progressively westward from the central backbone range to the west coast. The prevailing structural pattern is that of an elongate arc convexing to the west. The northern, shorter bend of the arc strikes east-north-east whereas the southern, major arm of the arc extends mainly north-south. All the major structures of Taiwan, including the important faults and fold axes, follow fairly closely this arcuate structure throughout the island.

13.3.1 Oligocene to Miocene Age Volcanism

In the western foothills, small-scale volcanism was active in the sedimentary basin from late Oligocene to Miocene times and ended in the Pliocene. In northern and north-central Taiwan, small lenses or irregular bodies of basaltic tuff, tuffaceous breccia and minor basaltic flows are present in all Miocene formations, but their distribution is limited to scattered local areas only.

The Coastal Range in eastern Taiwan is a volcanic arc containing a large amount of volcanic rocks and their pyroclastic derivatives. The most important igneous mass is the Chimei Igneous Complex, which outcrops also in a small exposure in Changyuan southeast of Chimei. The pyroclastic rocks of the Tuluanshan Formation are also associated with this phase of volcanism. The main constituents of the Chimei Igneous Complex are andesites of different mineralogical compositions and textures. These rocks are mainly of Miocene age. The two offshore islands in south eastern Taiwan, Green Island (Lutao) and Orchid Island (Lanyu), are composed of the same andesites and pyroclastic rocks. Their ages range from Miocene to Pliocene. All the igneous rocks in the Coastal Range are related to the Luzon Volcanic Arc, which is a chain of volcanoes in a north south line across the Luzon Strait from Taiwan to Luzon.

On the Penghu Island Group in the Taiwan Strait, fissure eruptions produced a widespread flood basalt that covers 64 islands and islets. This basaltic volcanic stage may occur at the end of the main orogeny in Taiwan and is of early to late Miocene age (17.4–8.2 million years ago). These basalts cover the entire island group except the westernmost island, Huahsu. They form flat-topped tablelands reaching a maximum elevation of some 50 m above sea level. Horizons of sand, clay, and carbonised wood fragments are inter-bedded with the basalt flows. The Penghu basalt erupted quietly, as pyroclastic deposits are comparatively limited. The eruption produced nine successive flows. The basalt is either dense and compact or amygdaloidal. Amygdules are generally filled with aragonite. These amygdaloidal basalts provide an ornamental stone for which Penghu is noted.

13.3.2 Pleistocene Age to Recent Volcanism

Large-scale andesitic eruptions took place in Pleistocene time in northern Taiwan and in some north-eastern offshore islands, depositing both lava and pyroclastic rocks. Two important Pleistocene andesitic volcano groups were formed, the Tatun Volcano Group and the Chilung Volcano Group, located on the west and the east side of the Chilung harbor respectively. No volcano is known to be active in Taiwan in the historical record, but geothermal evidence suggests that these volcanoes may not be entirely extinct.

The Tatun volcanoes are more mafic and contain some basaltic flows, whereas the Chilung volcanoes are more acid, composed mainly of dacite or quartz andesite. The five islands off north-eastern Taiwan (Pengchiayu, Mienhauyu, Huapingyu, Chilungyu, and Kueishantao) are all volcanic islands formed of either andesite or dacite. They are ascribed to the same volcanic belt and stage as those in northernmost Taiwan. These represent the western end of the Ryukyu Volcanic Arc. Fumarolic activity is present on at least two of these islands.

The Tatun (Datun) Volcano Group comprises a series of 20 volcanoes and volcanic cones in the northernmost part of Taiwan, the nearest being 15 km north of Taipei. Qixing Mountain (Chihsingshan) at an elevation of 1,120 m is the highest and youngest volcano in the group. Hot springs, fumaroles, and shallow volcanic craters are found at many places in this volcano group, mostly on the southeast side of the Chinshan longitudinal fault. Craters are preserved at the summit of some of the cones. A few have been eroded to craterless volcanoes or were erupted as high-viscosity volcanic domes. This group of volcanoes are mainly stratovolcanoes, built by successive eruptions of andesitic flows, volcanic ash, and coarse pyroclastic ejecta. Another peak, the Kuanyinshan volcano, rising to 612 m, is separated from the main volcanic chain by the Tanshui River. Guanyin Mountain (616 m high) is located at the most north-westerly edge of the Tatun Volcano Group.

Numerous quartz andesite or dacite flows are found on the north-eastern coast of Taiwan, east of Chilung harbour. These flows and associated pyroclastic rocks constitute the Chilung Volcano Group. Related gold and copper mineralisation is found in the Miocene host rocks or at the intrusive margins, forming the three principal deposits at Chinkuashih, Chiufen, and Wutankeng.

Pleistocene volcanism is represented at a location just south of the town of Tachi in Taoyuan-hsien in northern Taiwan by a conical volcanic hill rising to 347 m. It is made up of vesicular alkaline basalt (the Tsaolingshan Basalt) with a small amount of pyroclastic deposits exposed on the southern slope.

Turtle (Guishan) Island, located some 9 km east of Hengfang Harbour on Taiwan's north-eastern coast, represents the top of an andesite stratovolcano which rises from the seafloor, displaying active undersea fumaroles. The last recorded eruptions were during the period 1775–1795 A.D.

The Wandan mud volcano, one of Taiwan's 17 active mud volcanoes was dormant for 9 years between 1979 and 1988 but now usually erupts once or twice a year. There are two active mud volcanoes in South Taiwan, and several inactive ones. The Wushan mud volcanoes are located in the Yanchao District of Kaohsiung City, and are protected since 1992 as part of Taiwan's smallest nature reserve.

13.3.3 Taiwan's Geopark Network

Professor Jiun-Chuan Lin of the Department of Geography, National Taiwan University has advised that the Taiwan Geopark Network was launched in 2011. The purpose of Taiwan Geopark Network is to establish a national network for promoting the concept of a geopark, exchange the experience of management of the sites, to educate visitors to protect the landscape for the purpose of sustainable development during using or visiting the geological/geomorphological/ecological and cultural heritage, and to promote the local economic activities through geopark activities.

Six identified Taiwan geoparks (i.e. Northern Coast, Matsu, Penghu Marine, Liji, Yangchao, and Caoling) have different landscape characteristics, of which two—the Northern Coast and Penghu Marine comprise either part or full volcanic landforms (Fig. 13.2).

In 2009, Taiwan hosted the East Asian Geoparks Conference (Lin 2009).

13.4 Taiwan's National Treasures: Protected Areas in Taiwan—An Overview

Taiwan's natural heritage is protected in eight National Parks i.e. Kenting, Yushan, Yangmingshan (features the Tatun volcanoes), Taroko, Shei-Pa, Dongsha, Kinmen and Taijang as well as 13 designated scenic areas, with large areas of indigenous sub-tropical forest, particularly on the East Coast, and spectacular ocean scenery (National Treasure: Protected Areas in Taiwan 2007).

To supplement the main protected areas, Taiwan has identified and/or created a wide range of nature reserves, wildlife refuges, wildlife habitats, forest reserves. Bird watchers would be delighted to learn that these include

Fig. 13.2 Taiwan geoparks network

localities such as Guandu Nature Park in Taipei, the Mianhua (Mienhuahsu), and Huaping (Huapinghsu) Islands, the Penghu County Cat Islet Seabird Refuge, and the Matsu Islands Wildlife Refuge as well as an extensive array of wetland reserves.

Taiwan features some 18,400 species of wildlife of which about 20 % are considered rare or endangered.

13.5 Taiwan's Cultural Heritage

The aboriginal cultures (12 aboriginal tribes) of Taiwan represent 2 % of the population of the island. Taiwan also has more than 5,000 temples (Buddhist, Taoist and Confucius), ranging in size from single room shrines to vast multi-story complexes. Taiwan's cultural history includes elements resulting from previous Portugese and Japanese occupation.

The National Palace Museum in Taipei holds one of the largest collections of Chinese artifacts and artworks in the world, additionally most of the items were collected by China's ancient emperors. In addition, near Taipei, the excellent Shihsanhang Museum of Archaeology focuses on research, conservation, collection and display of cultural artifacts of Taiwan's indigenous people unearthed at the Shihsanhang excavation site and dating back some 500–1800 years to Taiwan's prehistoric time.

Fig. 13.3 Taiwan's spectacular Taroko Gorge, Taroko National Park

Taiwan is an excellent shopping destination, with everything from ultra-modern shopping malls to traditional markets (including 'night markets').

13.6 Summary: Taiwan's Experiential Tourism Features

1. Outstanding ecotourism and geotourism features related to Taiwan's recent volcanic and tectonic history, with particular reference to the following three key reference sites i.e.:

   - Yangmingshan National Park,
   - Turtle (Guishan) Island forming part of the Northeast and Yilan Coast National Scenic Area and designated as part of the Northern Coast Geopark,
   - Penghu Marine Geopark,

   and incorporated with visits to any of the following outstanding experiential attractions.

   - National Palace Museum in Taipei.
   - Other national parks, particularly the world-famous Taroko National Park (Fig. 13.3), renowned for its

Fig. 13.4 Yehliu Geopark's famous trademark Queen's Head—mushroom (wind erosion) of sedimentary rocks

nearly vertical marble gorge, one of the most astounding landscapes in the world. Taroko is also the home of the indigenous Truku people.

- Also Yushan National Park, the largest national park in Taiwan, located on the central part of the island. It contains Jade Mountain (Yushan literally means "Jade Mountain", 3,952 m high) which is the highest peak in East Asia.
- North Coast and Guanyinshan national scenic area, particularly the spectacular Yehliu coastal landforms within the Yehliu Geopark and the Shihsanhang Museum of Archaeology, and other parts of the Northeast and Yilan coastal landforms, which together form the balance of the designated Northern Coast Geopark (Fig. 13.4).
- Designated Liji, Yangchao, and Caoling Geoparks.
- Guandu Nature Park (Taipei).
- Jinguashi Gold Ecological Park (Rueifang, Northern Taiwan).
- Alishan National Scenic Area including the famous Alishan Forest Railway.
- Sun Moon Lake National Scenic Area.
- Green Island (Latou).
- Chung Tai Chan Monastery (Nantou), Fo Guang Shan Monastery (Kaohsiung), Dharms Drum Mountain (new Taipei City), Tzu Chi Temple (Kaohsiung), the world famous Juming Sculpture Museum (Jinshan).
- Night Markets of Taipei.
- Seasonal Lantern Festival, Aboriginal Celebrations, Matsu Festival, Taiwan International Orchid Show and other outstanding cultural events.

2. Cultural, religious and historical themes with indigenous, Chinese, Japanese and European influences. Taiwan has its own cuisine, but has also embraced the best of mainland Chinese cooking. Visitors to Taiwan can experience Chinese food from all of the key cuisine regions of Mainland China.

Getting around Taiwan is very easy. With excellent inter-city air, rail and bus services, Taiwan is also served by the Taiwan High Speed Rail service that runs approximately 345 km along the west coast of Taiwan, from the national capital of Taipei to the southern city of Kaohsiung. Taxi services are also excellent and very low cost by world standards. Taiwan's tourism infrastructure is robust with much detailed information including maps for the national Taiwan Tourism Bureau which is supported by visitor information centres in all regional centres.

13.7 Outstanding Ecotourism and Geotourism Features Related to Taiwan's Recent Volcanic and Tectonic History

13.7.1 Yangmingshan National Park

Located very close to Taipei, Yangmingshan National Park is the smallest and northernmost national park of Taiwan. Yangminshan is famous for its hot springs and geothermal activity. Each spring, Yangminshan also has a dazzling flower season (Fig. 13.5).

Whilst casual access into the tourist fumarole areas is easily achieved, the Yangmingshan National Park trail system can be utilised for a more extensive experience as the mountain peaks are accessible by a network of maintained hiking trails. For example, one long trail of some 8.5 km in length extends through Dingshan, Shihtiling, east peak, the main peak of Qixing Mountain (Chixingshan), the main peak, south peak and west peak of Datun Mountain (Datunshan), Miantianshan and Xiangtianshan among others. Enriched by andesite, Datun Mountain's conic or bell-shaped volcano body, crater and volcano lakes provide a unique geological terrain in the area. Hot springs and steam

Fig. 13.5 Steam vent area, Xiaoyoukeng, Yangmingshan National Park

Fig. 13.6 Aerial view of Guishan Island showing undersea fumaroles on the right hand side of the island

vents are mainly found along the Jinan Fault Zone between Baotou and Jinan, some 18 km long and 3 km wide. The most active steam vents are those at Tahuangtsui, Xiaoyoukeng, Macao, Dayoukeng, Ssuhuangtzuping and Kengtzuping. It is suggested that 1–2 days is allocated for the entire trip, which may be taken in different legs of 1/2 to 1 day each (Source: Yangmingshan National Park Headquarters website).

13.7.2 Turtle (Guishan) Island

Named for its similarity to the shape of a floating turtle, whilst Turtle Island is a well known landmark just off the coast of Taiwan, it is not a place well frequented by Taiwanese. The island is only open to tourists from 1 March to 30 November. The island is 3.3 km wide from east to west and 1.7 km long and reaches a maximum elevation of

Fig. 13.7 Profile of Guishan Island showing hydrothermal alteration of andesite

Fig. 13.8 Classic Penghu Basalt Columnar Jointing, Tongpan Islet

401 m. Comprised almost entirely of andesite, strongly altered by sulphurous hydrothermal activity, the island features steep cliffs, active hot springs, erosional caves, lakes, cold springs, and rare cliff vegetation surrounded by a rich ocean ecosystem situated in streaming warm water fed by the undersea fumeroles. Of particular interest is the opportunity to see spinner dolphins, bottle-nosed dolphins, small killer whales, and false killer whales—the best season for this activity is April through the end of August. Tours can be arranged to either engage in the wildlife watching and/or to actually visit the island by ferry and explore its landscape and its built heritage. There are restrictions on visitation because of its status as a marine ecological park and visitor applications need to be made in advance of any intended visit (Figs. 13.6 and 13.7).

13.7.3 Penghu Marine Geopark

The Penghu Marine Geopark, an archipelago comprising 64 islands and islets, lies off the western coast of Taiwan. These islands and islets are mainly formed by basalt lava flows and

Fig. 13.9 'Twin Hearts' Basalt Fish Trap, Cimei Island

Fig. 13.10 Sloping Basalt Columnar Jointing, Yuanbei Islet

a partly sedimentary rock sequence produced by intermittent eruption and weathering episodes (Lin 2009). The islands feature many attractive volcanic landforms including unusual columnar basalt jointing.

Lin (2009) has documented the key basalt geosites which include the following highlights:

- Jibei Island (Jibeiyu)—includes many fine stone weirs or fishing traps constructed from basalt and coral reef material that has become a distinguishing cultural feature of Penghu. There are some 590 stone weirs in Penghu, of which about 100 are located on this island.
- Siaomenyu—includes the unique 'whale cave', 10 beautiful stone weirs, and the Siaomen Geology Gallery (Tsai and Lin 2008).
- Kueibishan and Chihyu—tholeiite basalts with sea cave and sea stack erosional features.

- Tongpanyu—well known for its 'lotus platform' and basaltic columnar joint cliffs and platy jointing (Fig. 13.8).
- Wang-an—the largest island in Penghu features tholeiite xenoliths and outstanding coastal erosional features.
- Qimei (Cimei)—features pyroclastic deposits and iconic ancient fish traps (Fig. 13.9).

Various interesting geomorphological features can also be seen within islets forming part of the Penghu South Basalt Nature Reserve (Dongjiyu, Sijiu, Toujin and Tiejhan), the Penghu Columnar Basalt Nature Reserve (Chishanyu, Tingkoyu and Hsiaobaishayu), the North Sea islands (Yuanbeiyu, Siaobaisha and Niao) and others, notably Tongpan, Yuanbei, Hujing and Mao islets (Fig. 13.10).

The main island of Magong City/Huxi Township features the Penghu Marine Geopark Centre which exhibits both the diverse basalt landscapes and the cultural use of basalt in Penghu Island. Penghu also abounds in rich biodiversity including marine (including turtles), birdlife and unique plant species, information about which can be found in a wide range of visitor centres (Liu 2009).

Penghu can be accessed by daily flights from Taipei Songshan Airport for Magong City from where tourist boats leave daily (but not in the winter months) for visits to the various islands which usually include stops on Qimei, Wang-an and Tongpanyu islands.

13.8 Conclusions

Not well known for its nature-based tourism potential, Taiwan is one of the world's best kept secrets for experiential tourism and is an important destination in the global network of countries with volcanic experiences. The extensive and unusual geodiversity of volcanic landforms and the inter-relationship of this geology and geomorphology with cultural heritage in the Penghu Marine Geopark, as well as the undersea fumaroles at Turtle (Guishan) Island as a site for whale and dolphin watching are truly unique experiences. The proximity of 'calm' volcanic activity within Yangmingshan National Park located on the outskirts of a major global city is also cause for reflection.

Acknowledgments The support and contributions by way of supplementary photographs and diagrams by Professor Jiun-Chuan Lin of the Department of Geography, National Taiwan University and the Geographical Society of Taiwan is gratefully appreciated. Photos (Figs. 13.3, 13.4, 13.7 and 13.10) have been contributed by Angus M Robinson.

References

Central Geological Survey of the Ministry of Economic Affairs. http://www.moeacgs.gov.tw/english/twgeol/twgeol_generalgeol.jsp

Lin, J.-C. (2009a). East Asian geoparks—vision, problems and prospects. *Proceedings of the East Asian Geoparks Conference.*

Lin, J.-C. (2009b). The geodiversity of the Penghu Marine Geopark. *Proceedings of the East Asian Geoparks Conference* (pp. 315–326).

Lin, J.-C. (2009c). Geopark at Northeast and Yilan Coast national scenic area. *The Bitou Cape-Longdongwan Geopark Guidebook.* Department of Geography, National Taiwan University.

Liu, S.-L. (2009). Penghu Marine Geopark-culture & travel volume. *Agriculture and Fisheries Bureau, Penghu County Government.*

National Geographic (2005). Geotourism charter. www.nationalgeographic.com/travel/sustainable/pdf/geotourism_charter_template.pdf

National Treasure: Protected Areas in Taiwan (2007). *Forestry Bureau, Council of Agriculture, Executive Yuan.*

Taiwan Islands Net, a website for island studies at Graduate Institution of Environmental Education, National Taiwan Normal University.

The Beauties of Taiwan, Penghu National Scenic Area (2005). *Penghu National Scenic Area, Tourism Bureau, MOTC.*

Tsai, L.-C. & Lin, C.-H. (2008). The story of Siaomenyu. *Penghu Geopark Series, Agriculture and Fisheries Bureau, Penghu County Government.*

Galápagos: An Emerging Volcano Tourism Destination

14

Ross Dowling

Galápagos Islands

Geotour itinerary on Metropolitan Touring's *Isabella II*

Geographical location including map co-ordinates:
927 km off the west coast of South America in the eastern Pacific Ocean

Type of Volcanoes:
Oceanic Shield Volcanoes

Last known eruptions:
April 2009 (Fernandina Island)
GPS Location: 0° 22' 12" S, 91° 33' 0" W

Major attractions:
Galápagos National Park (70 sites),
Galápagos Marine Reserve (79 sites)

Status of protection:
UNESCO'S World Heritage Property
(land 1978, marine 2001)
National Park, Marine Reserve
Man & Biosphere Programme
Whale, Dolphins & Shark Sanctuary

Visitors per year:
125,059 International Visitors (2012)

14.1 Introduction

The Galápagos Islands of Ecuador are located in the eastern Pacific Ocean 927 km off the west coast of South America. The islands lie on the equator and cross into both the northern and southern hemispheres. The islands range from Darwin Island in the north to Española Island in the south, and they are spread out over a distance of 220 km. Several islands, Volcán Wolf and Volcán Ecuador on Isla Isabela, sit directly on the equator. The Galápagos Archipelago consists of 7,880 km^2 of land spread over 45,000 km^2 of ocean. The largest of the islands, Isabela, is 5,827 km^2 and makes up approximately three-quarters of the total land area of the Galápagos. Volcán Wolf on Isabela is the highest point, with an elevation of 1,707 m above sea level (www.wikipedia.com).

R. Dowling (✉)
Edith Cowan University, Joondalup, Australia
e-mail: r.dowling@ecu.edu.au

P. Erfurt-Cooper (ed.), *Volcanic Tourist Destinations*, Geoheritage, Geoparks and Geotourism,
DOI: 10.1007/978-3-642-16191-9_14,

14.2 The Natural Environment

The Galápagos Islands and the surrounding marine reserve have been called a unique 'living museum and showcase of evolution' (UNESCO 2013). They were inscribed on UNESCO's World Heritage List in 1978 and as a Marine Reserve in 2001. Located at the confluence of three ocean currents, the islands are a 'melting pot' of marine species. Ongoing seismic and volcanic activity reflect the processes that formed the islands. These processes, together with the extreme isolation of the islands, led to the development of unusual animal life such as the land iguana, the giant tortoise and the many species of finch that inspired Charles Darwin's theory of evolution following his visit in 1835.

The islands are situated on the Galápagos Submarine Platform and have been formed by volcanic processes. Most represent the summit of a volcano, some of which rise over 3,000 m from the Pacific floor. The archipelago is located on the Nazca Tectonic Plate which is moving in an east/southeast direction, diving under the South American Plate at a rate of about 6.4 cm per year (Pearson and Middleton 1999). It lies on top of the Galápagos hotspot, a place where the Earth's crust is being melted from below by a mantle plume, creating volcanoes.

The first islands formed here at least 8 million and possibly up to 90 million years ago (White 1997) and the western part of the archipelago experiences intense volcanic and seismic activity. While the older islands have disappeared below the sea as they moved away from the mantle plume, the youngest islands, Isabela and Fernandina, are still being formed. The most recent volcanic eruption occurred in April 2009 when lava from the volcanic island Fernandina started flowing both towards the island's shoreline and into the centre caldera.

The larger islands of Isabela, Santa Cruz, Fernandina, Santiago and San Cristobal typically comprise one or more gently sloping shield volcano, culminating in collapsed craters or calderas. Long stretches of shoreline are only slightly eroded, but in many places faulting and marine erosion have produced steep cliffs and lava, coral or shell sand beaches. Other noteworthy landscape features include crater lakes, fumaroles, lava tubes, sulphur fields and a great variety of lava and other ejects such as pumice, ash and tuff.

The islands vary greatly according to their altitude, area, and orientation and it is these differences combined with their separation that has contributed towards the species diversity and endemism that the islands are famous for (UNESCO 2009). They are relatively young in geologic terms and are the product of volcanoes that formed beneath the sea and which continued to rise until they broke through the surface of the sea and became islands. The islands on the east, especially the Island of Espanola, are much older than the western islands. Geologists suggest that the islands are situated over a 'hot spot', an area of extreme heat within the Earth's mantle. As the tectonic plates on which the islands are placed move, magma wells up causing earthquakes and volcanic eruptions.

14.3 Tourism

Global international tourist arrivals grew by 4 % in 2012 to reach 1.035 billion (UNWTO 2013). Emerging economies (4.1 %) regained the lead over advanced economies (3.6 %), with the UNWTO Region, *Asia and the Pacific,* showing the strongest results. The UNWTO Region *The Americas,* which includes South America and Ecuador, grew between 3 and 4 % following the *Asia Pacific* (5–6 %) and *Africa* (4–6 %). This growth is stronger than that for *Europe* (2–3 %) and the *Middle East* (0–5 %). Visitors to the Americas increased by 6 million arrivals, to reach 162 million in 2012.

In Ecuador the number of visitors is growing with the Galápagos being an iconic international tourist destination. Over the last thirty years the number of international visitors has grown by almost 100,000 per annum from 13,465 (1980) to 125,059 (2012). At the same time the number of national visitors has grown from 3,980 to 61,574. The combined number of domestic and international visitors to the region in 2010 was 173,296 (www.sangay.com).

This growth in tourism is reflected in the growing number of tourism companies opening for business there. Exploring the Galápagos just got more exciting. The luxury US based cruise line Silversea Cruises recently started cruises in the Galápagos archipelago with its ship *Silver Explorer*. In 2012 it purchased Canodros S.A., an Ecuadorian tourism company that operates in the Galápagos Islands, and their expedition ship, *Galápagos Explorer II*. In 2013 it re-branded this ship as the *Silver Galápagos*. The all-suite, 110-guest vessel offers 7-day expeditions providing guests with traditional wildlife watching opportunities as well as exploration of volcanic fields and natural lava tunnels.

However, the Ecuadorian company Metropolitan Touring has been operating in the Galápagos Islands for 44 years. During this time it has taken weekly departures depicting the stunning volcanic features of the Islands, and more recently it has introduced two specific geotours to the

Fig. 14.1 Metropolitan Touring's Yacht Isabela II photographed at Punta Suárez, Española (Hood) Island in the Galápagos Archipelago. Marine Iguanas are seen in the *foreground* of the picture. *Source* Metropolitan Touring, Ecuador

islands led by Dr Theofilos Toulkeridis from the Centro de Geologia, Universidad San Francisco de Quito (USFQ), with full commercial and operational support by Metropolitan Touring, Ecuador.

Metropolitan Touring's yachts *La Pinta* and *Isabela II* (Fig. 14.1) undertake low-impact geotours which are environmentally respectful and include the local people. A well-planned itinerary (see boxed map at start of chapter) also includes iconic wildlife sightings and a range of geological sites illustrating the understanding of how the former (fauna and flora) are dependent on the latter (geology and landforms). For example, the older islands (eastern side) are markedly different than the central or western islands. Thus understanding the diversity of the geological makeup of the islands is central to understanding the complex evolutionary processes occurring in the Galápagos Archipelago.

Wildlife encounters must be planned in a way that explorers know that, although wildlife is very easy to approach, we are still entering a wild habitat. This is where high skills of environmental interpretation are undertaken by the Naturalist Guides. The islands' wildlife is not a zoo, and the ultimate lesson here is the understanding of wild species interacting with each other, and being part of our complex web of life. Metropolitan Touring has fully trained Expedition Leaders and Naturalist Guides who understand the local environment and are experts in environmental interpretation. Their knowledge is supplemented by geological experts on their Galápagos geotours.

Thus tourism is growing and in 2012 new rules and regulations were instituted with regard to cruise routes and frequencies to and around the Islands. Tour operators now have to incorporate into their travel programs the Galápagos National Park Authority's new regulations, which are designed to protect the local animal and plant life (www.eturbonews.com). The new regulations allow guests of the on-site tour operators to stay a maximum of four nights and five days per ship, with a frequency of four landings within any 14-day period.

In the past the archipelago's annual visitors have been mainly concentrated on the three islands of Isabela, Santa Cruz and San Cristobal. The new visitor redistribution has been achieved by opening up other islands, and altering route frequencies. This also has helped to reduce fuel consumption and the overall amount of pollution emitted by the yachts, boats and ships working the islands. With the new system, the ports on Santa Fe and Tagus Cove are again accessible. In addition the islands of Española, Genovesa and Fernandina also now permit the arrival of smaller boats, allowing visitors to see rare species like the red-footed booby, the albatross, the flightless cormorant and various mimids (www.eturbonews.com). The new regulations and the protection of nature are designed to both grow tourism whilst being more environmentally respectful, by focusing on increased sustainability and more ethical, small-scale, low-impact tourism.

14.4 Geotourism and Volcano Tourism

Known the world over for its biological diversity, the Galápagos Islands have always been 'an essential destination' for dedicated ecotourists. However, the islands have recently witnessed the birth of geotourism (essentially *geological tourism*) and volcano tourism (where there is an emphasis on interacting with volcanoes, Erfurt-Cooper and Cooper 2010). A number of tourist companies are now offering geological tours of the islands as the Galápagos are one of the world's most active volcanic areas, with over 50 eruptions in the last 200 years. The most active volcanoes are the ones on Fernandina Island, which last erupted in April 2009, and Cerro Azul on Isabela (2008). All of the volcanoes are still active but most of the eruptions have been small. Other common geological activity includes uplifts, which are caused when magma beneath an island hardens and lifts the land through and above the water's surface, and large depressions called calderas, which happen when magma beneath the summit of a volcano cools and contracts, causing the entire peak to collapse inward.

The travel company Metropolitan Touring of Ecuador offers geological expeditions of the islands where explorers are offered the opportunity to examine its most significant geological sites and walk on the lava of the most recent eruptions (www.metropolitan-touring.com). Billed as 'Geology Expeditions in the Middle of the World', the tours are led by Dr Theofilos Toulkeridis, and guests learn the story of the formation of the islands through an understanding of the various forms of volcanism. He is the author of the book *Volcanic Galápagos* (Toulkeridis 2011). A typical expedition commences with a visit to North Seymour Island to observe inland gaunt marine skeletons of white coral that was once part of the ocean environment, and is exposed to air. This island is perhaps the best introduction to any Galápagos expedition since it has an extensive number of key species to Galápagos, as well as outstanding geological features.

North Seymour is an uplifted (as opposed to volcanic) island and is flat and strewn with boulders. A visit to Punta Suárez on Española (Hood) Island reveals the oldest rocks of the Galápagos group ranging from 3.2–3.5 million years old. This island is flat and without a volcanic crater. It is the oldest island and the southernmost in the chain. At Point Cormorant on Floreana (Charles) Island, visitors witness a myriad of parasitic cones, evidence of a violent volcanic past. The post-caldera collapse stage is clearly seen on this island as no caldera is present but what buried it is clearly obvious. Spatter cones, cinder cones, and tuff cones are seen from all angles especially when the vessel navigates along its coastline (Fig. 14.2).

Santa Cruz (Indefatigable) Island, Galápagos' third largest island, has a varied geologic history. Here tourists visit the twin pit craters geological formation, Los Gemelos. Genovesa (Tower) Island, is formed by the remaining edge of a large volcanic caldera that is now submerged. At anchor in the bay the walls of the caldera surround the tourist ship and many lava flows that make up this shield volcano can be seen.

In the western part of Galápagos are the younger shield volcanoes of the Isabela and Fernandina. Isabela is the largest island in the Archipelago and was formed by the fusing together of six large volcanic domes. One of the volcanoes, Ecuador Volcano, rises from the water line with one of its halves completely collapsed, and now rests underwater (Fig. 14.3). On this island volcanoes are still forming with eruptions occurring every 3 years or so. Tagus Cove is a natural harbor where centuries ago whalers and pirates left their ship's names painted or carved on the rocks. Here guests hike to the top of the Darwin Volcano which is another crater filled with salt water.

Fernandina (Narborough) Island is the youngest island and westernmost island in the Galápagos and in 1969 its caldera dropped about 300 m to its new position. It is an active volcano and eruptions occur intermittently (Fig. 14.4). Having no introduced mammals, the island is a unique environment with the abundant marine iguanas, sharing their habitat with sea lions, crabs, hawks, penguins and flightless cormorants. On this island volcanic processes happen regularly with eruptions occurring every 3 years or so.

On Santiago (James) Island explorers see evidence of violent eruptions which have occurred with tuff cones dominating the landscape. Lava flows which entered the ocean have left dozens of lava tubes and flat beds where lots of tidal pools form during low tide (Fig. 14.5). The geotour finishes on Bartolome Island which has been described as an open text book on geology due to its unique combination of features.

Fig. 14.2 The Galápagos Islands are probably one of the few volcanic places on Earth where almost all features of island volcanism can be witnessed as part of a regular expedition. Here we see a massive cinder cone in the distance, and lava flows that have entered the ocean. *Source* Francisco Dousdebés, Metropolitan Touring, Ecuador

Fig. 14.3 The Galápagos National Park Visitor Site, Punta Vicente Roca. Guests are shown the magnificent volcanic features of Ecuador Volcano on the northwestern corner of Isabela Island. Geological features such as sills, dikes, tuff-stone layers, volcanic bombs, and ash eruptions are clearly seen in the surrounding landscape. The Ecuador Volcano also includes a partially-collapsed caldera, and half of its dome has sunk into the Pacific Ocean. Life stages of newly-formed shield volcanoes are easily witnessed at this site. *Source* Francisco Dousdebés, Metropolitan Touring, Ecuador

Fig. 14.4 Volcanic eruption, Fernandina Island, Galápagos, April 2009. This impressive large-shield volcano has been erupting every 4–5 years, and its latest eruption was in April 2009. The eruption started off a radial fissure up in the slopes of the volcano and gradually the lava made it to the shoreline. The fissure was located about 14 km from the shoreline. As lava entered the water a tremendous amount of water vapor was released, and that is always exciting for geotourists to experience. *Source* Ramiro Jacome, Galápagos Naturalist, Metropolitan Touring, Ecuador

Fig. 14.5 The Galápagos National Park Site, Puerto Egas. Here geotourists explore the coastline of Santiago (San Salvador Island or James) Island. The lava flows are 3,000–5,000 years old and are topped with a large tuff cone known as Sugarloaf Hill formed by a more recent violent eruption with base surge and phreatomagmatic contact. Along the shores, marine iguanas graze on seaweed, and migrant shore birds, sea lions, and intertidal pools can be observed. *Source* Francisco Dousdebés, Metropolitan Touring, Ecuador

14.5 Conclusions

Thus the Galápagos Islands are now growing their tourism based on a more ethical platform that combines conservation and protection of the natural environment whilst fostering sustainable tourism through greater controls and dispersion of tourists. At the same time individual tour companies have realized the potential for geotourism to be added to their traditional wildlife tourism products and are thus opening up increased opportunities for their sustainable tours. This augers well for the future of the islands and is helping them to become one of the world's emerging geotourism destinations.

Acknowledgments This chapter is an expanded version of an original case study presented in the book *Global Geotourism Perspectives* (Dowling and Newsome 2010). The author also wishes to acknowledge the assistance of Francisco Dousdebés, Metropolitan Touring Ecuador, for providing information, photos and the map to support this chapter. His contribution was invaluable in preparing this chapter of the book.

References

Dowling, R. K., & Newsome, D. (Eds.). (2010). *Global geotourism perspectives*. Oxford: Goodfellow Publishers.

Erfurt-Cooper, P., & Cooper, M. (2010). *Volcano and geothermal tourism: Sustainable geo-resources for leisure and recreation*. London: Earthscan.

Pearson, D. L., & Middleton, D. W. (1999). *The new key to Ecuador and the Galápagos* (3rd ed.). Berkeley: Ulysses Press.

Toulkeridis, T. (2011). *Volcanic Galápagos*. Quito: Centro de Geología, Volcanología y Geodinámica, Escuela Politécnica del Ejército.

UNESCO (2009) The world's heritage: A complete guide to the most extraordinary places. United Nations Educational, Scientific and Cultural Organization and HarperCollins Publishers

UNESCO (2013) UNESCO World Heritage List: Galápagos Islands. http://whc.unesco.org/en/list/1. Accessed on 8 Feb 2013.

UNWTO (2013) International tourism to continue robust growth in 2013. UNWTO PR No: 13006, Madrid, 28 Jan 2013.

White, W. M. (1997) *A brief introduction to the geology of the Galápagos*. Cornell University Earth and Atmospheric Sciences. http://www.geo.cornell.edu/geology/GalápagosWWW/GalápagosGeology.html.

URL

www.cruisemates.com/articles/feature/Silversea-Galápagos-121101.cfm#ixzz2KHVQk7Kp.

Mount St. Helens, Washington, USA

15

Chris Newhall, Peter Frenzen and Carolyn Driedger

Mount St. Helens, southwest Washington State, USA -- 46°20'N, 122°18'W

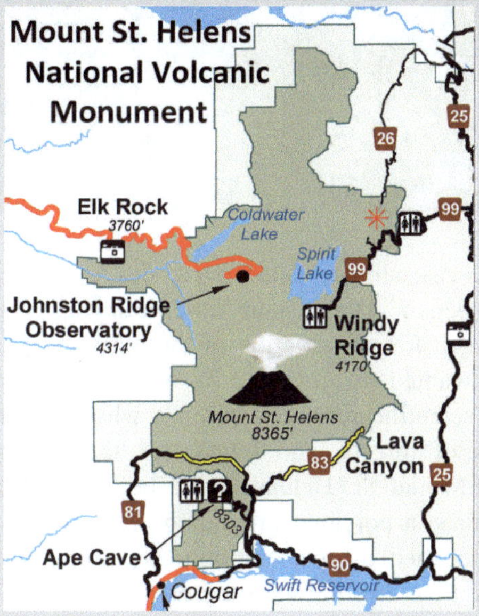

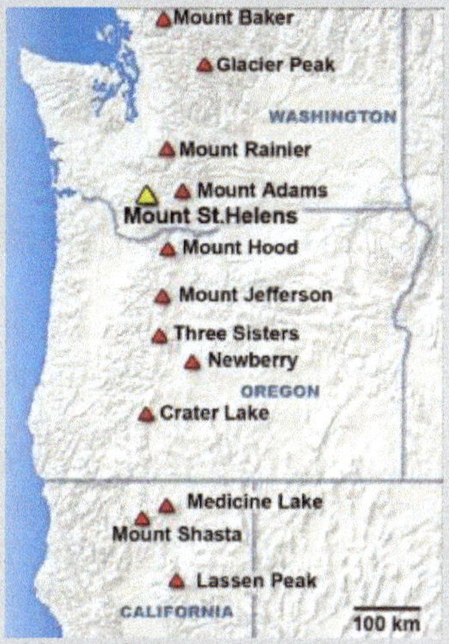

Maps of Mount St. Helens National Monument and WA State with Cascades

Composite volcano, with history of lava domes, plinian explosive eruptions, and 3 known giant landslides (sector collapses)

Latest eruption: dome-building eruption, 2004-2008

Site of largest landslide in recorded history, and a 300 mph laterally directed blast. These were seminal, paradigm changing events in volcanology and landscape ecology.

Status of protection: National Volcanic Monument administered by the U.S Forest Service.

Visitors per year: approximately 500,000, with peak of 1 million per year shortly after major visitor centres opened.

May 18, 1980. Sunrise and a crystal-clear morning. Mount St. Helens was a beautiful, peaceful, snow-capped mountain surrounded by pristine forests and lakes (Fig.15.1). Yes, it was known to be spitting ash and shaking, but it was still fundamentally the same Mount St. Helens that had for decades been a favorite camping and fishing destination of local residents. At 8:32 am Mount St. Helens turned ugly, unleashing a mind-boggling landslide that left a huge crater where the peak had been and a thick jumble of rock with an average thickness of 150-feet (45 m) filling the lush valley that once led to much-beloved Spirit Lake. Within minutes, a hurricane-force blast of hot steam and rock flattened mature forest up to 17 miles (27 km) away and killed everything except roots and snow-covered plants, rodents in their burrows and fish in protected ice-covered lakes. The eruption continued for nine hours, pumping pumice and ash high above the volcano where strong winds carried it eastward across the western U.S.turning day into night. Mudflows formed immediately and after a few hours delay, scoured valleys and carried away bridges, trucks, and homes like toys, and logs like matchsticks. 57 people died, mostly weekend campers and a few whose curiosity brought them to watch the volcano from what they thought was a safe distance.

C. Newhall (✉)
Earth Observatory of Singapore and formerly with US Geological Survey and University of Washington, Sto. Domingo, Albay, Philippines
e-mail: cnewhall@ntu.edu.sg

P. Frenzen
US Forest Service Mount St. Helens National Volcanic Monument, Amboy, WA, USA
e-mail: pfrenzen@fs.fed.us

C. Driedger
USGS Cascades Volcano Observatory, Vancouver, WA, USA
e-mail: driedger@usgs.gov

P. Erfurt-Cooper (ed.), *Volcanic Tourist Destinations*, Geoheritage, Geoparks and Geotourism,
DOI: 10.1007/978-3-642-16191-9_15, © Springer-Verlag Berlin Heidelberg 2014

Fig. 15.1 Before the devastating May 18, 1980 eruption, Mount St. Helens was considered to be one of the most beautiful and most frequently-climbed peaks in the Cascade Range. Spirit Lake was a vacation area offering hiking, camping, boating, and fishing. *Photo* US Forest Service, Gifford Pinchot NF (taken before May 18, 1980)

Because scientists had designated danger zones and public safety officials had closed most of them, many more people were saved than killed. Luck also was a factor. The eruption occurred early on a Sunday morning before many people had a chance to enter the 230-square mile blast area.

In the span of a few terrifying hours, Mount St. Helens was transformed from a symbol of the good life in the Pacific Northwest to a fearsome killer, and in succeeding months, to a worrisome threat, a mess that must be cleaned up, and eventually, to a world-class tourist attraction! In the next few pages, we'll recount how tourism was re-developed and has evolved at Mount St. Helens.

As soon as the search and rescue effort ended and residents, officials, and scientists pieced together what had happened—the curious public wanted to see the devastation as well (Fig. 15.2) Commercial flights from Seattle and Portland received permission to give passengers a better view of the incredible changes, albeit from frustratingly high altitude. At lower altitudes, light aircraft gave tourists closer views, but even with a birds-eye view it was hard to comprehend the scale.

Not long after the eruption, public pressure arose to reopen access to the area. Tourists would be excluded for now, but five billion board feet of timber had been felled in a few minutes and forest managers wanted to claim it before the insects. Public safety officials were cautious, as were loggers. Clearly the mountain could kill, but would it kill again? How dangerous would it be to enter the blast zone to recover the timber? After a few months of caution, land managers, scientists, and public safety officials sat down to quantify risk, and by that time the risk from the volcano was roughly equal to risk already taken every day by loggers. Timber salvage began. At the same time, lobbying of the US Congress began for preservation of the blast zone—by scientists who wanted to preserve a once-in-a-lifetime opportunity for scientific research, officials who promoted recreational use of forests, and businesses that foresaw tourism dollars. Mount St. Helens would become a laboratory for scientists, a poignant ground for those who lost relatives and cherished the pre-eruption landscape, and a huge draw for tourists from around the world.

Since many of these competing uses appeared to be incompatible, the only solution was to designate a core area for preservation (110,000-acre Mount St. Helens National Volcanic Monument, managed by the US Forest Service) with an outer area where timber salvage would be permitted. While each side grumbled a bit that the other got too much land, the eventual compromise was reasonable and worked well.

In the United States, National Parks and National Monuments give strong emphasis to preserving natural features but, at the same time providing access to visitors. The balance is remarkably successful. So, there was never a question about whether visitors would eventually be allowed back to Mount St. Helens. The only question was when? When would it be safe? And when could the Forest Service and others develop the infrastructure that would be needed to host hundreds of thousands, perhaps millions of visitors each year?

Fortunately, Mount St. Helens cooperated. After just a few months of explosive eruptions, it settled into a repeated

Fig. 15.2 Mount St. Helens and the devastated area are now within the 110,000-acre Mount St. Helens National Volcanic Monument, under jurisdiction of the United States Forest Service. Visitor centers, interpretive signs, and trails help serve the needs of the thousands of tourists, students, and scientists who visit the monument daily. Mount St. Helens is once again considered to be one of the most beautiful and interesting of the Cascade volcanic peaks. *Photo* USGS, Lyn Topinka (May 19, 1982)

pattern of months-long quiet broken by multi-day eruptions of mostly non-explosive, viscous magma. Most of the time, visitors could explore everywhere except the crater in reasonable safety, and needed to retreat only occasionally. A visitor's risk while visiting Mount St. Helens was little more than his or her risk of driving there. The first roads that reopened around the volcano supported timber salvage with 600 logging trucks exiting the area daily. These roads were closed to visitors. Construction crews did re-open a road from the northeast to provide access for the U.S. Army Corps of Engineers pumping station that prevented the now-impounded Spirit Lake from rising and overtopping the highly erodible natural landslide dam. Had that dam failed, several downstream towns would have been destroyed and the recently-dredged ship canal on the Columbia River might have refilled with sediment.

Perennial public debates involve land and lake use. While most land is open to visitors, some areas have been set aside for research of long-term ecological recovery. For now, the policy of the monument is to allow natural processes to continue substantially unimpeded by human activity. Recreational activities such as hunting, fishing and camping that are typically allowed in National Forests are not permitted within a 30,000-acre designated research area north of the crater (less than 1/3 of the Monument).

An amazing array of visitor facilities was constructed to support visitation. On the west side, five visitor centers were established: three by the Forest Service, one by the timber industry giant Weyerhaeuser Company, and one by Cowlitz County. Silver Lake Visitor Center, built by the Forest Service and located near Seaquest State Park about seven miles (11.2 km) east of Interstate 5—opened in 1987 and is now run by Washington State Parks. A new, US$150 million highway up the Toutle Valley (Spirit Lake Memorial Highway, WA 504) completed in 1993, enabled the Forest Service, Weyerhaeuser, and Cowlitz County to build additional visitor centers and allowed visitors to enter the blast zone. The Cowlitz County center focuses on tourism needs and includes a restaurant, gift-shop and a charter helicopter service, as well as interpretive displays. The Weyerhaeuser center presents a forestry perspective and tells the story of the salvage logging and replanting of their large-scale tree farm northwest of the volcano. Closer to the volcano, the Coldwater Ridge Visitor Center of the Forest Service was focused originally on the ecosystem impacts. While budget cuts forced closure of the building for several years, today, collaborating universities and community sponsors are working to repurpose the building as a Science and Learning Center that supports longer-term learning. The final leg of the highway, completed in 1997, reaches a high ridge looking straight across the Toutle valley and directly into the crater. Here, the Forest Service built the Johnston Ridge Observatory (JRO), named in memory of the late David Johnston, the USGS scientist on duty nearby on the fateful morning of May 18, 1980. As seen on approach, the crater is still about five miles (8 km) away, but it is so broad that it appears much closer. The Johnston Ridge Observatory emphasizes the dramatic impacts of 1980 eruption and ongoing geologic monitoring. A large outdoor patio and amphitheater provide dramatic settings for picture taking and interpretive presentations (Figs. 15.4 and 15.5) Each of the five visitor centers houses a small gift shop and several contain a theater.

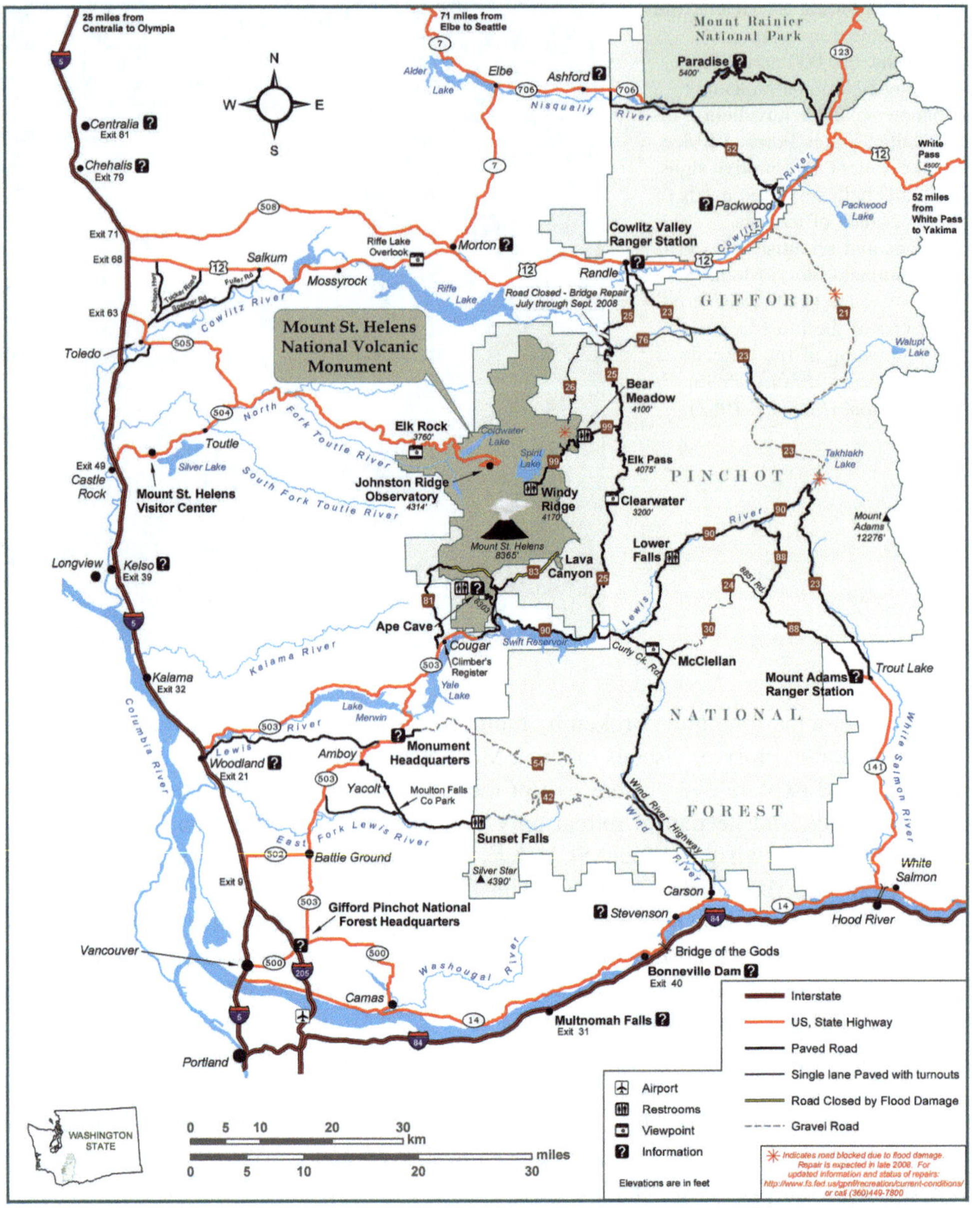

Fig. 15.3 Mount St. Helens and surroundings. Lightly shaded areas are National Forest for multiple uses, including sustainable logging and recreation. Dark shaded areas, the Mount St. Helens National Volcanic Monument and Mount Rainier National Park, emphasize conservation of nature and tourism. Source US Forest Service.

Much of the maintenance and staffing of visitor centers is paid for by visitor entrance fees. A daily fee for the Forest Service's visitor centers is nominal (USD $8/day), and entrance to the Weyerhaeuser center is free. Unfortunately, these fees do not cover the full costs of running all facilities. Budget pressures caused the Forest Service to turn the Mount St. Helens Visitor Center at Seaquest State Park over to the Washington State Parks, and temporary closure before eventual repurposing of the Coldwater Ridge Visitor Center.

Interpretation for visitors benefits from the close bond between the U.S. Geological Survey, Forest Service and universities including University of Washington, Oregon State University, and others. Researchers provide input for development and delivery of educational programs, exhibits and movies, and Forest Service rangers benefit from annual trainings and frequent research updates from scientists. Researchers benefit from the observations and help of monument staff. Together, they assemble interpretive messages of greatest importance to the interests and safety of visitors, not just at Mount St. Helens, but for the Cascade Range volcanoes.

Forest Service rangers express the significance of Mount St. Helens as a master teacher to a whole new generation of volcanologists. The 1980 eruption boosted interest in the study of explosive eruptions and monitoring efforts that improve warning systems and volcano-hazard mitigation around the world. Today, keeping watch over Mount St. Helens links scientists of many disciplines with one another, and with emergency and land-use managers. Out of the eruption of Mount St. Helens, scientists, officials, and the public have a new appreciation for potential effects of

Fig. 15.4 Planners had preservation of the landscape and aesthetics in mind when they designed Johnston Ridge Observatory to blend in with the surrounding hillside. Trail leads to a nearby viewpoint. *Photo* Harriet Kauffman (July, 2011)

Fig. 15.5 The deck at Johnston Ridge Observatory is the location for amazing views and inspiring interpretive talks. *Photo* US Forest Service, Scott Hinderman

volcanic eruptions throughout the Cascades and beyond. Mountains that we once thought were "stable forever" are now known to collapse and rebuild themselves repeatedly. Before 1980, only a handful of giant volcanic landslides were known. Now the list contains more than 400 and grows every year. Since 1980, research at Mount St. Helens has yielded quantum advances in the fields of volcanology and forest ecology, and even some insights into the origins of life!

Along the ridge top, the eruption obliterated an entire forest including old-growth trees, but visitors can still view evidence of the blast-shattered forest in preserved areas. Several trails lead hikers through the heart of the volcanic blast zone. The 2.3 mile (3.7 km) long Hummocks Trail loop offers the opportunity to explore hill-sized chunks of the former top and insides of the volcano that were carried into the Toutle valley by the massive landslide of May 18. In these hummocks, one can find some of the original layered deposits from past eruptions preserved despite their tumultuous transport. Based on their experience at Mount St. Helens geologists now recognize similar volcanic landslide and blast deposits elsewhere in the world.

Fig. 15.6 Visitors to Meta Lake learn about returning plant and animal life through Ranger-led hikes and roving interpretive talks offered during the summer season. *Photo* US Forest Service, Peter Frenzen

On the northeast side of Mount St. Helens, a pre-eruption logging road (now Forest Road 99) was widened and paved so that visitors can drive through spectacular standing dead and blown-down forests until they reach equally spectacular overlooks of the new crater and log-filled Spirit Lake. Because the northeast side is farther from the heavily traveled Interstate 5 highway, visitor numbers are lower than those on the west side, but on a nice summer day it is not uncommon to find full parking lots. On weekends during July and August, Forest Service rangers provide scheduled interpretive talks in an outdoor amphitheater at Windy Ridge.

Since the giant landslide and most of the eruptive activity was directed to the north, the south side of Mount St. Helens was less impacted. But even on the south, there are interesting features of both the 1980 and older eruptions, including some 1,900 year old basaltic lava flows complete with pahoehoe lava features and Ape Cave, a spectacular 2.5 mile (4 km) long lava tube. Southside visitors can also explore impressive andesite lava flows and waterfalls from the viewing decks and suspension bridges in Lava Canyon.

The climbing route on the south side also offers the opportunity to climb to the crater rim and gaze into the volcanic crater. The 10.5 mile (16.9 km) (4,450 foot [1,356 m] elevation gain) climb is arduous but is not technical (though ice axe and crampons may be necessary when there is snow on the route). Climbers should be aware that unstable cornices of snow and ice that line the rim in winter and spring often collapse. The US$22/person permits for the April through October climbing season are reserved online starting on February 1st of each year through the Mount St. Helens Institute (www.mshinstitute.org).

The 28-mile long Loowit Trail encircles Mount St. Helens, and feeder trails meet it at several locations. Two new trails are under consideration—one to the edge of the crater and another to a new close-up volcano viewpoint on the monument's south side.

The Mount St. Helens National Volcanic Monument (www.fs.usda.gov/mountsthelens/) employs permanent and seasonal staff to provide for public safety, facility maintenance, and education. Trained ranger interpreters answer visitor questions about volcanic events, forest recovery, eyewitness accounts, and recommended hikes and viewpoints (Figs. 15.6 and 15.7).

The Monument staff is supplemented by a large number of volunteers organized through the Mount St. Helens Institute, a non-profit organization devoted to helping people understand and protect the volcano. Monument staff members work closely with the Mount St. Helens Institute to offer classes in natural and cultural history, and interpretive talks, climbs, hikes and night sky-viewing within monument boundaries. The Mount St. Helens Institute also engages local communities by bringing volcano-themed talks and special events, and educational venues in nearby towns. In 2013, Forest Service and Institute educational programs reached more than 140,000 visitors.

Schoolchildren are an important component of visitation at the National Volcanic Monument. Today's schoolchildren (even many of their parents) were born after this world-renowned eruptive event that had such a profound effect on the landscape and psyche of the Pacific Northwest. Educational visits to Mount St. Helens encourage a greater understanding of Earth processes, about Cascade volcanoes and their hazards, and about the need for preparedness.

Fig. 15.7 Wildflowers thrive under the protection of a log in a forest blown down by the 1980 eruption. The rapid return of life has been one of many dramatic and inspiring stories to unfold in the Mount St. Helens National Volcanic Monument. *Photo* US Geological Survey, Carolyn Driedger

Can visitors expect to see an eruption? In the fall of 2004, Mount St. Helens did reawaken. Several small but spectacular explosions were witnessed by tens of thousands of people. Then, over a period of 3.5 years an exceptionally viscous lava dome—a 1,000-foot (304 m) tall mass of "rock toothpaste," was squeezed upward and sideways from a 100-foot (30.4 m) diameter vent deep in the crater behind the 1980–1986 lava dome. After several months of precautionary suspension of visitation, most areas were re-opened by the spring of 2005 so that visitors could come and watch the volcano rebuild. Much to the dismay of volcano watchers and to the relief of officials the eruption was largely gas-free and generally non-explosive. The only visible evidence was the growth of the new lava dome that grew at an average rate of 0.14 cubic yards (0.10 m^3) per second (nearly equivalent to filling four Olympic-sized swimming pools every day). The volcano thrilled visitors with minor steam plumes, and seismic traces of rock falls and tiny earthquakes that were recorded by seismometers in the crater and relayed to visitor centers. Has Mount St. Helens breathed its last for now? Scientists fully expect that future eruptions might occur even within our lifetime!

With many visitor facilities designed to accommodate hikers, cross-country skiers, and snowmobilers, the crater is the only part of the Monument generally off-limits to visitors, primarily because of rock fall from the crater walls and lava dome.

Establishment of the National Volcanic Monument, and the high quality of its facilities, has brought millions of visitors to southwest Washington State (as of 2012, a cumulative total of more than 30 million), with corresponding economic benefit to towns and businesses throughout the region. Except for an anticipated store with refreshments at the Science and Learning Center at Coldwater and at a restaurant at the Cowlitz County center at Hoffstadt Bluffs, food service and accommodation are only available in the surrounding communities. A few small-scale cottage industries have volcano themes. One might say that the eruption of Mount St. Helens spurred the economy just in time. While the timber industry is not dead, the days of easy access to old-growth forest are over and tourist businesses related to Mount St. Helens have helped to ease the transition away from what had been a timber-based economy.

Visitors to Mount St. Helens leave with a new appreciation for this amazing Cascade volcano and its neighbours, and a recognition that we live on an ever-changing planet (Figs. 15.8 and 15.9) increasingly via tourism. The living laboratory of Mount St. Helens is a place of wonder and continual change, a must-visit attraction to be added to anyone's future vacation and travel plans.

Fig. 15.8 Approximately 16,000 climbers ascend to the summit of Mount St. Helens annually. In clear weather they are rewarded with views of the volcano's crater, and other Cascade volcanoes. This photo shows members of a geologist-guided field seminar. *Photo* Jim Uyesugi

Fig. 15.9 Monument visitors explore the pyroclastic flow deposits north of the crater on the Truman Trail, part of a 300 km trail system surrounding the volcano. *Photo* Robert Hubner, Washington State University

Additional Resources

Dzurisin, D., Driedger, C., and Faust, L. (2013) *Mount St. Helens, 1980 to Now—What's Going On?* Fact Sheet 2013–3014, version 1.1. US Geological Survey, http://pubs.usgs.gov/fs/2013/3014/.

Cascades Volcano Observatory website: www.volcanoes.usgs.gov/observatories/cvo.

Discover your Northwest (2013) *Building community stewardship of Northwest public lands today and for generations to come.* Online: www.discovernw.org/.

Mount St. Helens National Volcanic Monument website: www.fs.usda.gov/mtsthelens.

Mount St. Helens Science and Learning Center website: www.mshslc.org.

Mount St. Helens Institute website: www.mshinstitute.org.

Volcano Tourism at Mount Batur in Bali, Indonesia

16

Patricia Erfurt-Cooper

16.1 Introduction

Bali is one of the Indonesian islands of the lesser Sunda group (Fig. 16.1) and is dominated by a chain of active and dormant volcanic mountain ranges including Mount Agung (3,143 m) and Mount Batur (1,717 m) as the highest mountains. Indonesia has around 500 volcanoes of which 129 are classed as active (Murdoch University 2012). Earthquakes are common in Bali due to its location in a highly active tectonic region. The main centres of activity are the volcanic complexes of Mount Batur and Mount Agung, which are separated by a narrow valley only 20 km apart (Seach 2013). Local authorities maintain that the Balinese volcanoes are safe for tourist activities unless they show significant signs of unrest. Volcano tourism in Bali takes place on and around active mountains such as Mount Batur and Mount Agung which can be reached from the capital Denpasar in a few hours. Many tourist brochures

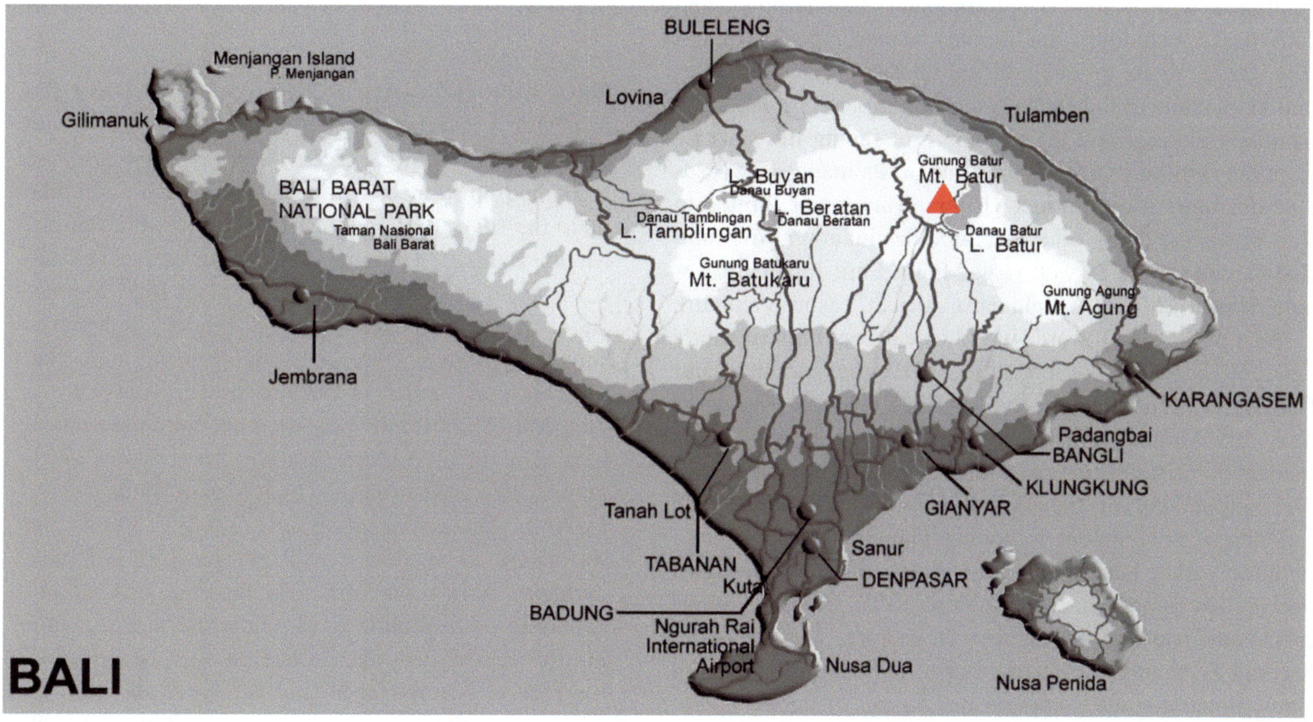

Fig. 16.1 The Indonesian island of Bali with Mount Batur. Map modified by Patricia Erfurt-Cooper. *Source* OrangeSmile Tours (2013)

P. Erfurt-Cooper (✉)
James Cook University, Australia, P.O. Box 5016 Torquay,
QLD 4655, Australia
e-mail: Patricia.Erfurt@my.jcu.edu.au

P. Erfurt-Cooper (ed.), *Volcanic Tourist Destinations*, Geoheritage, Geoparks and Geotourism,
DOI: 10.1007/978-3-642-16191-9_16, © Springer-Verlag Berlin Heidelberg 2014

Fig. 16.2 Tourist brochures include volcanoes as main attractions together with local hot springs, cultural performances, traditional architecture and unique agricultural sites like rice terraces. *Photo* Patricia Erfurt-Cooper

offer volcano tours as one of the main attractions usually combined with visits to cultural sites like temples and activities including traditional dance performances (Fig. 16.2). Visits to local volcanic hot springs are another option on some trip agendas. Volcano tours are promoted as 'sunrise trekking' and the adventure of seeing the sun rise from the summit of a volcano is one of the major drawcards tour operators use to attract visitors. This is not limited to Bali however; many other volcanic destinations offer sunrise viewing tours as well (e.g. Mount Fuji, Japan; Haleakala, Hawaii; Mount Kilimanjaro, Africa; Yasur, Vanuatu).

As a major step towards worldwide recognition of Bali's volcanic heritage in September 2012 the Mount Batur Caldera was included in the UNESCO Global Geoparks Network. Mount Rinjani and Lake Toba are two further volcanic tourist destinations currently among the candidates for membership of the Global Geoparks Network. The Indonesian government is hoping that the UNESCO recognition will be beneficial for the local economies through a geotourism approach (Jakarta Post 2013; United Nations 2013) and increase the number of visitors. The following account of a sunrise trekking tour to the summit of Mount Batur illustrates the complete integration of active volcanoes as well as volcanic hot springs into Bali's tourism industry. For people interested in further information about Bali's geological heritage and the cultural significance of Mount Batur the Ministry of Mining and Energy (Directorate of Volcanology and Geology) has established a four-story volcano museum in Kintamani.

16.2 Sunrise Trekking

Nusa Dua, Bali—April 2010—ready to leave the hotel for a day trip. It is 2 a.m. in the morning and our driver is waiting outside the hotel. By 4 a.m. we are supposed to meet with a tour guide who will take us to the summit of Mount Batur to view the sunrise. Mount Batur or Gunung Batur is an active volcano 2.5 h north of Denpasar, the capital of Bali, Indonesia. The driver speaks excellent English and makes sure his passengers are comfortable in his new van. The air conditioner is on full power—outside the temperature is over 30 °C—no relief from the tropical heat at night time. After 2 h we arrive at our destination; the caldera out of which Mount Batur rises to 1,720 m. The actual climb covers only 1,300 m, but we all are aware that it will take place in complete darkness. When we arrive in the car park at our departure point a number of guides are already waiting for us. The driver introduces us to our tour guide, we receive a torch each and off we walk into the darkness. At the beginning it is easy walking on a well used track. According to our guide it may only take an hour to reach the summit. This is not even close to the truth as we are about to discover, although it could be clever psychology to keep us walking. The slope becomes steeper and after a while we sit on a tree to

Fig. 16.3 Mount Rinjani on the island Lombok erupting at sunrise (24 April 2010). *Photo* Patricia Erfurt-Cooper

catch our breath; not our guide though as he is only 20 years old and does this tour at least three times a week he tells us. However, he is very caring and patient and tries to keep us comfortable. While we are sitting on our tree we can see the lights of Kintamani in the distance and it is now obvious that we have left the forest at the foot hills of Mount Batur behind us. We can see more torches shining through the trees of the next group following us and by the time we reach the summit, there will be over 50 volcano tourists with 15 tour guides.

The trek up the mountain becomes slower due to the loose scoria on the slopes, which is now replacing the sandy paths further down. As we are climbing higher and higher towards the summit crater, the lights of Kintamani disappear in the early morning mist. At times it is hard to continue on the steep track which none of us can see; only our guide knows exactly where to go and we follow him blindly into the unknown. It is certainly not recommended for people with health problems or low levels of physical fitness. There are moments when it feels impossible to reach the summit in safety and one wonders, why we are doing this and where we all will end up. When we finally reach the top, we are welcomed into a mountain hut built from volcanic rocks where everybody is offered breakfast. Eggs steamed in fissure vents, sandwiches with fresh bananas as sweet as jam and tea or coffee add a pleasant note to the unique experience of watching the sun rising from the top of an active volcano. Any exhaustion fades away quickly and everybody starts to take photos. Then for a short time another volcano is visible in the distance—the guides inform their groups that this is Mount Rinjani on the island of Lombok, which lies across the Strait of Lombok to the east of Bali. While the sun is rising and Mount Rinjani is clearly visible above the cloud layers we suddenly notice that this volcano had started erupting (Fig. 16.3). According to the tour guides this was the first morning without rain and with a clear view in two months! Such luck, as the view of the sunrise and Rinjani's eruption was truly amazing and made the difficult climb worthwhile.

A short time later a group of monkeys appeared and waited for people to throw them pieces of food. In return they entertained the crowds by presenting an additional unique photo opportunity (Fig. 16.4). So much for a 'balanced' ecology, where volcano tourists meet wildlife that is used to regular and unrestricted interaction and handouts from tourists.

Once the sun is up and breakfast is finished, the different groups and their guides move on to the actual crater of the main volcano, which was last active in 2000, although it was erupting from a vent on the lower flanks (Tour guide 2010). From the crater walls higher up fumaroles emit hissing steam while the crater itself is an intimidating hole with a vertical drop of around 150 m (Tour guide 2010). Up close the crater rim looks even more forbidding with only a few strands of rusty barbed wire as protection from the fall to death (Fig. 16.5). No warning signs anywhere. This hopefully has changed since 2010 to provide better protection for volcano tourists. However, despite the repeated warnings from the tour

Fig. 16.4 Local monkey community on Mount Batur after sharing breakfast with the visitors. *Photo* Patricia Erfurt-Cooper

Fig. 16.5 In March 2010 a Swedish tourist died after falling in Mount Batur's crater. Two weeks later the only barrier consisted of these rusted strands of barbed wire, which ended after a few metres. In wet weather this is not a safe option, although it may have changed since 2010. *Photo* Patricia Erfurt-Cooper

guides to stay away from the crater rim, some tourists cannot resist the urge to go as close as possible.

For the majority of our climbing companions watching the sun rise from a volcano is a challenge and offers something different. While climbing a volcano classed as active presents a challenge for the more adventurous minded traveller, unfortunately this can have fatal consequences. Only 2 weeks prior to our ascent a young tourist from Sweden died when he fell 150 m into the steep sided crater we were looking

Fig. 16.6 Mount Abang in front of Mount Agung viewed from the slopes of Mount Batur. *Photo* Patricia Erfurt-Cooper

Fig. 16.7 Lava flows from the last eruption of Mount Batur. *Photo* Patricia Erfurt-Cooper

at in front of us (Tour Guide 2010; Sydney Morning Herald 2013). As a consequence the government has banned hiking around the crater rim beyond the viewing area were steaming vents are indicative of the continuous state of activity and act as a reminder of the November 2009 unrest of Mount Batur (Seach 2013). After these sobering observations everybody is slowly heading back downhill—easier in daylight, but not less treacherous due to the now visible surroundings distracting from watching one's step on the steep path strewn with loose black scoria.

Our guide tells us that although not many accidents happen, they do occur and every guide hopes it is not a tourist in their group. Consequently they take their responsibility very serious and even lend a helping hand to prevent us from falling down the steep track when they notice us stumble. During daylight the full size of the caldera becomes apparent with Lake Batur in front of Mount Abang and Mount Agung offering an amazing view. The Balinese people are right when they say they live in paradise (Fig. 16.6).

Looking down into the caldera the ground is covered with black lava flows from the eruption in 2000. Only a small wooded mound stands out of the black mass (Fig. 16.7). Leaving the summit behind us on the way down the full beauty of the mountain becomes apparent. The trail that was not noticeable in the dark is now visible, zigzagging up the steep sides

Fig. 16.8 Looking back at the winding track that leads up to the summit of Mount Batur. *Photo* Patricia Erfurt-Cooper

of the volcano (Fig. 16.8). In hindsight we think our trek was quite an achievement.

At the foothills of Mount Batur the fertile volcanic soils are supporting extensive vegetable plantations, were tomatoes, beans, peppers, peanuts and corn are grown and which are easy to access by motorbikes and 4WD vehicles. At the end of our hike to the crater our guide explains that to overcome 1,300 m in height we had to walk approximately 4 km in the dark; a challenge for most of us who were carrying cameras and water bottles. Without our tour guide's encouragement and considerate manner not all of us would have made it to the top.

Tourism in Bali is well organised. Tours are usually booked at the hotels and guests are picked up by their drivers at the entrance. Drivers and guides take their responsibility serious and aim to provide a safe experience as their livelihoods depend on pleased customers. Nevertheless, volcano tourism in Indonesia presents various challenges as can be seen from the Mount Batur case study. It is not necessary for a volcano to be in eruption to contribute to accidents and even fatalities as the case of the Swedish tourist shows, who died from severe injuries to his head and body. According to the police the fatality occurred as a result of the tourist falling on the sharp rocks after losing his footings on the slippery rim of the crater (Kencana 2010; Standora 2010). Adventure tourism always includes certain risks and accidents remain an unfortunate possibility. It is the degree of personal risk perception and risk acceptance however, that determines the actual risk level for health and safety of the individual volcano tourist.

As a result of the inclusion of Mount Batur in the Global Geopark network the region is receiving special attention from investors and developers. While Bali's overall visitor numbers in 2011 reached 2.9 million tourists, the Kintamani/ Mt. Batur area was visited by 400,000 tourists. With the new Global Geopark label the target for 2014 was set at 1.5 million visitors (Jakarta Post 2012), although this requires major investments into the local infrastructure. Despite the volcanic attractions Bali's focus officially is on ecotourism, not volcano tourism; however, with a UNESCO Geopark this trend may change in the future and include geotourism based on the volcanic heritage as well.

References

Jakarta Post. (2011). Sam Ratulangi Airport closed after volcanic eruption. Retrieved March 14, 2013, from www.thejakartapost.com/news/2011/07/03/sam-ratulangi-airport-closed-after-volcanic-eruption.html.

Jakarta Post. (2012). Investors race to develop Batur Geopark. Retrieved March 14, 2013, from www.thejakartapost.com/bali-daily/2012-11-29/investors-race-develop-batur-geopark.html.

Jakarta Post. (2013). Mt. Rinjani proposed as geopark, again. Retrieved January28, 2013, from www.thejakartapost.com/news/2012/12/08/mt-rinjani-proposed-geopark-again.html.

Kencana, M. A. (1 April 2010). Swedish tourist falls to his death in Bali volcano crater. Retrieved March 14, 2013, from www.thejakartaglobe.com/home/swedish-tourist-falls-to-his-death-in-bali-volcano-crater/366818.

Murdoch University. (2012). Bali case study: Physical setting. Retrieved March 14, 2013, from www.istp.murdoch.edu.au/ISTP/casestudies/Case_Studies_Asia/bali/case01.html.

OrangeSmile Tours. (2013). High-resolution Maps of Bali, Indonesia. Retrieved July 21, 2013, from www.orangesmile.com/travelguide/bali/high-resolution-maps.htm.

Seach, J. (2013). Batur volcano. Retrieved February 2, 2013, from www.volcanolive.com/batur.html.

Standora, L. (1 April 2010). Swedish tourist Daniel Fetersan falls to death in Bali volcano. Retrieved March 14, 2013, from www.nydailynews.com/news/world/2010/04/01/2010-04-01_hiker_falls_to_death_in_bali_volcano.html.

Sydney Morning Herald. (2013). Young tourist slips into volcano crater. Retrieved February 2, 2013, from www.smh.com.au/travel/travel-news/young-tourist-slips-into-volcano-crater-20100401-rf8v.html.

Tour Guide. (2010). Conversation between Tour Guide and Author during Sunrise Trekking Tour, Mount Batur.

United Nations. (2013). Earth Sciences for Society: UNESCO Geopark Members. Retrieved January 28, 2013, from www.unesco.org/new/en/natural-sciences/environment/earth-sciences/geoparks/members/.

17 Krakatau: Tourism and the Recovery of a Volcanic Rainforest

Richard Field and David Newsome

17.1 Introduction

One of the most famous volcanoes in the world, Krakatau (often referred to as 'Krakatoa'), is in a group of four uninhabited islands situated in the Sunda Strait between Java and Sumatra (Fig. 17.1). The islands comprise Rakata, Sertung, Panjang (sometimes known as Rakata Kecil, meaning 'little Rakata') and the currently active Anak Krakatau ('child of Krakatau') that emerged from the sea in the 1920s, becoming a permanent island in 1930. A series of massive eruptions in 1883 blew the volcano apart and effectively destroyed all life on what was left of the remaining islands following the explosion. The 1883 eruptions are notable for many reasons, including the catastrophic tsunamis that are thought to have killed over 36,000 people along the coastlines of Java and Sumatra. This 'event' in 1883 has been studied by an extraordinary range of researchers, including those from the fields of geology, biogeography, ecology, climatology, oceanography, geomorphology, history and sociology (Thornton 1996). According to Thornton (1996, p. 4), the event "still has a special place in volcanological research".

The fame of, and widespread interest in, Krakatau probably stems from its status as the first global news story about a natural disaster. Winchester (2003) argued that the 1883 eruptions of Krakatau changed the world's view of itself, giving birth to ideas associated with the 'Global Village'. This was due to various factors, most notably the recently laid undersea telegraph wires that allowed news of the events in Indonesia to reach a global audience within hours of their occurrence (and also led to the common mis-spelling 'Krakatoa'). Eye-catching photographs were taken, bringing the events to life: of some of the eruptions, before-and-after shots of coastal areas in Java and Sumatra that were devastated, and so on. Photography was still quite new at the time. The Krakatau story, in various forms, ran for months and allowed people all around the world to share some of the hopes and despairs of individuals caught up in the events. It was prominent in the rise of Reuter's newsagency, which found itself perfectly placed to do the reporting.

Krakatau is notable in other ways, also. One of the explosions remains the loudest sound recorded, being heard by people as far away as Perth, Australia and the island of Rodriguez (4,700 km away) and being recorded (by instruments) travelling around the world three and a half times—helping meteorologists understand how air waves travel through the atmosphere (Thornton 1996). One of the tsunamis was registered and measured in both Alaska and the English Channel. The great volume of volcanic ash and dust ejected into the atmosphere caused spectacular sunsets (similar to the eruption of Tambora in 1815 that featured famously in some of Turner's paintings); two thirds of the 494-page report of the Royal Society's 'Krakatoa Committee' (Symons 1888) were devoted to optical phenomena in the atmosphere (Winchester 2003). Indeed, Olson et al. (2004) suggested that the very red sky in the painting 'The Scream' by Edvard Munch was caused by the Krakatau eruptions of 1883.

From an ecological point of view, Krakatau is highly significant in being an example of the recovery of rainforest from a sterilised platform. Today the vegetation consists of tropical rain forest re-growth and this is best developed on the largest of the islands, Rakata. The islands of Sertung and Panjang (Fig. 17.1) also contain coastal and secondary lowland rainforest while Anak Krakatau, one of the most active volcanoes in the world, shows earlier stages of vegetation development. The ecological and biogeographic significance of these islands is mainly that the archipelago

R. Field (✉)
School of Geography, University of Nottingham, Nottingham, UK
e-mail: richard.field@nottingham.ac.uk

D. Newsome (✉)
School of Environmental Science, Murdoch University, Perth, WA, Australia
e-mail: D.Newsome@murdoch.edu.au

P. Erfurt-Cooper (ed.), *Volcanic Tourist Destinations*, Geoheritage, Geoparks and Geotourism,
DOI: 10.1007/978-3-642-16191-9_17,

Fig. 17.1 Map of Krakatau

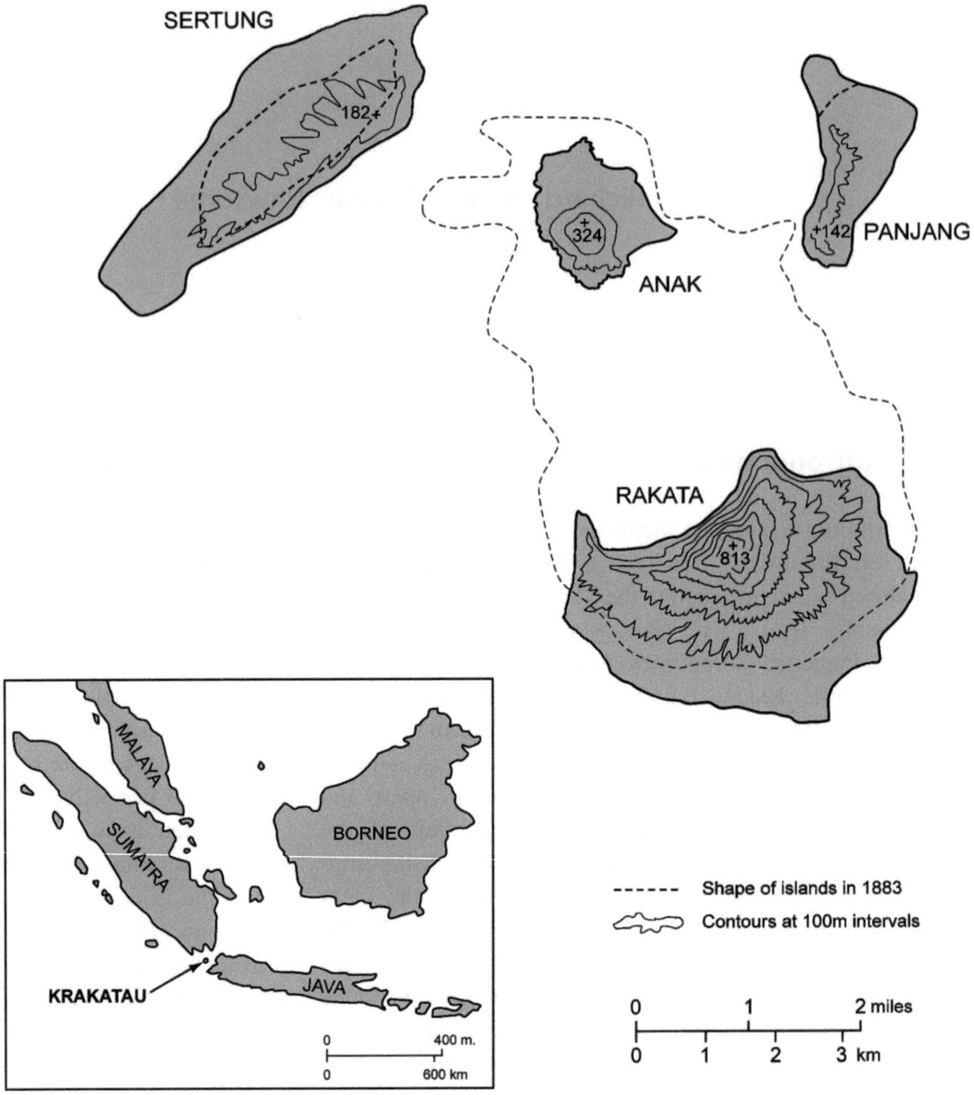

represents a natural experiment of rare size and quality—rarer still given that there is even today a global scarcity of tropical ecological data on the destruction of rainforest by volcanoes. Rakata, Sertung and Panjang islands were all sterilised at the same time (August 1883); this gives replication. Their ecosystems have been rebuilding ever since, almost completely free from human disturbance. There is also a remarkably good record of the colonisation of plants and various groups of animals since the 1880s, thanks in large part to the Herculean efforts, early in the 20th century, of the Dutch scientists Willem Docters van Leeuwen and Karl Dammerman. Their monographs on the plants (Docters van Leeuwen 1936) and animals (Dammerman 1948) of Krakatau during the first 50 years after the 1883 eruptions have greatly enhanced biogeographical and ecological research ever since—particularly the many dozens of papers produced by both Robert J. Whittaker and his colleagues (Oxford University, UK) and Ian W. B. Thornton and his colleagues (La Trobe University, Melbourne, Australia). In addition, the Krakatau bird data were famously used by MacArthur and Wilson (1967) in their classic Equilibrium Theory of Island Biogeography, which revolutionised the science of biogeography (and strongly affected related sciences) and remains very influential today, including in conservation theory. Krakatau, then, provides an extremely important case study of tropical rain forest succession and dynamics, which has become a mainstay of most biogeography textbooks (e.g. Cox and Moore 2010).

The volcanic activity of Anak Krakatau, coupled with its iconic status and the archipelago's natural beauty (Fig. 17.2), has provided a possible international tourism experience for over three decades (since Indonesia opened up to the wider world), but until recently only small numbers of westerners and international travellers visited the islands. In recent years, local people from the surrounding coastlines of Java and Sumatra have been visiting Krakatau,

Fig. 17.2 View from Owl Bay, northern Rakata, at sunrise. The islands visible, from *left* to *right* are Panjang, Sebesi (with Sebuku just visible behind it) and Sumatra (Gunung Rajabasa). © Richard Field, 2002

Fig. 17.3 Local people visiting Anak. © Richard Field, 2007

in increasing numbers (Fig. 17.3). What does this increasing interest and changing demographic mean for Krakatau as a tourism destination in the future and how does it affect its status as a protected area? After giving some background on the geology and vegetation of the islands, this chapter aims to explore the potential of Krakatau as a tourist destination and then examine the strengths and weaknesses of Krakatau's current tourism set-up—both with particular reference to the potential educational and interpretive resources associated with its geology and, particularly, its ecology. The chapter concludes with a look towards the future.

People are always seeking out new experiences and interesting stories. In the modern world of risk assessments and accident-chasing lawyers, many feel stifled and like to seek out thrills and danger, as indicated by the rapid rise in extreme sports (Buckley 2006). For these reasons and more, Krakatau provides an extraordinary opportunity for tourists. In one trip, they can visit an active volcano, see how a forest

Fig. 17.4 The charred remains of trees from 1883. **a** Burnt 1883 tree trunks uncovered by coastal erosion at the bottom of ash cliffs at the top of the beach at Zwarte Hoek, Rakata; © Richard Field, 2002. **b** A burnt 1883 tree trunk exposed by gully erosion inland on Rakata; © Richard Field, 2007

can recover from damage, learn about geology and the nature of Asian rain forest, 'feel' history (e.g. see and touch the charred remains of 1883 plants at the bottoms of ash cliffs; Fig. 17.4), witness the scale and raw power of 'nature', experience danger and excitement (especially when climbing the active crater) and be physically challenged (e.g. by climbing Rakata). Many of the views are spectacular and the islands remain almost completely free from signs of human influence. Authentic ecotourism can provide the framework for sustainable tourism, provided that appropriate planning, management and tour guiding are employed.

17.2 Geology and Eruption History

Southern and western Indonesia lie on a very active plate margin where the Indo-Australian Plate is moving towards the Eurasian Plate and being subducted under it. This forms the deep Java Trench in the ocean just to the south of Java and Sumatra. The rising magma resulting from the subduction reaches the surface in a long string of volcanoes stretching from north west Sumatra across to Flores, Komodo and other islands in the east; in terms of the volcanic legacy of this part of Indonesia, Java and Sumatra can be viewed as joined-up strings of volcanoes. Krakatau lies at a particularly unstable point in this, one of the world's most active plate boundaries. Notice in Fig. 17.1 (inset) the change in orientation of the islands (and hence the plate boundary) from northwest–southeast (Sumatra) to east–west (Java and islands to its east). This is because Sumatra is rotating clockwise, and has been doing so for millions of years: perhaps 20° in the last 2 million years (Nishimura et al. 1986). Associated with that rotation, there is thought to be considerable stretching of the plate at the location of the Sunda Strait (Huchon and Le Pichon 1984), which may explain the existence of the Sunda Strait itself. The stretching has been estimated at 70–100 km over the last 2–3 million years and is associated with a complex fault system (Sieh and Natawidjaja 2000). As a result, the Sunda Strait is very active geologically, with much seismicity and the potential for much vulcanism (Harjono et al. 1991).

Consequently, Krakatau is a very active volcano! Sometimes, however, the vent gets blocked and the volcano stays quiet for a few hundred years, during which time the material builds up in the magma chamber underneath and the pressure mounts until it explodes spectacularly, as in the 1883 eruptions. This probably also happened in the much bigger explosion of 'ancient Krakatau', an island volcano about 10–15 km in diameter and 2,000 m in altitude (Ōba et al. 1982; Williumsen 1997) that is thought to have blown up in the fifth century AD (Javanese Book of Kings, cited in Simkin and Fiske 1983). Both sets of eruptions resulted in much of the volume of the pre-existing islands being blown into the sky (Fig. 17.1), leaving large undersea calderas. Rakata, Panjang and Sertung are the surviving remnants of ancient Krakatau. Following the 1883 eruptions, about half

Table 17.1 Eruption history of Krakatau and Anak Krakatau

Krakatau		Anak Krakatau					
250AD	1150	1927–1930	1938–1940	1946	1955	1979	1996
416	1350	1931–1932	1941	1946–1947	1959	1980	1997
850	1550	1932–1934	1942	1949	1959–1963	1981	1999
950	1680–1681	1935	1943	1950	1972–1973	1988	2000
1050	1883	1936	1944	1952	1975	1992–1993	2001
		1937	1945	1953	1978	1994–1995	2007–ongoing

of the Rakata volcano survived and the submarine crater was measured as approximately 9–10 km in length and more than 250 m in depth (Ōba et al. 1982). Volcanic activity resumed in about 1919, rapidly filling up the caldera and in 1927 this material reached the sea surface and above (Ōba et al. 1982). For a few years erosion by the sea restricted the formation of the new island, but in 1930 new material was added faster than the sea could remove it and the new island, named 'Anak Krakatau' (Krakatau's child) became permanent.

The islands themselves are mostly basalt and andesite, with considerable deposits of acid pyroclastic rocks, mainly dacitic ignimbrite (Self and Rampino 1981). It has been estimated that 18–21 km^3 of pyroclastic deposits, including ash, were deposited as a result of the 1883 eruptions (Verbeek 1885; Self and Rampino 1981), much of it on and around the islands themselves.

Since 1930, Anak Krakatau has tended to erupt in a series of very frequent, relatively minor eruptions (Table 17.1). For example, throughout July 1994 there were eruptions of ash (as 'mushroom clouds') and volcanic bombs approximately every 10 min (e.g. Fig. 17.5). In 2009, from 1–25 April, 4060 explosions were recorded (Seach 2013). There have been quiet periods, such as during most of the 1980s and the period 2001–2007, when there was only minor fumerolic activity. The most recent eruptive phase began in late 2007 and has once again radically changed the appearance of the volcano's cone. The level of alert at Krakatau has been fluctuating during this time, often being at 'Waspada' (Level 2) or 'Siaga' (Level 3), out of a maximum level 4. Heavy ash-fall can be experienced on the other Krakatau islands, 5 km or more away (personally experienced many times by R.F. in July 1994). Pyroclastic flows may regularly reach 1 km or more from the crater and the volcanic bombs quite often land in the sea 1.5 km or so from the crater (both witnessed personally by R.F.). During the recent eruptive period, the Indonesian Directorate of Volcanology and Geological Hazard Mitigation has requested that people stay at least 2 km from the volcano, while local communities have still been allowed to fish outside the 2 km danger zone radius.

Explosion-type earthquakes measured during the 2009 eruptive period totalled 4,311 in August, 541 in September and 34 in October; in November 2009 the alert level was lowered from Siaga (Level III) to Waspada (Level II).

Fig. 17.5 Two eruptions of Anak within 10 min in July 1994. The second eruption is just starting, while the ash from the first is still moving up and away towards Sertung. © Richard Field, 1994

Eruptive activity commenced again in 2010 and on 29th October 2010 there were 136 eruptions at the volcano, compared to a preceding average of 10 per day, and the volcano was raised to alert level 2, again. Eruptions forced the evacuation of tens of thousands of residents on 11th January 2011. Districts affected included Kalianda, Rajabasa, Katibung, Sidomulyo, Ketapang, Sragi and Palas, all lying in Lampung province at the extreme southeast tip of Sumatra (Fig. 17.1). Ash from Anak Krakatau covered

Fig. 17.6 Two types of eruption. An ash and bomb eruption is taking place in the background, while in the foreground a lava flow from an eruption a year or two previously has split the young Casuarina forest into two patches. © Richard Field, 1994

Fig. 17.7 Both ash and volcanic bombs being ejected from Anak. The impacts of numerous bombs can be seen as puffs of ash over most of the island. © Richard Field, 1994

residential areas in parts of Banten province (western end of Java) and South Lampung and fisherman and tourists were again advised to stay at least 2 km from the volcano.

The recent volcanic activity clearly raises the question of safety for tourists visiting active volcanoes and this issue has been recently been discussed by Karkut (2010), who considered volcano hazard management and emergency planning in a tourism context. But, as in the case of many volcanoes, frequent eruptive activity is a major strength, particularly in the case of Krakatau where tourists can get views of current activity such as spectacular ash and volcanic bomb emissions, lava flows and the destructive impacts on vegetation (Figs. 17.6 and 17.7).

17.3 The Colonization of the Islands by Vegetation: Rainforest Assembly

Following the catastrophic eruption of 1883 the new substrate almost entirely comprised unconsolidated ash and was over 50 m deep over much of the archipelago (Fig. 17.8). There was extensive gully formation, resulting from rapid

Fig. 17.8 Thick ash deposits on Sertung with deeply incised gullies, stabilised by vegetation. Some of the gullies have been left hanging by coastal erosion. View of the west coast of the island. © Richard Field, 2002

Fig. 17.9 A selection of buoyant seeds of sea-dispersed plants collected in a few minutes on a beach on Panjang. For scale, an Indonesian parang is shown—an essential tool for the walk up Rakata. © Richard Field, 2007

erosion of the loose ash, before vegetation largely stabilised the gullies. There was also extensive coastal erosion; where this was not too severe, strandline communities could develop, dominated by sea-dispersed plants, mostly with large, buoyant seeds (Fig. 17.9) and this marked the onset of what Whittaker (1998, mainly drawing on Docters van Leeuwen 1936; see also Whittaker and Jones 1994) has called 'Phase 1' of plant colonisation of the islands. At the same time, open ground-dwelling ferns, grasses and composites with small, wind-dispersed seeds/spores started to grow in the interior and these plants' roots held together and stabilised the ash. Species that became prominent were *Saccharum spontaneum* (a wild sugar cane relative) and *Casuarina equisetifolia*, a tree that typifies early succession and colonises, very disturbed areas and the landward side of beaches. This species arrived in great numbers during this period and was recorded from around the islands by the mid 1890s. A very similar phase of vegetation development can be observed today on parts of Anak.

In Phase 2, which can also be observed on Anak, the same processes as in Phase 1 continued to operate, but the number of new species arriving in these ways declined because many suitable species were already present. The main feature of Phase 2 was that there was now suitable vegetation present to attract frugivores (fruit-eating birds and bats). This was partly because of stems or branches for perching and roosting, as well as some fruit becoming available from plants arriving during Phase 1. The most important frugivores were/are fruit bats (particularly *Pteropus vampyrus*, the world's largest bat, and other species in its family—the so-called 'flying foxes') and the large fruit pigeons (particularly the pied imperial pigeon *Ducula bicolor* and the green imperial pigeon *Ducula aena*). Both groups are very strong fliers and regularly fly between Krakatau and other islands, including Java and Sumatra, as well as between the Krakatau islands themselves, carrying seeds in their excreta. The imperial pigeons may also have imported moderate-sized seeds in their crops (Bush et al. 1992).

Of particular interest for attracting frugivores in the first place are species with two primary dispersal modes: sea and animal ('diplochorous' species). Their fruits or seeds can float on the sea and survive the journey to form part of the strandline vegetation, but they are also fleshy and attractive to fruit-eaters. So once they have established on the strandline and started fruiting, they attract fruit eaters (mostly birds and bats; also crabs), which disperse them inland. Typically the fruits are quite large (allowing sea dispersal), and can be carried by an animal to somewhere safe, where the fleshy parts are eaten and the (usually large) seed dropped. A good example is *Terminalia catappa* ('sea almond'), which is a large tree that is very common on the strandlines of Krakatau, noticeable to tourists because of the distinctive layered patterning of the branches and the often red leaves (during the drier season).

Key fleshy fruit-bearing plants arriving in Phase 2 included fig (*Ficus*) species and the shrub *Cyrtandra sulcata*. In the early 1900s, *Cyrtandra* developed large areas of 'wilderness' in which it completely dominated the mid-to-higher altitudes of Rakata. The many species of fig on Krakatau (at least 23 recorded on the Krakataus to date) can be considered 'keystone' species, as elsewhere (Shanahan et al. 2001). This is because figs fruit both profusely and at any time of year. Most tropical rainforest tree species fruit synchronously (all together), thus satiating seed predators during fruiting, but limiting their numbers because of starvation for the rest of the year. Fruit eating species that survive do so in large part because of the figs. Thirteen fig species were recorded on the Krakatau islands by 1920 (approximately the end of Phase 2), with some establishing as early in the colonisation process as the late 1890s. *Ficus septica* for example, was there by about 1900, which was approximately the start of Phase 2, and remains very common today, including on Anak.

Phase 3 of plant colonisation was the development of young forest covering large parts of the islands. This made the islands even more attractive to frugivores, which brought more and more animal-dispersed forest plants in their droppings—allowing the development of increasingly complex forest. There was also a new wave of establishment of wind-dispersed species. These were again typically small-seeded/spored plants, and their seeds and spores would have been arriving from the start, although there had previously been no habitat for many species such as epiphytes including arboreal ferns and orchids.

Phase 3 was really about the building-up of 'proper' forest, complete with the epiphytes and emergent trees you expect in a tropical rainforest (Fig. 17.10). However, it was and remains relatively low in biodiversity. This is partly because a substantial proportion of the plant species have gone extinct on the islands (Whittaker et al. 2000), but mainly because many plant species have still not arrived. Many more will almost certainly keep on arriving and establishing, but for some it is very difficult. Of three main types that remain absent, or nearly so, first are those dispersed by terrestrial mammals (e.g. including primates), which are themselves absent from the islands because they do not disperse well over the sea (the metabolic requirements of mammals are too high). Second are the plants dispersed by bats, but with seeds too big to swallow: bats carry them in their feet, eat the flesh and drop the seeds. The bats, however, will not carry them far in this way. The final group is wind-dispersed plants with relatively big, heavy seeds. Typically these seeds have wings helping them to be blown short distances, but very rarely as far as Krakatau. Most notable here are the dipterocarps: a family of trees that typically dominate South East Asian rainforests. These are the typical canopy emergents, sometimes reaching 90 m in places like Borneo. No dipterocarp has yet been recorded on the Krakataus. The canopy emergents on Krakatau tend instead to be figs (sometimes 60 m tall or more, now) and the dominant canopy tree over large parts of the islands is currently *Dysoxylum gaudichaudianum*, which is more of a secondary forest tree elsewhere in South East Asia. Thus, the forests of Krakatau contain some of the same species as nearby places like Java and Sumatra, but are missing whole groups—a phenomenon known as 'disharmony'.

Anak Krakatau shows elements of the above scheme, but keeps being reset by periods of volcanic activity, with the island growing at about 7 m a year. Anak comprises mostly lava which is hard and impenetrable compared to the volcanic ash that dominates the other islands and as a consequence only a few early-successional plant species are present, usually growing in cracks which allow some build-up of water and some shade from the intense heat of the sun on the black rock. The east coast of Anak has rather more mature vegetation where not covered by recent lava—the most mature being young woodland dominated by *Casuarina* and *Ficus*, plus some quite mature strandline vegetation.

Fig. 17.10 Characteristics of 'proper' forest. **a** A giant, emergent fig tree (*Ficus pubinervis*) on Rakata; Dr Harriet Hawkins for scale. **b** Epiphytic ferns in forest near the summit of Rakata. Both images © Richard Field, 2007

17.4 The Impact of Recent and Current Volcanic Activity on the Vegetation of Krakatau

The continued volcanism of Anak keeps 'resetting the system' on Anak itself by covering the existing vegetation with lava, and heavy bombardment with ash and larger ejecta. It also creates spatial patterning in the vegetation types (Thornton 2007), with young forest in patches around the east coast, far enough from the main rain of ash and boulders from the eruptions and dissected by the odd lava flow (Fig. 17.6). Ash from Anak impacts on the vegetation of the other islands, to differing degrees, both directly and indirectly. Ash-fall is a direct effect and when Anak is active, some parts of the forests can look grey, because several mm of ash covers the surfaces of the leaves. This adds a lot of weight to the branches, causing branch or tree death and also reduces photosynthetic activity, stressing the plants and can lead to widespread mortality, especially if combined with a dry spell. The build-up of ash on the ground can also encourage landslips, which seem to be very common on steeper slopes when Anak is active. It may be that there are many more lightning strikes during eruptive phases of Anak, probably resulting from ash particles in the air causing electrical charges to build up more. The lightning strikes can trigger more landslides, and also often directly kill large trees, creating sizeable gaps in the forest, which are very rapidly filled by fast-growing pioneer trees.

The way Anak influences the growth and survival of vegetation leads to many questions such as a consideration of the importance of wind direction. Certainly there are reasonably distinct forest types on the three islands – yet they are all very close together and were all sterilised at exactly the same time. Sertung and Panjang tend to receive much greater ashfall than Rakata (Whittaker et al. 1989, 1992), and there may be a seasonal bias, with wind blowing ash more over Sertung in the drier season and more over Panjang in the wetter season. This has led to speculation by researchers that differential ash-fall on the three islands has resulted in distinctly different ecological communities (Bush et al. 1992), but this is still a matter of debate (Ward and Thornton 2000).

17.5 Krakatau as a Tourism Destination

From a tourism perspective the eruptive history and geological setting of Krakatau, the colonisation of the islands by vegetation and its subsequent modification by Anak Krakatau and the prominent place of Krakatau in biogeography and conservation represent rich sources of information, that can be built into tour guide commentary and inform the content of interpretive media such as interpretive panels, pamphlets and guidebooks. Visitors who learn interesting facts and become engaged at an educational level typically show increased satisfaction with their visit, which in turn can

Fig. 17.11 A group of tourists hiking up Anak. © Richard Field, 2007

lead to further tourism, increased funding and related economic benefits (Newsome et al. 2002; Dowling and Newsome 2006; Newsome and Dowling 2010). Importantly, Krakatau serves to illustrate the close connection between geology and biology and is a model of rain forest recovery following catastrophic damage. Scientists are often accused of not sharing their findings and thoughts with the general public but educational tourism can help achieve these aims. Given the massive losses of rainforests in Asia and other equatorial regions it is possible to use Krakatau very tangibly in asking questions such as:

- Does lost tropical rain forest regenerate quickly?
- Where do the plants and animals come from and how do they arrive?
- Is the succession of the forest always different or often very similar?
- What will the re-built rain forest look like?

This can be achieved in a simple and accessible way via tourism and such questions engage the visitor more deeply with the experience of the visit. Such an approach would also add value to Krakatau's status as part of Ujung Kulon National Park in highlighting the scientific, educational and tourism importance of the islands.

This discussion raises the question of what the existing tourism profile of Krakatau is and how it is conducted. Official figures on tourist numbers do exist but are almost certainly highly inaccurate. Anecdotal evidence suggests that the current visitation to Krakatau, although increasing, is relatively low, with tourists mainly visiting beaches for swimming, snorkelling, picnicking and viewing and climbing Anak. Nonetheless, the islands are now visited on a daily basis, especially at the weekends, when as many as 50 people have been encountered on Anak at the same time (Quinn 2012). Tourists who arrive at the islands typically hike up Anak (Fig. 17.11 and see www.gunungbagging.com/anak-krakatau/), when safe to do so, and sometimes when not. Very rarely some take on the much more arduous but very rewarding climb up Rakata (see www.gunungbagging.com/rakata/)—a climb that offers the best views (Fig. 17.12) but requires a knowledgeable guide, of which there are very few.

Tour operators mostly run from Java (Anyer and Carita) and less so from Sumatra (Canti Pier, Kalianda). Bookings can be made to visit Krakatau from hotels and guesthouses (there is also promotion via the internet) but local people often visit the islands as a private arrangement with local boat owners, especially from the island of Sebesi (Dean 2012; Quinn 2012). According to both Dean (2012) and Quinn (2012) the quality of guiding is highly variable and mostly of low quality—paralleling the findings of Newsome et al. (2002) in the context of wildlife tourism in Sumatra. It is not a requirement to be knowledgeable about the islands in order to offer trips and few operators tend to be competent or offer high-quality guiding services. Most 'operators' appear to be local people who have the means to get to Krakatau (a boat), know where to land a boat and locate the start of the Anak walk trail.

Fig. 17.12 Views from the summit of Rakata. **a** View of Anak; © Richard Field, 1994. **b** View of Panjang with Sebesi, Sebuku and Gunung Rajabasa on Sumatra behind; © Richard Field, 2002

Fig. 17.13 a–c All the tourist information boards on Anak. © Richard Field, 2007

Given the lack of guidance in the form of education and sustainable tourism, and as more and more people visit Krakatau, there is a danger of degradation in the form of littering, damage to colonising vegetation, unauthorised camping and attendant impacts such as graffiti and waste disposal problems (see Newsome 2010). Quinn (2012) and Dean (2012) have noticed litter problems associated with picnicking on Anak. Bush et al. (1986) reported similar problems three decades ago, along with collection of green turtle eggs (*Chelonia midas*), the deliberate killing of these turtles, dynamite fishing, crab hunting, logging of *Casuarina* trees and the suspicion of hunting of reticulated pythons (*Python reticulatus*). Many of these latter problems—common to most protected areas in Asia (Corlett 2009)—seem to have reduced, probably as a result of increased ranger activity. There is also a strong risk of introduction of new species, affecting the natural experiment in progress. Bush et al. (1986), for example, found both chickens and pigs on the islands, and several species of fruit tree are known to have colonised the Krakatau islands via human means (Whittaker et al. 2000). Recently there has also been some illegal sand mining on Anak (*Python reticulatus*) (Jakarta Post 2009), an issue possibly connected with bureaucracy and politics: Krakatau is officially part of Lampung province (Sumatra), which therefore foots the bill for maintaining it, but most of the tourism to Krakatau, especially from the more profitable international tourists, is run from western Java.

Fig. 17.14 Buildings for use by rangers on **a** Panjang (background) and **b** Sertung. These are the only buildings on Krakatau, except for a small shelter at the trailhead on Anak. The boat in **a** is the rangers' patrol boat, with PHPA rangers on board. Both images © Richard Field, 2007

As with many other Indonesian volcanos (Newsome 2010) education/interpretation is largely missing from the tourism activity on Krakatau, with a few information boards on Anak being the sum total for most tourists (Fig. 17.13). Although there is a ranger presence on Krakatau (Fig. 17.14) this is small and mainly connected with ostensibly assessing the activity of the volcano rather than the status of Krakatau as a protected area.

Krakatau would benefit from an overall tourism strategy which addresses a number of critical areas which includes the positioning of fixed and managed landing points, stronger ranger presence and environmental management e.g. controls on random, informal camping, waste disposal and rubbish collection (Newsome et al. 2002; Smith and Newsome 2002). Most important of all, a formal system of guide training would be highly beneficial. If guides were trained in all aspects of ecotourism (volcano tourism), including safety and visitor management, and in educating tourists about the rich history, scientific significance and overall value of Krakatau to the world, many tourists would find a trip to Krakatau to be tremendously rewarding. With increased popularity, further measures would be needed, such as defined walking trails on Anak to reduce the degradation of natural values, and for safety reasons. Developing a hiking trail to the summit of Rakata could attract additional interest. Such trails need to be marked and managed according to the principles of sustainable trail management (Mende and Newsome 2006; Randall and Newsome 2008; Marion and Leung 2004).

17.6 Conclusion

Krakatau is an iconic volcano, but even so, its significance remains poorly appreciated by tourists and the general public alike. Besides the obvious volcanic activity there is a story to be told of how life has returned to the islands since the catastrophic eruptions of 1883. There is the opportunity to view the volcano as a living laboratory, as well as to indulge in the rich history of the place. Existing tourism is increasing, particularly amongst Indonesians, but the quality of this tourism activity is dubious. Krakatau has the capacity to be an iconic tourism destination, delivering high levels of tourist satisfaction, but this is dependent on appropriate management according to the principles of sustainable tourism. The island group, although a protected area, remains under threat from sand mining and the negative effects of visitation by tourists. As in the case of many natural attractions in Indonesia, environmental management and tourism activities are under-funded and poorly resourced and subject to ad-hoc interests and arrangements. The vision of Krakatau being an iconic, well-managed tourism destination is particularly dependent on well-trained guides who understand the values the volcano holds as a geo-ecological system.

Acknowledgments We would like to thank Gill Dean and Daniel Quinn for useful insights into the nature of current tourism activity on Krakatau.

References

Buckley, R. (2006). *Adventure tourism*. Oxford: CABI Publishing.

Bush, M., Jones, P. A. & Richards, K. (1986). Conservation of Krakatau. In M. Bush, P. A. Jones & K. Richards (Eds.), *Krakatoa centenary expedition: Final report*. University of Hull, Department of Geography, Miscellaneous Series No. 33.

Bush, M. B., Whittaker, R. J., & Partomihardjo, T. (1992). Forest development on Rakata, Panjang and Sertung: Contemporary dynamics (1979–1989). *GeoJournal, 28*, 185–199.

Corlett, R. (2009). *The ecology of Tropical East Asia*. Oxford: Oxford University Press.

Cox, C. B., & Moore, P. D. (2010). *Biogeography: An ecological and evolutionary approach* (8th ed.). USA: John Wiley & Sons Inc.

Dammerman, K. W. (1948). The fauna of Krakatau, 1883–1933. Kon. Ned. Akad. Wet., *Verhandelingen* (12th Section), *44*, 1–594.

Dean, G. (2012). *Personal communication*. Canada: Research associate, Botany Department, University of British Columbia.

Docters van Leeuwen, W. M. (1936). Krakatau 1883–1933. *Annales du Jardin Botanique de Buitenzorg, 46–47*, 1–506.

Dowling, R., & Newsome, D. (Eds.). (2006) *Geotourism*. Oxford: Elsevier/Heineman Publishers.

Harjono, H., Diament, M., Dubois, J., Larue, M., & Zen, M. T. (1991). Seismicity of the Sunda Strait: Evidence for crustal extension and volcanological implications. *Tectonics, 10*, 17–30.

Huchon, P., & Le Pichon, X. (1984). Sunda Strait and central Sumatra fault. *Geology, 12*, 668–672.

Jakarta Post. (2009). Sand quarrying on Mount Anak Krakatau protested. Retrieved April 12, 2012, from www.thejakartapost.com/news/2009/10/23/sand-quarrying-mount-anak-krakatau-protested.html.

Karkut, J. (2010). Reconsidering the boundaries and applications of geotourism—lessons learnt from tourism at Mount Vesuvius. In D. Newsome & R. Dowling (Eds.), *Geotourism: The tourism of geology and landscape*. Oxford: Goodfellow Publishers Limited.

MacArthur, R. H., & Wilson, E. O. (1967). *The theory of island biogeography*. Princeton: Princeton University Press.

Marion, J. L., & Leung, Y. F. (2004). Environmentally sustainable trail management. In R. Buckley (Ed.), *Environmental impacts of tourism*. Canberra, ACT: CAB International.

Mende, P., & Newsome, D. (2006). The assessment, monitoring and management of hiking trails: a case study from the Stirling Range National Park, Western Australia. *Conservation Science Western Australia, 5*(3), 27–37.

Newsome, D. (2010). The need for a planning framework to preserve the wildness of Sibayak Volcano, north Sumatra. In P. Erfurt-Cooper & M. Cooper (Eds.), *Volcano and geothermal tourism: Sustainable geo-resources for leisure and recreation*. London: Earthscan.

Newsome, D., Moore, S., & Dowling, R. (2002). *Natural area tourism: Ecology, impacts and management*. Clevedon: Channel View Publications.

Newsome, D., & Dowling, R. (Eds.). (2010). *Geotourism: The tourism of geology and landscape*. Oxford: Goodfellow Publishers Limited.

Nishimura, S., Nisida, J., Yokoyama, T., & Hehuwat, F. (1986). Neotectonics of the Strait of Sunda, Indonesia. *Journal of Southeast Asian Earth Sciences, 1*, 81–91.

Ōba, N., Tomita, K., Yamamoto, M., Istidjab, M., Badruddin, M., & Parlin, M. et al. (1982). Geochemical study of lava flows, ejecta and pyroclastic flow from the Krakatau Group, Indonesia. Reports of the Faculty of Science, Kagoshima University (Earth Sciences and Biology) 15, 41–76.

Olson, D. W., Doescher, R. L., & Olson, M. S. (2004). When the sky ran red: the story behind the "Scream". *Sky & Telescope, 107*, 28–35.

Quinn, D. P. (2012) Personal communication. Lead writer, Gunung Bagging. www.gunungbagging.com.

Randall, M., & Newsome, D. (2008). Assessment, evaluation and a comparison of planned and unplanned walk trails in coastal south-western Australia. *Conservation Science Western Australia, 7*(1), 19–34.

Seach, J. (2013). Krakatau Volcano—2009 Eruptions. Retrieved August 5, 2013, from www.volcanolive.com/krakatau.html.

Self, S., & Rampino, M. R. (1981). The 1883 eruption of Krakatau. *Nature, 294*, 699–704.

Shanahan, M., So, S., Compton, S. G., & Corlett, R. (2001). Fig-eating by vertebrate frugivores: A global review. *Biological Reviews, 76*, 529–572.

Sieh, K., & Natawidjaja, D. (2000). Neotectonics of the Sumatran fault, Indonesia. *Journal of Geophysical Research, 105*, 28295–28326.

Simkin, T., & Fiske, R. S. (1983). *Krakatau 1883: The volcanic eruption and its effects*. Washington: Smithsonian Institution Press.

Smith, A. J., & Newsome, D. (2002). An integrated approach to assessing, managing and monitoring campsite impacts in Warren National Park, Western Australia. *Journal of Sustainable Tourism, 10*(4), 343–359.

Symons, G. J. (Ed.). (1888). *The eruption of Krakatoa and subsequent phenomena*. Report of the Krakatoa Committee of the Royal Society. London: Trubner & Co.

Thornton, I. W. B. (1996). *Krakatau: The destruction and reassembly of an island ecosystem*. Cambridge: Harvard University Press.

Thornton, I. W. B. (2007). In T. New (Ed.), *Island colonization: The origin and development of island communities*. Cambridge: Cambridge University Press.

Verbeek, R. D. M. (1885). *Krakatau*. Batavia, Java: Lansdrukkerij.

Ward, S. A., & Thornton, I. W. B. (2000). Chance and determinism in the development of isolated communities. *Global Ecology and Biogeography Letters, 9*, 7–18.

Whittaker, R. J. (1998). *Island biogeography: Ecology, evolution, and conservation*. Oxford: Oxford University Press.

Whittaker, R. J., Bush, M. B., & Richards, K. (1989). Plant recolonization and vegetation succession on the Krakatau Islands, Indonesia. *Ecological Monographs, 59*, 59–123.

Whittaker, R. J., Field, R., & Partomihardjo, T. (2000). How to go extinct: Lessons from the lost plants of Krakatau. *Journal of Biogeography, 27*, 1049–1064.

Whittaker, R. J., & Jones, S. H. (1994). Structure in re-building insular ecosystems: an empirically derived model. *Oikos, 69*, 524–530.

Whittaker, R. J., Walden, J., & Hill, J. (1992). Post-1883 ash fall on Panjang and Sertung and its ecological impact. *GeoJournal, 28*, 153–171.

Williumsen, P. (1997). *Krakatau—Events and geology. A practical guide to Krakatau and surroundings*. Indonesia: Indonesian Petroleum Association.

Winchester, S. (2003). *Krakatau. The day the world exploded*. London: Viking (Penguin Books, Ltd).

18 Volcanic National Parks in Japan

Malcolm J. M. Cooper

18.1 Introduction

Volcanic tourism destinations are the theme of this book, and the National Parks on the islands of Japan are a tourist mecca in this regard, with many active sites, geothermal fields and their related phenomena such as volcanic hot springs as attractions. While the central focus of this chapter is the island of Kyushu it also makes reference to the more important volcanic tourism destinations on the islands of Honshu and Hokkaido. These destinations are found in National Parks in Japan, which have a history of over 80 years and were created to preserve the nation's representative natural environments. They are also intended as recreational resources for tourists and locals, as educative centres, and as risk management research centres. 29 have been designated to date.

Parks as destinations are managed through a system of zoning based on the characteristics of their natural environments and scenic values; where the most strictly protected area is a 'Special Protection Zone', the main protected area is labelled 'Special Zone', and there is usually a buffer zone known as the 'Ordinary Zone' (Ministry of the Environment 2013). Tourist activities and infrastructure that might affect these zones require permits, from Local Government or from the Ministry of the Environment. A Park Ranger system deals with day-to-day activities such as interpretation for visitors, promoting public awareness, and risk management.

M. J. M. Cooper (✉)
Ritsumeikan Asia Pacific University, Beppu, Japan
e-mail: cooperm@apu.ac.jp

18.2 Geology and Tourism

18.2.1 The Geological Background

Japan is an archipelago of 6,852 islands with a total land area of 377,944 km^2, geographically located from 24 to 46° north latitude and from 123 to 146° east longitude in East Asia. Hokkaido is the most northerly of Japan's main islands, followed by Honshu, Shikoku and Kyushu. The country is tectonically very active due to extreme crustal instability and is home to approximately 108 active volcanoes (Japan Meteorological Agency (JMA) 2010), and many more dormant ones (Fig. 18.1). This single fact means that the countryside of Japan is an important geotourism destination (Dowling and Newsome 2006), which generates some 150 million, mainly domestic, visits per year (JNTO 2012) to the volcanoes, hot springs and related geothermal activity and facilities. On Honshu, the largest island, volcanic landforms make up much of the environment of the mountainous areas known as the Japan Alps, while hot springs and other geothermal phenomena are to be found all over the Island. Central Japan is in fact marked by the convergence of three mountain chains that form the Japan Alps, several of whose peaks are higher than 3,000 m. The highest point in the country is Mount Fuji (Fujisan), a volcano dormant since 1707 that rises to 3,776 m above sea level just 70 km from Tokyo. Shikoku, Kyushu and many of the outlying islands are equally mountainous and geothermally active.

Of the by-products of this geological situation, earthquakes and hot springs are the most common. Japan has experienced about 50,000 earthquakes since 1997 (Table 18.1), and magnitudes of 1–3 on the Shindo Scale (www.hinet.bosai.go.jp/-about_earthquake/part1.html) are common. Minor tremors occur daily in one part of the country or another (up to 400/day) but major earthquakes do fortunately occur infrequently. The most well-known in the twentieth century were the great Kantō (Tokyo) earthquake of 1923, in which 130,000 people died and the Great Hanshin (Kobe) earthquake of 1995 where the death toll was about 6,500.

P. Erfurt-Cooper (ed.), *Volcanic Tourist Destinations*, Geoheritage, Geoparks and Geotourism,
DOI: 10.1007/978-3-642-16191-9_18,

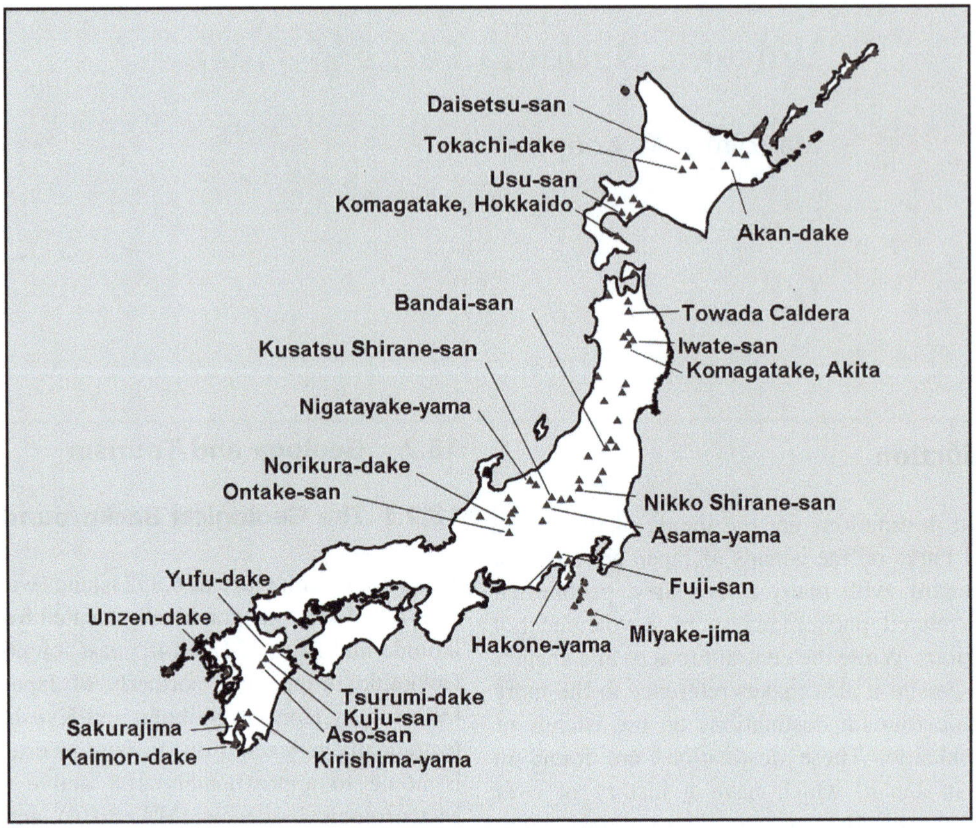

Fig. 18.1 Distribution of the major active volcanoes in Japan. *Source* Geological Survey of Japan/AIST—Map modified by Patricia Erfurt-Cooper

Table 18.1 The number of earthquakes of magnitude 1–7 on the Japanese Shindo Scale 1997–2011 (those of 5 and 6 are divided into weak/strong)

震度	1	2	3	4	5弱	5強	6弱	6強	7	計
1997年	1,188	434	129	30	5	4	1	–	–	1,791
1998年	999	374	118	27	1	–	1	–	–	1,520
1999年	641	278	81	20	3	–	–	–	–	1,023
2000年	12,211	3,975	1,133	312	30	8	6	1	–	17,676
2001年	1,014	352	110	28	5	3	1	–	–	1,513
2002年	821	309	95	24	4	–	–	–	–	1,253
2003年	1,344	573	191	63	2	–	5	1	–	2,179
2004年	1,317	612	212	66	11	11	2	2	1	2,234
2005年	1,055	486	121	38	4	5	2	–	–	1,711
2006年	862	340	113	25	3	–	–	–	–	1,343
2007年	1,351	520	170	48	4	2	1	2	–	2,098
2008年	1,216	483	163	34	6	–	1	1	–	1,904
2009年	1,066	399	124	36	3	–	1	–	–	1,629
2010年	883	294	99	32	5	–	–	–	–	1,313
2011年	6,365	2,833	998	255	45	17	4	4	1	10,522
2012年	2,010	816	232	65	12	4	–	–	–	3,139
計	34,343	13,078	4,089	1,103	143	54	25	11	2	52,848

Source Japan Meteorological Agency (www.jma.go.jp/en/quake/ and www.hinet.bosai.go.jp/-about.earthquake/part1.html)
The Shindo scale is considered to be more precise as additional factors are taken into account compared to the Richter scale or the Mercalli scale

Undersea earthquakes also expose the Japanese coastline and coastal parks to danger from tsunamis. On March 11, 2011 the country was subject to a devastating magnitude 9+ (Richter scale) earthquake followed by a massive tsunami. This earthquake was the largest ever recorded in Japan and was the world's fourth largest earthquake since 1900 (U.S. Geological Service 2011). At least 100 aftershocks registering at magnitude 6.0 or higher followed the main earthquake, and at least 16,000 people died as a result of it, mainly in the tsunami that travelled up to 10 km inland. As with the on-shore ones, undersea earthquakes cannot be accurately predicted, but as a result of these disasters Japan has become a world leader in research on the causes and prediction of earthquakes, as well as on risk management policies and practices for them. The development of advanced technology has permitted the construction of buildings even in earthquake-prone areas, and extensive civil defence efforts focus on training in protection against earthquakes and tsunamis. These are some of the important tourism considerations and resources that this chapter discusses.

18.2.2 The Tourism Background

On the positive side there is a geothermal by-product that has made Japan famous in geotourism terms, its *onsen*. Contributing probably at least 90 % of the recorded annual tourism trips in the Country, the onsen, or volcanic hot springs, have a myriad of uses ranging from simple bathing (the majority) to an important resource in the field of health and wellness and medical tourism (Erfurt-Cooper and Cooper 2009). Traditional bathhouses are frequented regularly as a means of socialising with friends, family and neighbours, while soaking in natural hot spring water and relaxing, and the tourist has similar experiences and access. Preferred for accommodation in association with many of the more famous hot springs are the traditional Japanese inns called *Ryokan*, which use their onsen as the main attraction, combined with beautiful garden settings, regional culinary specialties, local attractions and unique landscapes. This combination of volcanic hot springs with the natural environment draws domestic visitors from the corporate sector (collective getaways for social bonding) to families and couples of all ages.

Tracing the development of tourism and of the National Parks system for this discussion shows that, at the beginning of the Meiji era modern tourism became possible when the travel restrictions of the preceding Edo period (Cooper et al. 2008) were lifted. In 1869 the Meiji administrators declared that passports were no longer necessary for domestic travel by Japanese citizens (Ashiba 1994), even though they remained obligatory for foreigners, since they were a convenient means of controlling and monitoring the movement and duration of stay in Japan. In addition, it was recognized that the building of infrastructure for domestic and international travel was a priority.

As with European and North American tourism, the railways contributed much to the growth of this infrastructure in Japan—in the nineteenth century, private railways based their routes on existing tourist attractions like shrines, temples and *onsen* to secure basic demand. In the early years of the twentieth century, railways were associated with hotels and department stores at their urban termini, suburbanisation along the line, and recreational facilities and tourist attractions at the rural end of the line. A third phase of development coincided with the establishment of the National Park system in 1931–1932. Tourism to areas of outstanding natural beauty became popular and required the large-scale development of cottages, hotels and leisure facilities (Funck and Cooper 2013) at entrances or within them. In the Tokyo area in the 1920s the railway companies of Seibu and Tôkyû fought a development war over the Hakone (Mt Fuji) area, a major national park and tourist destination in close proximity to the capital (Havens 1994, p. 26).

The advent of the National Park system and the ease of using rail to travel throughout Japan also attracted international visitors (from North America and Europe). The famous Fujiya Hotel in Hakone had up to 19,000 guests per year by 1915 (Funck and Cooper 2013). By 1936, spending by foreign visitors represented Japan's fourth largest source of foreign-exchange revenue after cotton, raw silk and silk products (Leheny 2003, p. 61). When it became publicly known that the government was preparing a national park system during the 1920s, regions across the country competed to suggest sites, just as they tried to attract railway lines, as it was thought both would bring gains to regional economies from tourism. All in all, sixty sites were proposed to Parliament between 1920 and 1931; famous sites like Mount Fuji several times. However, the initial twelve sites eventually chosen in 1932 (and many of the subsequently added National Parks and Quasi-National Parks) concentrated on mountains, gorges and forests, leaving out famous coastal landscapes except for the Seto Inland Sea. This was because of the fascination with geothermal landscapes in the form of volcanoes and onsen—and subsequent concern that such destinations were in danger of being affected by too many visitors. Although the numbers of international tourists from all sources declined after World War II, domestic tourism to these geotourism destinations remained very high (Erfurt-Cooper and Cooper 2010).

In recent years though Japanese National Parks, hot springs and their volcanic environments have again developed into a major tourist attraction for overseas visitors, but now those from China, Taiwan and Korea (Funck and Cooper 2013). This source of visitors is on a rapid growth curve at the moment; in November 2013 the monthly total of Chinese visitors was 102,000, and the total number of visitors to Japan reached 10 million in 2013, considerably above the previous record of 8.6 million set in 2008 (Japan Tourism Agency 2013). The decline in the value of the Yen and the close proximity of China and Korea to Japan have proved to be stronger attractors than any detracting disputes over territory or resources in the seas between them.

18.2.3 The Volcanoes

The volcanoes and their associated National Parks (over 5 % of the Country's land area) are the tourism resource that a majority of both domestic and in-bound travellers come to visit. There are over 1,500 volcanoes classed as active in the world at the present time (Siebert et al. 2010), and Japan is home to nearly 10 % of them. The reason for the high activity in the Japanese archipelago is that four tectonic plates meet each other, the Eurasian Plate, the Philippine Sea Plate, the Pacific Plate and the North American Plate. The Pacific Plate is subducted underneath the Philippine Plate and the North American Plate (convergent boundaries), while the Philippine Plate margins are subducted under the Eurasian Plate (GLGArcs 2013). This means that the country is prone to volcanic eruptions and earthquakes as a result of tectonic plate movement and shifting boundaries. Volcanoes in Japan are distributed in the Western Japan volcanic belt (runs from Kyushu to Chugoku (near Kyoto), and along the Tokara Islands towards Okinawa) and the Eastern Japan volcanic belt (includes the Kuril Arc towards Kamchatka, Hokkaido and NE Honshu and south out along the Izu-Bonin Arc towards Guam). There is a gap of some 150 km between the two major arcs around the central Honshu area of Osaka-Kyoto and Lake Biwa.

The JMA defines active volcanoes as those that have erupted in the past 10,000 years or currently have fumarolic activity, and is continuously monitoring the activities of at least 20 of these using seismic and visual observations. These more active volcanoes include Meakandake, Tokachidake, Tarumae, Usu, Hokkaido-Komagatake, Azuma, Adatara, Bandai, Nasu, Kusatsu-Shirane, Asama, Ontake, Izu-Tobu, Izu-Oshima, Miyakejima, Kuju, Aso, Unzen, Kirishima, and Sakurajima (see Table 18.2 and Fig. 18.1). International recording and monitoring systems such as the *Global Volcanism Program* database built by the Smithsonian Institute of the USA for Japan currently include 20 volcanoes with thermal activity on Hokkaido, 47 on Honshu, 9 on Kyushu, and 43 for the rest of the Japanese archipelago (Smithsonian Institution 2013).

The other volcanoes listed are surveyed regularly but not continuously by the JMA, and such surveys are now carried out at the Iwatesan, Akita-Komagadake, Fuji, Satsuma-Iwojima, Kuchinoerabujima, and Suwanosejima volcanoes, in recognition that even those volcanoes in Japan that have not shown signs of activity in the recent past are merely dormant, and local communities and their visitors require protection should a major event occur. If some abnormal phenomena are detected, temporary observation stations are installed and the data are constantly monitored.

18.3 The Major Volcanoes of Hokkaido and Honshu

This section covers the volcanoes of Hokkaido and Honshu and their associated national parks (Table 18.2). The following section covers Kyushu. For each island in this section a single case study provides sufficient information to elucidate the main features of volcano tourism to National Parks in that area.

18.3.1 Mt Usu, Shikotsu-Toya NP, Hokkaido

Shikotsu-Toya (Fig. 18.2) was designated a national park (983 km^2) in 1949. Principal attractions are the scenery created by caldera lakes and active volcanoes. Its location close to Sapporo means popularity as a recreational destination; for some 15 million tourists annually. Mt Usu and Showashinzan are the most active of the volcanoes, with 4 major eruptions in the last 100 years (last in 2000). There is a crater trail walk that has as an added attraction in the form of the results of eruptions—buried houses and onsen facilities, damaged cars and buses, and displaced bridges. There is also a cable car to the rim on the east side of Mt Usu, and one of the most famous onsen in Hokkaido, Noboribetsu is found in this area. The inclusion of the Toya Caldera and Usu Volcano into the Global Geoparks Network was officially announced in 2009.

18.3.2 Mt Fuji, Fuji-Hakone-Izu NP, Honshu

Mount Fuji is Japan's highest mountain at 3,776 m and most frequented tourist attraction (Fig. 18.3). The last eruption of this volcano was in 1707, but the view of the National Research Institute for Earth Science and Disaster Prevention (Japan Times, 7 September 2012) is that there may be some more volcanic activity in the next few years since the March 11, 2011 Tohoku earthquake has affected the pressure in Fuji's magma chamber.

Table 18.2 Some examples of popular volcanic tourist attractions in Japanese National Parks

Volcano	Type	Location	Last known eruption
Hokkaido			
Usu	Stratovolcano	Shikotsu-Toya NP, GGP	2001
Showa-Shinzan	Lava Dome	Shikotsu-Toya NP, GGP	1944/45
Daisetsuzan (Mt Asahi)	Stratovolcanoes	Daisetsu NP	1739
Meakan	Akan Stratovolcano	Akan NP	2006
Oakan	Caldera	Akan NP	2008
Rausu	Stratovolcano	Shiretoko NP, WHA	1800
Honshu			
Fuji-san	Stratovolcano	Fuji-Hakone-Izu NP, WHA	1708
Kiso Ontake	Stratovolcano	Nth Japan Alps (Pref Park)	1980
Haku-san	Stratovolcano	Hakusan NP	1659
Yakeyama	Stratovolcano	Itoigawa Geopark	1997
Tateyama	Stratovolcano	Chūbu-Sangaku NP	1839
Asama	Complex volcano	Joshin'etsukogen NP	2009
Kusatsu-Shirane	Stratovolcano	Joshin'etsukogen NP	1983
Bandai	Stratovolcano	Bandai-Asahi NP	1888
Chokai	Stratovolcanoes	Chokai Quasi-NP	1974
Akita-Komaga-Take	Stratovolcanoes	Towada-Hachimantai NP	1971
Iwate	Complex volcano	Towada-Hachimantai NP	1919
Akita-Yake-Yama	Stratovolcano	Towada-Hachimantai NP	1997
Iwaki	Stratovolcano	Tsugaru Quasi-NP	1863
Hakkōda Volcanic Group	Stratovolcanoes	Towada-Hachimantai NP	1550
Osore-Yama	Stratovolcano	Towada-Hachimantai NP	1787
Kyushu			
Aso	Stratovolcano	Aso-Kuju NP	Continuous
Kuju	Stratovolcano	Aso-Kuju NP	1996
Tsurumi	Lava domes	Aso-Kuju NP	867
Unzen	Stratovolcano	Unzen-Amakusa NP, GGP	1997
Sakurajima	Stratovolcano	Kirishima-Yaku NP	Continuous
Kirishima (Shinmoedake)	Shield Volcano	Kirishima-Yaku NP	2011
Kaimondake-Ibusuki	Volcanic Field	Kirishima-Yaku NP	885

Source compiled by the Author from various sources

Mt Fuji is considered one of the 3 Holy Mountains (along with Mount Tate and Mount Haku) of Japan and is a stratovolcano about 70 km south-west of Tokyo, with an exceptionally symmetrical shape making it into a famous symbol of Japan and an important element in Japanese art. It is a popular destination for excursions. More than 300,000 people climb to the top every year, while some 105 million visitors to the mountain and its surrounding region (Hakone) are recorded annually (Funck and Cooper 2013).

18.3.2.1 Risk Management for Mt Fuji

It is estimated by the National Research Institute for Earth Science and Disaster Prevention that a new eruption of Mount Fuji could force some 567,000 people to evacuate their homes (in the worst case scenario). Two of the Prefectures surrounding the mountain are working on evacuation plans; the Shizuoka Prefectural Government estimates that more than 130,000 people from approximately 50,000 households would have to relocate if lava were to reach the city of Fuji located just south of the 3,776 m mountain. Depending on the number and locations of vents, evacuation from wider areas could be needed. Under this evacuation plan, 10 zones have been designated at the foot of the mountain based on the expected direction of lava flows, and the plan specifies evacuation zones in four grades. The Yamanashi Prefectural Government is preparing a similar plan, and each will be finalized at a consultation forum involving Shizuoka, Yamanashi and Kanagawa Prefectures

Fig. 18.2 Lake Tōya is part of the volcanic Shikotsu-Toya National Park. In the background covered in clouds the volcano Mount Yotai is just visible. The lake can be best viewed from the active volcano Mount Usu which is accessible by a ropeway to the summit. *Photo* Patricia Erfurt-Cooper

Fig. 18.3 View of Mt Fuji from Hakone-En. *Photo* Patricia Erfurt-Cooper

and the Central Government. Joint evacuation drills in the three prefectures will also be conducted (Japan Times, 9 September 2012), although the impact on the large number of tourists usually in the area is not formally being discussed.

18.4 The Major Volcanoes of Kyushu

Kyushu has a number of active volcanoes (Table 18.3 and Fig. 18.4), some of which are constantly erupting but generally maintain a low level of activity except for Sakurajima's series of stronger eruptions. All of these volcanoes and volcanic complexes are tourist destinations, are surrounded by National or Prefectural parklands, and have more or less developed facilities for access and accommodation in close proximity to the volcanic area. All of these active volcanoes are also located close to large cities (compare Table 1.9) and have a rather recent eruption history as Table 18.3 shows. Mt Unzen and Mt Sakurajima were included in the list of 'decade volcanoes' some years ago to be specifically studied and monitored through the International Association of Volcanology and Chemistry of the Earth's Interior (IAVCEI).

18.4.1 Mt Unzen and the Western Area

Mount Unzen is one of the 16 decade volcanoes across the world identified by the IAVCEI as being worthy of

Table 18.3 Active volcanic complexes in Kyushu

Volcano	Elevation (m)	Last erupted	Nearby cities
Unzen Stratovolcano complex	1,500	1997	Kumamoto (670,000 inhabitants)
Mount Aso, Aso-Kuju national park	1,529	Continuous	Kumamoto (670,000 inhabitants) and many villages in and around the caldera
Mount Kuju, Aso-Kuju national park	1,788	1996	Kumamoto, Oita, Kokonoe
Tsurumi Lava dome complex	1,584	867	Beppu, Oita
Kirishima shield volcano	1,700	2011	Kagoshima (600,000 inhabitants)
Shinmoedake shield volcano		2011	Kagoshima
Sakurajima	1,117	Continuous	Kagoshima
Kaimondake-Ibusuki volcanic field	922	885	Kagoshima
Fukujima stratovolcano	Undersea	550 BCE	–

Source Smithsonian Institute and the Japan meteorological agency

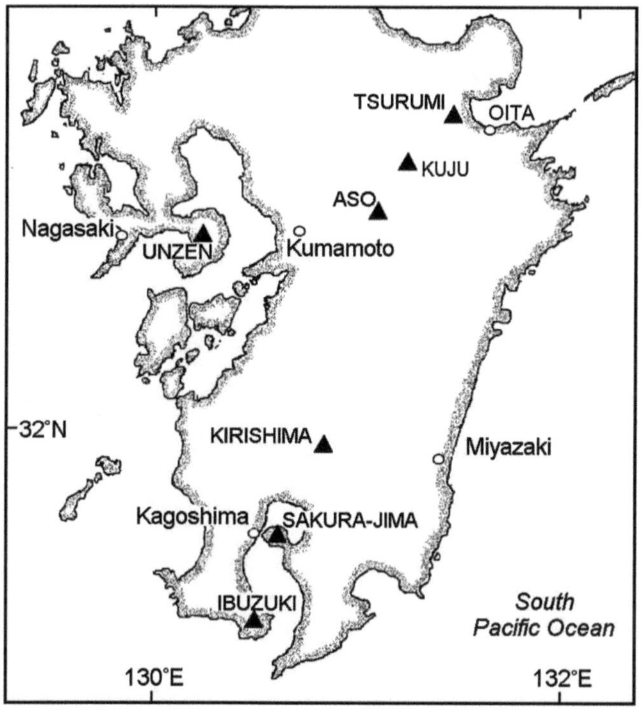

Fig. 18.4 Kyushu's Active Volcanic Centres. *Source* www.volcano.si.edu, modified by Patricia Erfurt-Cooper

particular study in light of their history of large, destructive eruptions and proximity to populated areas, and is part of the Unzen-Yamakusa National Park (Yamamoto et al. 1993; Newhall 1996). Since August 2009 it has also been recognized by UNESCO as the Unzen Global Geopark.

Mount Unzen is part of the Shimabara Peninsula and close to the city of Nagasaki, and been the site of extensive volcanism over thousands of years (Uto et al. 2001). Activity from 150,000 years ago to the present has occurred at a number of sites around the volcanic complex of four main domes at different times: Nodake (70–150,000 years old), Myōkendake (25–40,000 years old), Fugendake (<25,000 years old) and Mayuyama (4,000 years old). Fugendake has been the site of most of the eruptions during the past 20,000 years and lies about 6 km from the center of the city of Shimabara. Unzen's deadliest eruption occurred in 1792, with a large dacitic lava flow coming from Fugendake and a post-eruption earthquake, which caused a mega-tsunami that reached a height of 100 m, and killed an estimated 15,000 people. As of 2013 this is the highest volcanic related death toll in Japan.

After 1792, the volcano remained dormant until 1990 when the threat of further disastrous events prompted the authorities to evacuate 12,000 local residents from their homes. On June 3, 1991, the volcano erupted violently. A pyroclastic flow reached 4.5 km from the crater and claimed the lives of 43 scientists and journalists (Erfurt-Cooper 2010). Between 1991 and 1994 the volcano generated at least 10,000 small pyroclastic flows, destroying about 2,000 houses. Since then heavy rains have frequently re-mobilized pyroclastic material, generating lahars. Dikes have been constructed in several river valleys to channel lahar flows away from vulnerable areas, and warning systems and evacuation plans have been developed and deployed.

18.4.1.1 The Mt Unzen Geopark

The theme chosen by the managers of the Unzen Volcanic Area Global Geopark (460 km^2, designated in 2009) is "the coexistence of an active volcano and human beings" (JNTO 2012). The area surrounding Unzen volcano was designated as the first national park in Japan in 1934 and its natural environment has been conserved since then (Fig. 18.5). The Unzen Geopark contains 150,000 inhabitants, outstanding landscapes, volcanic hot springs, fresh spring water, fertile agricultural soil and supports many types of tourism.

On the other hand, Unzen volcano has erupted in the past and has caused serious disasters, which have damaged both local tourism and the economy. However, the residents of the area have shown the ability to recover from these volcanic disasters, as has tourism.

Fig. 18.5 Mt Unzen and its lava dome viewed from downtown Shimabara. *Photo* Patricia Erfurt-Cooper

Fig. 18.6 The Aso Caldera. *Source* Google Earth

18.4.2 Mt Aso and the Central-Northern Area

One of the most popular volcano complex destinations in Japan is the Mt Aso Caldera (Fig. 18.6), which is located in the centre of Kyushu in the Prefecture of Kumamoto, and is part of the Aso-Kuju National Park. There are 5 peaks within the caldera (Nekodake, Takadake, Nakadake, Eboshidake, and Kishimadake) and Mt Kuju lies just to the north. The highest point is Mt. Taka, at 1592 m above sea level, and the presently active crater is located on Mt. Naka. This vent continuously emits gas and has occasional eruptions. The area of Mt Aso is continuously monitored in real-time at the Aso Volcano Observatory, which shares a building with the Aso Volcano Museum and a tourist complex with shops and restaurants located a little over 2 km away. Hazard warnings are continuously broadcast in several languages to warn visitors of any changes in the conditions of the active crater.

The caldera contains the city of Aso (population 29,000) as well as agricultural land and onsen resources. It has a circumference of around 120 km, with a diameter of 25 km north to south. Nakadake Crater is one of the few places in the world where tourists can drive up to the summit and approach the edge of an active crater.

Mt Aso draws large crowds on any day with reasonable weather and favourable winds that keep the toxic gas

Fig. 18.7 The active crater (Mt Nakadake) at Mt Aso. *Photo* Patricia Erfurt-Cooper

Fig. 18.8 The Kuju Dome Complex, a group of active stratovolcanoes is located within the boundaries of the Aso-Kuju National Park. *Photo* Patricia Erfurt-Cooper

emissions from the crater away from the visitors (JNTO 2012). This volcano is unique as it is not just classed as one of the most dangerous volcanoes in the world, but it also has access to the summit not just via a toll road, but also by ropeway to cater for the mass tourism that takes place here on a regular basis. A number of other attractions such as hot spring resorts, are linked to the volcano, and recreational activities are combined with visits to the crater, for example horse riding, hiking, trekking, and bird watching. Souvenir shops, restaurants and the Aso Volcano Museum are located at a short distance from the volcano and can be accessed even if the crater area is closed (Fig. 18.7).

The Mt. Kuju volcano complex (Fig. 18.8) to the north is also part of the Aso-Kuju National Park and consists of 20 lava domes and cones. The Mt Kuju area also has one of the few geothermal power stations in Japan (Hatchobaru, the largest plant (110 MW) in Japan). It is also a site where sulphur used to be mined (Nabakoh et al. 2003).

18.4.2.1 The Aso-Kuju Geopark

Mt Aso became a member of the Japanese Geoparks Network (JGN) in October 2009. It was designated a National Park in 1934. An application to the Global Geoparks Network has been made for global status (2013). The major landform, the caldera was created between 270,000 and 90,000 years ago and more than 50,000 people now live within it, while approximately 1 million people visit the area in the course of a year. 33 specific Geosites form the main reason for the area becoming a Geopark (Table 18.4). Several onsen complexes are included in these sites, the largest being Uchinomaki with some 80 springs and more than 20 accommodation facilities. The agriculture of the caldera is grassland farming, in fact the largest continuous area of such farming in Japan. Horse riding schools, walking trails, and a local agricultural zoo also complement the geotourism resources. These are supplemented by additional resources based on the town of Kokonoe in the foothills of Mt. Kuju, which is a popular place for weekend houses and country-side retreats.

People come from all over Japan to the area around Mt. Kuju and Mt. Aso to enjoy a different lifestyle for either a few days or weeks or buy a second home to retire to. An active volcano is no deterrent at all; the many volcanic hot springs with their waters rich in beneficial minerals hold an enormous attraction for domestic tourists. The Ryokan owners in the area often use one of the nearby volcanoes in the background of their hot spring inn or resort marketing material, and also use the volcano research facilities to verify the quality of the local onsen water.

18.4.3 The Tsurumi Dome Complex and the Eastern Area

The Tsurumi dome complex is also part of the Aso-Kuju National Park, and an active volcano (1,375 m) located on the east coast of Oita Prefecture. The mountain was formed in the Beppu-Shimabara graben zone in the same manner as Mt. Unzen, Mt. Kuju, and Mt. Yufudake. In this area both

Table 18.4 Geosites in the Aso-Kuju geopark

Geosite	Theme	Geosite resources
Daikanbo caldera Geosite	Caldera of world scale	Aso Caldera, Daikanbo, Akahoya volcanic ash layer
Nakadake Geosite	The active volcano	Nakadake Crater, Sunasenri-ga-hama
Kusasenri Geosite	Grazing land and volcanic activity remnants	Aso Volcanic Museum, Kusasenri-gahama observation station, Mt. Eboshi
Komezuka Geosite	2,000 yr old cinder cone	Komezuka, Kami-Komezuka
Furubochu Geosite	Volcanic activity and sacred mountain	Shrine
Sensuikyo Gorge Geosite	Rock peak and alpine plants	Sensuikyo, Takaoka
Kishimadake Geosite	Last activity 3000 to 4000 years ago	Kishimadake (Mt. Death)
Fumotobochu Geosite	Cityscape of the history and culture	West Iwao temple, Museum of Aso rural district
Ogidake Geosite	Isolated peak in pyroclastic flow plateau	Ogitake
Koganotaki Falls Geosite	Waterfall in the outer rim	Waterfall
The Deity of the Volcano Geosite	The roots of volcano-based Aso culture	Aso-jinja Shrine, Kokuzoshinsha old burial mound group, Frost Shinto shrine
Cluster of Springs in Aso Valley District Geosite	Alluvial fan and pressured underground water	Miyaji, Yakuinbaru district springs
Futaenotouge Mountain Pass Geosite	Myth and history	Double peak daimyo's alternate-year residence, Matoishi Ochiyaya trace artificial pond
Aso Yellow Soil (Aso Odo) Geosite	Culture, blessing of great volcano to environment	Aso yellow ocher
Uchinomaki Hot Spring Geosite	Hot springs and regional history	Uchinomaki hot spring
Milk road grassland Geosite	Very large grassy plain which opens to north outer ring	Grassy plain scenery
Oshito-ishi Geosite	Fusion with nature and culture	Oshitoishi

(continued)

Table 18.4 (continued)

Geosite	Theme	Geosite resources
North Outer Rim Pyroclastic Geosite	Huge pyroclastic flow	Panka waterfall, Yusuikyo, Nanadaru, sedimentation side of Aso pyroclastic flow (Manami road)
Oguni-go Area Hot Spring Geosite	Hot-spring village	South small country hot-spring village, Tsuetate Onsen, Sparrow Jigoku, Waita Onsen
Ikeyama and Yamabuki Geosite	Spring water	Ikeyama water source, Japanese mountain rose water source
Sankinkotai Road Geosite	Road which utilized nature	Daimyo's alternate-year residence to Edo stay requirement (Funck and Cooper 2013)
Kusakabe Geosite	Shrine dedicated to God of Aso	Yoshimi Shrine
Camel Mountain (Rakudayama) Geosite	Scenery which intruded rocks created	Camel mountain, Minamiaso visitor center, Aso wild grass garden
Cluster of Springs in Minami-Aso District Geosite	Water resources from the volcano	Minamiaso-mura spring group, Takamori Yusui Tunnel
Jigoku and Tarutama Hot Springs Geosite	Terrestrial heat zone and hot spring, the explosion crater	Ikenokubo Jigoku, Tarutama Onsen
Nango Valley Geosite	Regional characteristic topography	The bench topography of Nango Valley
Mt. Rakanyama Geosite	Scenery and historical resources	Mt. Rakan stone group, guardian deity of children at the mountain pass
Tawarayama Mountain Pass Caldera Geosite	Caldera of world scale	Aso Caldera, Tawarayama observation station
Tateno Gorge Geosite	Place where landform dislocation and volcanic activity are observed	Tateno gorge, Sugarugataki waterfall, sweetfish, columnar basalt
Shiraitonotaki Falls Geosite	Waterfall on the outer rim	Shiraito Falls
Omine Volcano Geosite	Active fault evidence	Daihokasan
Soyokyo Gorge Geosite	Valley beauty that volcano and water created	Soyokyo
Heitate Shrine Geosite	Connection with religious world	Heiritsukyu

Source Aso geopark promotion committee, modified by the Author

these stratovolcanoes and the different monogenetic (single eruption) volcanoes, for example Mt. Onimi exist, revealing that the geological structure of the area is complicated. Lava domes made of highly viscous andesitic-dacitic magma, thick lava flows and pyroclastic debris form the Tsurumi complex with widely spread volcanic alluvial fans around the mountain (Siebert et al. 2010).

The area's pre-eminent tourism destination, Beppu City, has developed on one of these volcanic alluvial fans. Beppu has become one of the largest tourism cities in Japan because of its rich resources of hot springs; "Onsen tourism" is not only the main characteristic, but also the basic industry. This was confirmed after WWII when in 1950 (Showa 25), a law entitled *Beppu kokusai kanko onsen bunka toshi kensetsu ho* (the law for the construction of Beppu international tourism and an onsen culture) was adopted (Beppu City 1985, p. 244), making this city the only city in the country apart from Nagasaki and Hiroshima to have its own development legislation passed by the national government. Uniquely, the development of Beppu tourism was promoted as a national project. The approximately 2,600 onsen, the Jigoku (Hells), and the 11.5 million tourists that visit Beppu thus show the importance of the volcanic resource in Japanese tourism (Funck and Cooper 2013).

A special attraction is the *jigoku meguri* (Erfurt-Cooper and Cooper 2009). Jigoku is the Japanese word for hell and meguri refers to a tour or pilgrimage to certain specific locations. This 'pilgrimage' however is not strictly religious, although each of these 'hells' have several sacred shrines on site. The Jigoku of Beppu are a group of 10 small geothermal parks with different themes utilising the individual variance in hot springs like boiling ponds, mud pools and geysers to their full potential, attracting high numbers of visitors on a daily basis (Fig. 18.9). Several jigoku also offer pleasantly warm foot and hand spas for their visitors and hot spring cooked food as well as the visual geothermal attractions. The tour of the Beppu hells is one of the most popular tourist attractions of the city and as with many locations throughout Japan; repeat visits during different seasons are encouraged (Erfurt-Cooper 2010).

Fig. 18.9 Umi Jigoku, one of the 'hells' in Beppu, which attracts tourists all year round to the geothermal and volcanic attractions. *Photo* Patricia Erfurt-Cooper

Fig. 18.10 Sakurajima Island is located close to the large city of Kagoshima separated only by Kinko Bay. The volcano is erupting on a daily basis with activity increasing in recent years. *Photo* Patricia Erfurt-Cooper

18.4.4 Kirishima, Sakurajima and the Southern Islands Area

Sakurajima, one of Japan's most active volcanoes, was an island situated opposite the city of Kagoshima in the south of Kyushu until 1914, but during the great *Taisho* eruption in that year was connected to the mainland by extensive lava flows (Fig. 1.22), and is now considered a peninsula within the Aira caldera which includes Kagoshima (*Kinko*) Bay. Sakurajima erupts constantly on a daily basis and attracts many visitors to the island and the surrounding mainland with striking displays of rising ash columns and Strombolian eruptions (USGS 2013). Sakurajima is part of the Kirishima-Yaku National Park. In 2013 500+ eruptions were recorded up to August, following a year in 2012 that experienced the highest ever number of eruptions (1200+). The latest series have left the city of Kagoshima (600,000 people, 8 km away) blanketed in ash from the 5,000 m plume and experiencing disruptions in public transport and goods delivery services (Fig. 18.10). A small outflow of lava has also been reported.

To the north of Kagoshima Bay is Kirishima, a group of 20 volcanoes including Takachihonomine, Nakadake,

Fig. 18.11 Kaimondake viewed from the Nishi-Ōyama Railway Station in Ibusuki. *Photo* Malcolm Cooper

Ohatayama, Karakunidake, Tairoike, Ohachi, and Shinmoedake. These volcanoes cover an area occupying an area of about 20 × 30 km elongated in the NW-SE direction. An eruption occurred at Kirishima volcano, Japan on 26 January 2011. Some residents evacuated their homes from a town located 7 km from the crater at Shinmoedake. The eruption forced the cancellation of train services on several lines, including the JR Nippo Main Line between Oita and Miyazaki cities. Part of the Miyazaki Expressway remained closed off due to poor visibility. The eruption produced lava fountains, lava flows and ash emissions. Ash emissions reached a height of 7,700 m.

To the south of Kagoshima is the volcanic field of Kaimondake-Ibusuki. The central point, Kaimondake is a 924 m tall volcano located at the southern end of the Satsuma Peninsula in Kagoshima Prefecture (Fig. 18.11). It is considered to be one of the 100 famous mountains in Japan for its shape. The northeastern half of its base faces land while the southwestern half faces the ocean. A spiral hiking trail takes visitors to the top of the mountain in about 3 h from where a panoramic view from Sakurajima to Mt. Kirishima can be enjoyed in good weather conditions.

18.4.4.1 Tourism

Tourism is concentrated in Kirishima City and in the Kirishima-Yaku National Park, Mount Sakurajima in Kagoshima City, and in Yakushima Island (JNTO 2012). Kagoshima City, the administrative, economic, and cultural center of Kagoshima Prefecture, is a castle town dating back to the fourteenth century, and the volcano is located only about 4 km from its center. Lava fields can be seen everywhere on the island, testifying to the history of its violent eruptions. The Sakurajima Visitor Center has a museum that showcases a detailed history of the eruptions of the mountain and its relationship with the local people. The Yunohira Observation area on the slopes of Sakurajima at 373 m above sea level, is nearest to the active Minamidake crater. There are other observation points such as the Arimura Lava Observation facility. Sakurajima is also known for the Sakurajima Radish, the largest variety in the world, and the Sakurajima Mandarin Orange the smallest variety in the world, as well as Sakurajima Marine Park where visitors can observe the seabed of the Bay, its coral reefs and sea life.

The red-painted main building of Kirishima shrine, designated as an 'Important Cultural Property' of Japan, is a place where music and festivals are held as offering for the gods, including the 'Kagura' dance, the Minami-Kyushu Kagura festival in summer, the 'Tenson Korin Gojinkasai' fire festival in autumn, and the performance of Kumendaiko drums in winter. There are ten crater lakes in the Kirishima mountain range, located at 1,239 m above sea level—the highest elevation for such lakes throughout Japan. Hot spring villages such as Myoban, Yunotani and Shin-yu are found on the southwestern slope of the Kirishima mountain range. Accommodation facilities are available visitors who wish to stay overnight, but it is also possible to make a day trip just to enjoy the hot springs. The Shinmoedake crater, which erupted in 2011, was the site of filming for an early James Bond film.

The most important tourist destination of the islands off the south coast of Kyushu is Yakushima, one of the Ōsumi Islands, which has an area of about 500 km^2 and a population of roughly 15,000 and attracts some 90,000 visitors each year. The highest point on the island is Miyanouradake at 1,935 m. It is covered in dense forest noted especially for old growth cedar trees known as *sugi* in Japan. The island forms part of Kirishima-Yaku National Park. In 1980 an area of 18,958 ha was designated a UNESCO Man and the Biosphere Reserve, and in 1993 10 ha of wetland at

Nagata-hama were designated a Ramsar Site (Ministry of the Environment 2011). Yakushima is Japan's wettest place; total precipitation is one of the world's highest at 4,000–10,000 mm. The huge Yaku Cedar trees growing on the island's mountains were placed on the World Heritage (Natural Property) list in 1993 (Funck 2009; UNESCO 2013).

18.5 Risk Management

As noted by Erfurt-Cooper and Cooper (2010), despite the continuous activity of Kyushu's major volcanoes, all of them are main attractions for visitors in their particular area. The attraction and activities of volcanic areas in Japan are very similar to those in other parts of the world, with the difference that for millions of Japanese the use of natural hot springs plays a large role in their everyday life. So, while the tourism infrastructure in active volcanic areas generally includes a typical *Onsen Ryokan* or hot spring resort as well as historical sites, shrines and temples and restaurants, it also includes many thousands of small baths (onsen). As a result, the inherent dangers of an active geothermal site may not be readily apparent to foreign or domestic visitors because the local population has learnt to live with them over centuries of use.

Nevertheless, as a result of the disasters that have occurred, and the cultural values placed on geothermal activity in Japan, the country is very well organized when it comes to safety issues in public geothermal places, including national parks. However, with the growing numbers of foreign tourists in rural areas as well as the big cities more volcanic areas need to include signage in languages other than Japanese to warn visitors of any potential danger. In this regard one of the prime volcanic sites in Kyushu, *Mt Aso*, is a very good exemplar of risk management for the visitor (Erfurt-Cooper 2010). The Aso Volcano Disaster Prevention Council maintains a website in English which contains valuable information for tourists not familiar with Mt Aso and its immediate surroundings. Maps and other useful information are also made available in English, Chinese and Korean for the increasing numbers of foreign visitors to Kyushu. The best advice on this website are the real-time updates of crater status as well as a live webcam. Every volcano visited by tourists should have a website like this to provide useful information for visitors before they arrive at their destination. From a safety point of view the audibility of the PA system at the crater is also very good while the volcano is quiet, and evacuations happen swiftly as soon as the wind direction changes and toxic fumes from the crater are blown towards the viewing areas and the walkways (pers. observation). If that is not possible for all, there are several concrete shelters which are replaced on a regular basis as they deteriorate due to the emissions from the crater (Fig. 18.12) (Erfurt-Cooper 2010).

In the areas surrounding *Yufudake* and *Tsurumidake,* the central government and Oita Prefecture have carried out afforestation and sabo (channeling pyroclastic flows and lahars) works for many years. However, it is predicted that very serious damage will be caused if a volcanic eruption occurs, because main transportation networks, such as the Trans Kyushu Expressway, national roads and JR lines cross these areas, and because the urban district of Beppu is not far from Mt Tsurumi's summit. Thus, volcanic eruption warning and evacuation works have also been promoting a

Fig. 18.12 The viewing platform on the summit of Mt Aso with tourists lining up along the Crater Rim to take photos. To the *right* are several concrete shelters. Photo: Patricia Erfurt-Cooper

range of non-structural measures since 1994—hazard maps of potential eruption-damage areas and plans for installing a volcanic disaster monitoring system have been prepared for the purpose of establishing a warning and evacuation system for emergencies.

At *Sakurajima*, the visitor centre is a reinforced concrete structure, built to withstand at least the average eruptions of the volcano, and offers a good selection of information for tourists about the volcanic history of the mountain, the local ecology and about disaster prevention measures. Many sightseeing spots can be found all over the island and these areas have concrete shelters constructed of reinforced cement walls with rounded roofs, all facing away from the volcano. Even public phones are surrounded by concrete to protect this important lifeline for emergencies (Erfurt-Cooper and Cooper 2010), and on the slopes of the mountain diversion channels leading lahars and mudflows to the ocean are constantly undergoing either upgrading or maintenance.

With the approval by UNESCO of its Volcanic Geopark status (UNESCO 2013), *Mt. Unzen* now has a heightened tourism marketing ability, but one which will attract even more visitors to the area surrounding this potentially dangerous mountain. That is why a school building that burned out in a pyroclastic flow as well as houses destroyed in a post eruption lahar at the time of the last eruption, have been preserved just as they were. They are now major attractions in the Geopark and have been utilized widely for local disaster prevention educational programs throughout the country. At Mt Unzen the visitor information and sign boards are usually in at least four different languages. Hazard maps and leaflets with essential information are generally available and extra information can be found on local government websites. These websites usually also have English translations for foreign residents and visitors (Erfurt-Cooper 2010).

A different form of risk management is apparent on *Yakushima Island* of the Osumi Island Archipelago. Yakushima Members of UNESCO's World Heritage Committee visited the area in 1997 and pointed out that the roots of the Jomon cedar were more exposed than when the tree was designated as a World Heritage site in 1993. Acting on the committee members' warning, local authorities established wooden steps at several points along the route to the Jomon Sugi (cedar) tree and also set up a viewing deck at a point some 12 m away to prevent visitors from stepping on the tree roots. But visitors often trample the ground when they pass each other on the steps, and a shortage of toilets has also become a serious problem. As a result, the Yakushima municipal government is currently studying capping the number of daily visitors to the Jomon Sugi tree at 420 (Japan Times, Sunday Jan 1 2012).

18.6 Conclusions

This chapter has discussed some representative tourism destinations of Japan that are in the form of active volcanoes with associated National Parks. The island of Kyushu, for example, has nine major volcanic areas. The most important are Mt Unzen, Mt Kirishima, Mt Sakurajima and Mt Aso, each of which is a major active site and location for considerable tourism activity, while Mt Tsurumi supplies the City of Beppu with the onsen water that attracts 11.5 million visitors each year (Ura 2002). Risk management is a significant concern as these resources are both highly active and a magnet for tourists, both domestic and foreign. As might be expected then, considerable research has been undertaken on volcanoes as the potential and actual source of natural disasters in Japan. This has ranged from research on the source process of large earthquakes, through gaining a greater understanding of hazard mechanisms caused by earthquakes and volcanic eruptions, to the development of disaster mitigation technologies through observations, experiments and field surveys. The fruits of this research can be seen at volcano sites such as Mt Aso in Kyushu, with a geotourism experience that is as safe as possible as a result.

References

Ashiba, (1994). *Shin Kankogaku Gairon*. Kyoto: Mineruva Shobo.

Beppu City Council. (1985). *Beppu shishi* [Beppu history]. Kyushu: Beppu City.

Cooper, M., Eades, J. S., & Ogata, M. (2008). Heritage tourism in Japan—a synthesis and comment. In B. Prideaux, D. Timothy & K. Chon (Eds.), *Culture and heritage tourism in the Asia Pacific* (pp. 107–117). London: Routledge.

Dowling, R. K., & Newsome, D. (2006). *Geotourism*. Oxford: Elsevier Butterworth-Heinemann.

Erfurt-Cooper, P. (2010). Volcano and geothermal tourism in Kyushu, Japan. In P. Erfurt-Cooper & M. Cooper (Eds.), *Volcano and geothermal tourism: Sustainable geo-resources for leisure and recreation*. London: Earthscan.

Erfurt-Cooper, P., & Cooper, M. (Eds.). (2009). *Health and wellness tourism: Spas and hot springs*. London: Channel View Press.

Erfurt-Cooper, P., & Cooper, M. (2010). *Volcanic tourism worldwide—geo-resources for leisure and recreation*. London: Earthscan.

Funck, C. (2009, July 11–13). Ecotourism in Yakushima Island: Mass or niche tourism. *Sustainable and Alternative Tourism: An International Conference*, Yangshuo (Guilin), China.

Funck, C., & Cooper, M. (2013). *Japanese tourism: Spaces, places and structures*. Oxford: Berghahn.

GLGArcs. (2013). *Introduction to the landforms and geology of Japan—Japan in a subduction zone*. Retrieved November 2, 2013, from http://www.glgarcs.net/intro/subduction.html

Havens, T. R. (1994). *Architects of affluence*. Cambridge, MA: Harvard University Press.

Japan Meteorological Agency (JMA). (2010). *Volcanic information*. Retrieved November 2, 2013, from www.seisvol.kishou.go.jp/tokyo/STOCK/kaisetsu/English/level.html

Japan Times. (2012, September 7). *Quakes added to eruption risk on Mount Fuji*.

Japan Times. (2012, September 9) *Mt Fuji Volcano (Japan): increase of pressure in the magma chamber as a sign of an impending eruption*.

Japan Times. (2012, January 1). *Yakushima seeks environment-tourism balance*.

Japan Tourism Agency. (2013). *White paper on tourism*. Tokyo: MLIT.

Leheny, D. (2003). *The rules of play: National identity and the shaping of Japanese leisure*. Ithaca, NY: Cornell University Press.

Ministry of the Environment. (2011). *Ramsar sites in Japan—Yakushima Nagata-hama*. Retrieved November 12, 2012, from http://www.env.go.jp/en/nature/npr/ramsar_wetland/pamph/yakushima_33.pdf

Ministry of the Environment. (2013). *The National park profile*. Retrieved December 14, 2013, from http://hokkaido.env.go.jp/nature/mat/park/en/shikotsu/p14.pdf

Nakaboh, M., Ono, H., Sako, M., Sudo, Y., Hashimoto, T., & Hurst, A.W. (2003). Continuing deflation by fumaroles at Kuju volcano, Japan. *Geophysical Research Letters*, 30(7), 49.1–49.4.

Newhall, C. (1996). *IAVCEI/International Council of Scientific Unions' Decade Volcano projects: Reducing volcanic disasters, status report*. Washington, D.C: United States Geological Survey.

Organisation, Japan National Tourism. (2012). *Practical Travel Guides*. Tokyo: JNTO.

Siebert, L., Simkin, T., & Kimberley, P. (2010). *Volcanoes of the world* (3rd ed.). Washington DC: Smithsonian Institution.

Smithsonian Institution. (2013). *Global volcanism program*. Retrieved November 10, 2013, from www.volcano.si.edu/

Uto, K., Hoshizumi H., Matsumoto A., Oguri K., & Nguyen H. (2001). *Volcano tectonic history of Shimabara Peninsula and the evolution of Unzen volcano in Southwest Japan*. Presented to the fall meeting of the American Geophysical Union. Washington: American Geophysical Union.

UNESCO. (2013). *Unzen volcanic area Geopark (Japan)*. Retrieved November 12,2012, from http://www.unesco.org/new/en/natural-sciences/environment/earth-sciences/geoparks/members/japan/unzen-volcanic-area-geopark/

Ura, T. (2002). *Beppu Onsen niokeru Atarashii Kanko no Doko (New Trend of Tourism in Beppu Onsen)*. Retrieved November 12, 2012, from http://www2.tourism.ac.jp/mei-ura/beppu021120.htm

U.S. Geological Service. (2011). *Largest earthquakes in the world since 1900*. Retrieved November 12, 2012, from http://earthquake.usgs.gov/earthquakes/world/10_largest_world.php

U.S. Geological Service. (2013, August 27). *Big Blast at Sakaurajima, Japan*. Washington: USGS. Retrieved December 14, 2013, from http://earthobservatory.nasa.gov/IOTD/view.php?id=81891

Yamamoto, T., Takarada, S., & Suto, S. (1993). Pyroclastic flows from the 1991 eruption of Unzen volcano. *Japanese Bulletin of Volcanology, 55*, 166–175.

Additional Websites Accessed

www.hinet.bosai.go.jp/about_earthquake/part1.html.

Rabaul: Papua New Guinea

19

Bruce Prideaux

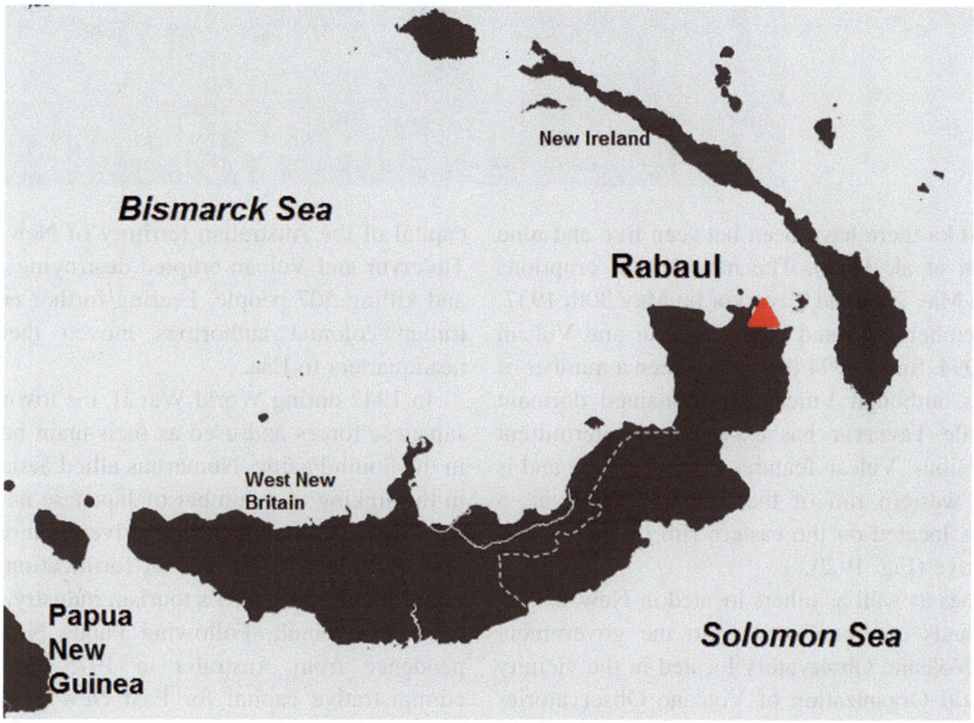

Fig. 19.1 Location of Rabaul in the East New Britain Province of Papua New Guinea. Map modified by Editor

19.1 Rabaul Volcano

Rabaul volcano, located on the island of New Britain, Papua New Guinea (PNG) features a breached flooded caldera open to the sea (Fig. 19.1). The volcano is located proximate to the junction of the South Bismarck, Solomon Sea and South Bismarck tectonic plates, is tectonically active and one of the most active and most dangerous volcanoes in PNG. Past effusive and violent eruptions resulted in a beached sea flooded caldera about 6000 BP (Williams 2013). The volcano has a history of effusive basaltic eruptions, dacitic and rhyolitic explosive eruptions and strombolian eruptions. The caldera has a number of volcanoes including Tovanumbatir, Kabiu, Turangunan, Tavurvur and Vulcan (Green et al. 1986). The highest point of the Rabaul caldera rises 688 m above the sea and features remnants of a shield formed by past pyroclastic flow deposits.

B. Prideaux (✉)
James Cook University, Cairns, QLD, Australia
e-mail: Bruce.Prideaux@jcu.edu.au

P. Erfurt-Cooper (ed.), *Volcanic Tourist Destinations*, Geoheritage, Geoparks and Geotourism,
DOI: 10.1007/978-3-642-16191-9_19, © Springer-Verlag Berlin Heidelberg 2014

Fig. 19.2 Minor eruption of Tavurvur, an active stratovolcano near Rabaul in 2006. Tavurvur and Vulcan were responsible for the destruction of Rabaul in 1994. *Photo* Bruce Prideaux

In the last 20 ka there have been between five and nine eruptions (Nairn et al. 1995). The most recent eruptions were Vulcan on May 29th and Tavurvur on May 30th 1937, Vulcan in September 1944 and both Tavurvur and Vulcan in September 1994. Since 1994 there have been a number of minor eruptions, although Vulcan has remained dormant after 1944, while Tavurvur has experienced intermittent small-scale eruptions. Vulcan features a pumice cone and is located on the western rim of the Caldera. Tavurvur, a stratovolcano, is located on the eastern rim of the caldera and remains active (Fig. 19.2).

Both volcanoes as well as others located in New Britain and nearby islands are monitored from the government funded Rabaul Volcano Observatory located in the vicinity of Rabaul (World Organization of Volcano Observatories 2013). Established after the 1937 volcanic eruption the Rabaul Volcanic Observatory monitors 14 active and 23 dominate volcanoes in PNG. A detailed record of recent eruptions can be found at the Global Volcanism Program of the Smithsonian Institution (Global Volcanism Program 2013). A live view of the volcano can be viewed on http://webcams.volcanodiscovery.com/rabaul while aerial and space views can be found on the Michigan Technological University's website, which covers the Rabaul Caldera (MTU Volcanoes Page 2013).

19.2 The Town of Rabaul

Rabaul became the colonial headquarters for the German colony of New Guinea in 1910, in part because its location on the rim of the Rabaul caldera provided access to a large deep water port. Rabaul was captured by Australian forces in 1914 during World War I and subsequently became the capital of the Australian territory of New Guinea. In 1937 Tavurvur and Vulcan erupted destroying most of the town and killing 507 people. Fearing further eruptions the Australian colonial authorities moved their administrative headquarters to Lae.

In 1942 during World War II, the town was captured by Japanese forces and used as their main base for operations in the South Pacific. Numerous allied aerial attacks resulted in the sinking of a number of Japanese naval vessels which later became popular wreck drive locations. The Japanese also built a large number of fortifications that are now a major focus of the area's tourism industry. After the war the town was rebuilt. Following Papua New Guinea's independence from Australia in 1975 Rabaul became the administrative capital for East New Britain province. By 1990 the population had grown to over 17,000 and the town had emerged as the main urban centre of the province and a major exporter of Copra. The town had also become the focus of a growing dive tourism industry offering excellent diving on nearby coral reefs and the wrecks of numerous World War II ships.

The town was largely destroyed by a volcanic eruption in 1994 when Tavurvur and Vulcan again erupted. Compared to the previous eruption the town was better prepared through evacuation planning and drills. Only 5 persons were killed however about 80 % of the buildings in the town were destroyed when they were covered in ash (Fig. 19.3) and collapsed. The airport was also destroyed. The government of Papua New Guinea decided to move the capital to the former German settlement of Herbertshöh, renamed Kokopo. A new international standard airport was also built at Tokau to the south east of the volcanoes however occasional small-scale volcanic activity that produces ash clouds periodically closes the airport.

Fig. 19.3 This photo taken in 2006 illustrates how the volcanic eruption buried much of the original town of Rabaul with volcanic ash in 1994. The haze in the centre of the photo is caused by volcanic fumes. *Photo* Bruce Prideaux

Despite the heavy damage caused by previous eruptions Rabaul has been slowly rebuilt although it has not been able to regain its position as a major dive tourism centre despite the high quality of the town's marine resources including coral reefs and ship wrecks. This is in part due to lack of investment by the government and a similar lack of interest by the tourism industry. PNGs reputation as a country with a high crime rate and the perceived vulnerability of tourists to robbery and assault has been a major factor in lack of interest in the area by international tourists. Current tourism activity is small scale and focuses on diving and expeditions to the region's volcanoes. In recent years Rabaul has begun to attract a growing number of cruise ship visits, however, compared to land based tourism facilities such as hotels, cruise passengers generate very limited income for the community. There is significant opportunity for the reestablishment of tourism as a major economic sector with the relocated Tojau airport enabling direct flights to Port Moresby, the capital of PNG, and Cairns in Australia.

The current status of Rabaul's inventory of tourism related resources is quite small and comprises two hotels, several lodges, several land tour operators, a small museum, at least one dive operator and a number of restaurants that mainly cater for locals and visiting business people. The tourism related site of the East New Britain government website lists a comprehensive record of tourism attractions, many related to the Japanese occupation (East New Britain 2013).

From a tourism perspective Rabaul is a good example of a destination that is in serious decline. In most instances a major disaster such as the volcanic eruption would be overcome through a partnership between the private sector and the public sector directed towards re-establishing the infrastructure needed to support the tourism industry and to stimulate marketing particularly by national marketing bodies. This has not been the case in Rabaul and the small number of operators that currently constitute the region's tourism supply have found it difficult to reinvigorate tourism.

From a theoretical perspective the post 1994 decline in Rabaul follows the path of decline predicted in Butler's (1980) resort cycle. In terms of comparative and competitive advantage (Fig. 19.4) Rabaul has a significant comparative advantage over many other destinations in PNG based on its indigenous culture, marine resources include diving wrecks, beaches, military heritage and rainforests.

However the failure of the private sector to take advantage of these resources and the failure of the public sector to assist the private sector, allied with PNGs discouraging image as an unsafe destination has stifled investment. Rabaul is unfortunately a realistic example of a destination that has failed to recover from a major crisis despite having a significant comparative advantage based on natural and cultural resources. Figure 19.4 based on an idea proposed by Prideaux (2009) illustrates the downward trajectory of post eruption tourism that has occurred in Rabaul.

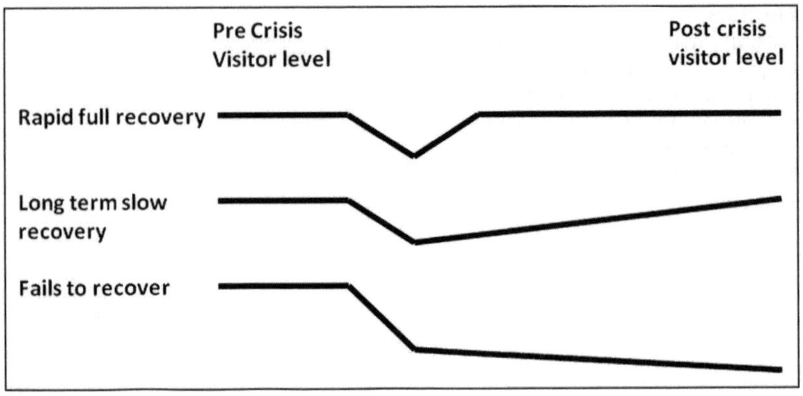

Fig. 19.4 Possible visitor trajectories in the post crisis phase. *Source* Prideaux 2009

Recovering its pre 1994 image as a 'must visit' dive site will require long term planning by the public sector with considerable assistance from the PNG Tourism Promotional Authority, the regional and national governments and Airlines of PNG and Air Niugini.

Note See Journal of Volcanology and Geochemical Research for numerous articles on the Rabaul volcanoes.

References

Butler, R. (1980). The concept of a tourist area cycle of evolution: Implications for management of resources. *The Canadian Geographer, 24*, 5–12.

East New Britain. (2013). Tourism. Retrieved September 27, 2013, from http://www.eastnewbritain.gov.pg/tourism.htm.

Global Volcanism Program. (2013). Rabaul. Smithsonian Institution. Retrieved August 10, 2013, from http://www.volcano.si.edu/volcano.cfm?vnum=0502-14=.

Green, G., Fiffin, D., & McKee, C. (1986). Structural deformation and sedimentation in an Active Caldera, Rabaul, Papua New Guine. *Journal of Volcanology and Geochemical Research, 30*, 327–357.

MTU Volcanoes Page. (2013). Rabaul Caldera, Papua New Guinea. Michigan Technological University. Retrieved September 27, 2013, from www.geo.mtu.edu/volcanoes/rabaul/.

Nairn, I., McKee, B., Talai, B., & Wood, C. (1995). Geology and eruptive history of the Rabaul Volcano area, Papua New Guinea. *Journal of Volcanology and Geochemical Research, 69*, 255–284.

Prideaux, B. (2009). *Resort destinations: Evolution, management and development* (pp. 1–310). Oxford: Butterworth Heinemann.

Williams, I. (2013). Rabaul Volcanic Caldera, PNG. Retrieved September 27, 2013, from www.anaspides.net/earth_life_sciences/rabaul_volcanic_caldera.html.

World Organization of Volcano Observatories. (2013). Rabaul Volcanology Observatory. Retrieved August 10, 2013, from www.wovo.org/0500_0504.html.

Part IV
Volcanic Destinations Worldwide

Fig. 1 Hexagonal volcanic columns at East Dam of High Island Reservoir, Sai Kung Volcanic Rock Region, Hong Kong. *Photo Martin Ng* (Minghong)

While many active volcanoes located along the Ring of Fire are a resource for tourism, dormant volcanic regions also attract visitors with their unique landforms, which often include volcanic hot springs. The volcanic World Heritage listed Korean Jeju Island for example, also known as the '*Island of the Gods*', is popular with young couples as a destination to spend their honeymoon. Famous for its lava tubes and the volcano Mount Halla (1,950 m), Jeju Island is located approximately 154 km southwest of Korea's mainland with nature-based tourism as the main industry. In China, several geoparks include volcanic landforms as well as many geothermal features, which attract visitors with extreme hot springs and steaming fissure vents. Countless volcanic hot springs in China are also utilised for recreational and health purposes. South of mainland China, the large city of Hong Kong is mainly known as a business empire with a distinct skyline of office towers and apartment blocks. However, despite the population density large areas have been set aside as nature reserves and the island region of Hong Kong with its extensive geological heritage was recognized in 2011 as a member of the UNESCO Global Geopark Network (GGN). Included in the unique landforms is a considerable volcanic legacy, which has become a major tourist destination. The diversity of Hong Kong's Global Geopark can be explored by boat to reach places like the Sai Kung volcanic rock region, where volcanic remnants include thick lava layers cooled into giant columns up to 100 m high (Fig. 1). The official website describes the Hong Kong Global Geopark as 'invaluable geo-tourism resources that draw great interest' (Hong Kong Geopark 2013).

Part 4 (Chaps. 20–27) consists of a selection of volcanic areas worldwide. Not located along the 'Ring of Fire', they are either the result of intraplate volcanism connected to hot spots or they are located on plate boundaries such as continental rift valleys and oceanic spreading ridges. Included are regions known for their active volcanoes as well as areas with a rich, but currently dormant volcanic heritage.

In Chap. 20 Patricia Erfurt-Cooper reviews the volcanic heritage of Australia and looks at its potential for geotourism. Isolated in the centre of the Pacific Plate, surrounded by the RoF, the Hawaiian Islands are an individual cluster of hot-spot volcanoes. Chapter 21 by Lisa M. King and Ted A. Bratstrom analyses the famous Hawai'i Volcanoes National Park and its visitor attractions. Volcano tourism in Iran is illustrated by Alireza Amrikazemi in Chap. 22 and tourism at the active volcano Erta Ale in Ethiopia is described by Rosaly Lopes and Jani Radebaugh in Chap. 23. Jasmine Cardozo Moreira and Gilson Burigo Guimarães investigate the visitor attractions of the Brazilian Fernando de Noronha Archipelago in Chap. 24 and in Chap. 25 Ross Dowling and Julie Margueritte draw attention to Reunion Island's active volcanic attractions, an area not well covered in the tourism literature.

Chapter 26 by Henry Gaudru reports on the impact of volcanic disasters and their attraction for tourism in countries such as Chile, the Philippines, Iceland and the Caribbean island of Montserrat. The last chapter raises key issues in relation to risk perception and risk management, which are explored and debated by leading volcanologist Chris Newhall in Chap. 27, who offers valuable advice on decision making based on risk assessment.

References

Hong Kong Geopark (2013). Hong Kong Global Geopark of China. Online: www.geopark.gov.hk/. Accessed 8 Oct 2013.

Lockwood, J.P., & Hazlett, R.W. (2010). *Volcanoes: Global perspectives*. Chicester, UK: Wiley-Blackwell.

Schmincke, H.U. (2006). *Vulkanism*. Berlin: Springer.

Sigurdsson, H., Houghton, B., McNutt, S., Rhymer, H., Stix, J. (2014). *Encyclopedia of volcanoes* (2nd ed.). San Diego: Academic Press, An Imprint of Elsevier.

Australia's Volcanic Heritage

20

Patricia Erfurt-Cooper

20.1 Introduction

In tourism terms Australia is a country perhaps better known for its unique wildlife than its geological heritage. Nevertheless there are unusual rock formations such as Uluru (Ayers Rock), the Bungle Bungle Ranges, and Wave Rock in Western Australia, as well a surprising abundance of remnant volcanic landscapes (Fig. 20.1). More than 500 national parks and over 28 million hectares of land designated as national parkland represent almost four per cent of Australia's land area (Australian Government 2012). Apart from national parks and other protected sites Australia has

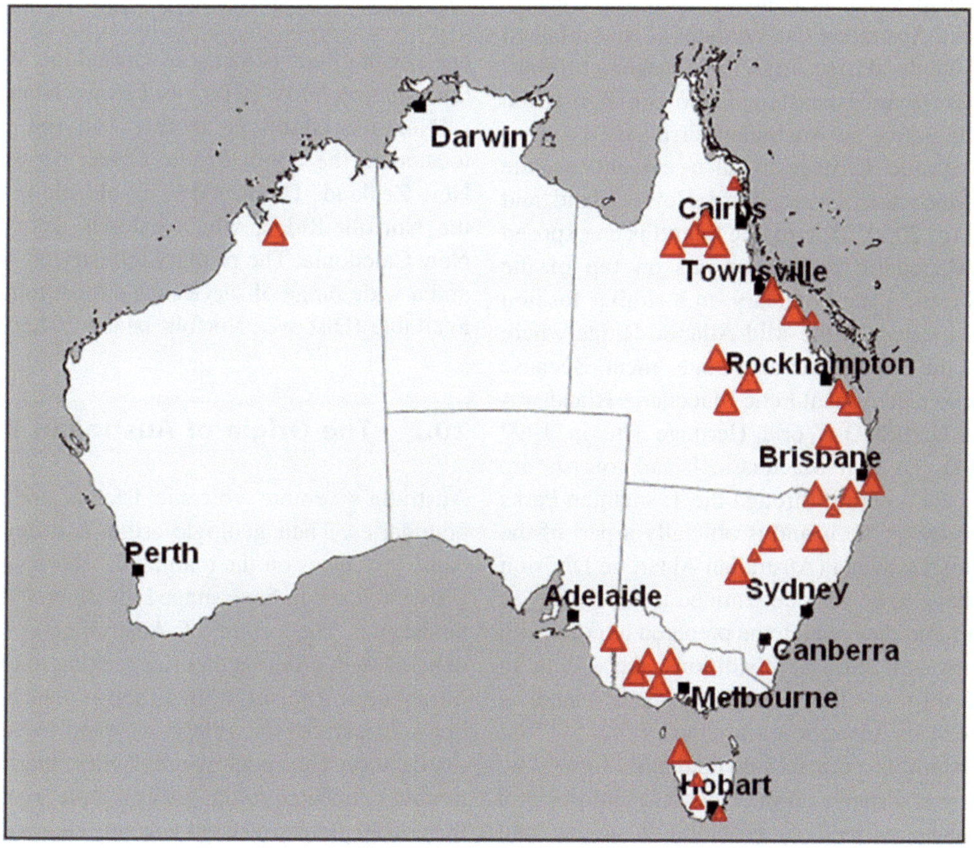

Fig. 20.1 Australia's main volcanic areas. Map modified by Patricia Erfurt-Cooper

P. Erfurt-Cooper (✉)
James Cook University, Cairns, Australia
e-mail: Patricia.Erfurt@my.jcu.edu.au

P. Erfurt-Cooper (ed.), *Volcanic Tourist Destinations*, Geoheritage, Geoparks and Geotourism,
DOI: 10.1007/978-3-642-16191-9_20, © Springer-Verlag Berlin Heidelberg 2014

Table 20.1 Australian World Heritage sites which include volcanic landforms

Volcanic World Heritage Areas in Australia
Heard and McDonald Islands
Lord Howe Island Group
Macquarie Island
Gondwana Rainforests of Australia
Tasmanian Wilderness
Wet Tropics of Queensland

nineteen (19) UNESCO World Heritage listed properties of which at least ten (10) contain volcanic heritage (Table 20.1). Not all of them are accessible for tourists though; for example **Heard and McDonald Islands** are mainly visited by scientists and their rare pristine ecosystems are therefore largely preserved from human impact. While Heard Island can be accessed by tourists, a permit is required and visitors must agree to stay within specified zones and the island group is managed as a strict nature reserve through the Australian Government's 'Heard Island and McDonald Islands Marine Reserve Management Plan' (UNESCO[1] 2012). Heard Island's volcano Big Ben is the only one currently active on Australian territory.

Australia's volcanic heritage is also evident on the Pacific islands Macquarie Island, Lord Howe Island and Norfolk Island (Fig. 20.2). **Macquarie Island** is an exposed section of the Macquarie Ridge and sits on top of the Australian and Pacific plate boundary, in a similar fashion to Iceland being located on the Mid Atlantic Ridge where the Eurasian and the North American plates meet. Because of its unique features the subantarctic Macquarie Island was recognised as a UNESCO World Heritage site in 1997 (UNESCO[2] 2012). To visit Macquarie Island tourists are required to apply for a permit through the Tasmanian Parks and Wildlife Service as the island is officially a part of the Australian State of Tasmania (Australian Antarctic Division 2012). Tourists have to be very determined to set their eyes on these locations and they need to be prepared to cope with unexpected events and extreme conditions. According to demand the option to visit these remote volcanic islands is frequently included in Antarctic excursions.

Lord Howe Island is a remarkable land mass formed by volcanic activity and displays all the emergent features of a large shield volcano as well as providing a unique and spectacular exposure of the Lord Howe Rise (Australian Government 2005). The eroded remnants of this ancient shield volcano with their impressive topography and local wildlife make it a popular tourist attraction, although visitor numbers are limited to control the environmental impact. One of the local volcanic attractions is Ball's Pyramid, a remainder of a volcanic shield that is located about 20 km southeast of Lord Howe Island. Ball's Pyramid belongs to the

Fig. 20.2 The approximate location of Australia's volcanic islands. Map modified by Patricia Erfurt-Cooper

Lord Howe Island Marine Park and at 562 m height is considered to be the tallest volcanic stack in the world. Mount Gower (875 m) is a remnant volcanic plug with guided tours as the preferred option to access the summit. A ranger with extensive knowledge about the natural history of Lord Howe ensures that the visitors stay safe on the Mount Gower Trek (Destination NSW 2012; Lord Howe Island Tours 2014).

Norfolk Island is another outlying part of Australia located in the South Pacific Ocean between Australia and New Zealand. The island is a volcanic outcrop and part of the Norfolk Ridge, which extends from New Zealand to New Caledonia. The primary industry on Norfolk is tourism and a wide range of services and most modern comforts are available (Discover Norfolk Island 2012).

20.2 The Origin of Australian Volcanism

Australia's remote volcanic islands are located on plate boundaries. Their geologic origin is different from the volcanic provinces on the mainland, where various episodes of active volcanism have shaped the diversity of the Australian landscape. The origin of Australian volcanoes has been debated with different theories seeking to explain Australia's intraplate volcanism. With no active plate boundaries in sight hot spot activity is widely accepted as the most plausible explanation. Hot spots are areas where thermal plumes, which remain connected to the lower mantle, rise up from beneath the crustal plates and form volcanic chains during continental drift (Ferrett 2005; Lockwood and Hazlett 2010). For more detailed information about the geophysical aspects of Australia's volcanoes a text by Johnson, Knutson and Taylor (1989) holds comprehensive information about intraplate volcanism in Australia.

Some of the more well known volcanic destinations of Australia were covered in a previous book by Erfurt-Cooper and Cooper (2010) and therefore this chapter will look at a number of remarkable volcanic areas along the Australian eastern seaboard that ought to be mentioned in greater detail. While some areas are well recognised for their volcanic features, others are more popular for their recreational potential with a minor focus on the geological heritage. Interesting geosites such as the Capricorn Coast volcanics, the Peak Range volcanoes in Central Queensland, the rather unusual Mt Hypipamee National Park and the Undara Volcanic National Park in the tropical north are gradually becoming centres for the growing interest in geotourism in Queensland and are highlighted as representative case studies in this chapter. The volcanic heritage of the southern states from New South Wales to Victoria and South Australia, although to a lesser extent, is covered to give a picture of the diversity and the scale of volcanic destinations on the Australian continent.

20.3 Volcanic Geotourism in Australia

Despite the fact that Queensland has some of the most amazing geological heritage of Australia (Figs. 20.3 and 20.4), the absence of geotourism as an additional important nature based tourism sector in this State to date is noticeable.

The Capricorn Coast of central Queensland is the first region in the State that actively promotes geotourism in a first tentative attempt to provide visitors not just with information about rainforests, reefs and wildlife, but also with education about the surrounding landforms and their geological heritage.

> Geotourism is rapidly being recognised as an exciting new direction for tourism surrounding geological attractions and destinations (Capricorn Online 2012).

The 'Dig the Tropics' project aims to position Central Queensland as a 'world leader' in geotourism by including all aspects that enhance the geographical character of this region from the coast to the outback (Dig the Tropics 2013). The local volcanic attractions, the numerous national parks and other protected sites are promoted as part of a geo-journey which has put Capricorn Tourism on the forefront of a developing geotourism industry sector that is drawing increasing numbers of visitors. It may follow in time that other Queensland regions recognise their exceptional geological attractions and combine them with the cultural heritage, following the lead of the Capricorn Coast by adding geotourism to showcase all of their available tourism resources. In north Queensland another geotourism development project is currently at planning stage with a focus on geological interpretation of the rich volcanic heritage of the Tablelands region and Undara (pers. com. with Tourism Coordinator, January 2014).

Other states of Australia are slowly accepting the concept of geotourism with Western Australia gradually being followed by Tasmania, New South Wales and Victoria. Considering that the State of Queensland has more than 900 protected parks and forests the adoption of geotourism could be a logical consequence for the near future. Based entirely on natural resources geotourism is a value adding

Fig. 20.3 The remnant volcanic plugs of the Glasshouse Mountains in South East Queensland are outstanding examples of Australia's volcanic geoheritage. *Photo* Patricia Erfurt-Cooper

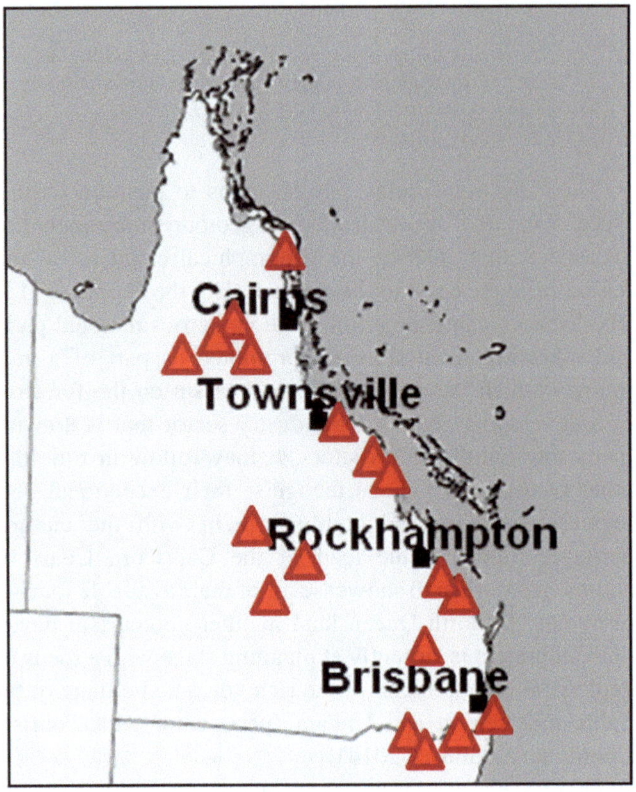

Fig. 20.4 Queensland's major volcanic areas. Map modified by Patricia Erfurt-Cooper

market sector which complements ecotourism by providing additional information about the natural and cultural geosphere that surrounds us.

20.4 Central Queensland Coastal Region

At the *Capricorn Coast* of Central Queensland we find an abundance of interesting volcanic features. In Rockhampton the Berserker Ranges and Mount Archer National Park are based on past volcanism with ancient igneous rock formations eroded and weathered into their current shape. Facing north from the road between Rockhampton and the coastal town of Yeppoon an impressive group of twelve volcanic plugs is scattered over the rural landscape with Mt Jim Crow towering next to the highway. Separated from the group by the road is Mount Wheeler, another remnant plug of an ancient volcanic vent. The eroded volcanoes belong to the Mount Hedlow trachyte group and are located on land mainly used for cattle grazing (Fig. 20.5). For those willing to climb to the summit of one of the volcanic plugs Mt Jim Crow (Fig. 20.6) with its surrounding small national park is easy to access from a car park adjacent to the road. However, due to the steep cliffs and scree slopes climbing is considered dangerous and not recommended by the Department in charge of Queensland's national parks. For the experienced and adventurous climber reaching the peak of the 221 m high volcanic remnant is not too difficult to manage, although it may be treacherous in wet conditions. The Department of National Parks, Recreation, Sport and Racing (2012) has provided some safety advice (Table 20.2) for visitors of the Jim Crow National Park, despite the fact that climbing is not encouraged. For the local Aboriginal people Mount Jim Crow is a site of cultural significance. The area of the national park is largely undeveloped without any facilities other than a car park and some signs.

From Mt Jim Crow and the Hedlow Trachyte Group it is only a short distance to the Capricorn Coast. The Double Head section of the Capricorn Coast National Park (Figs. 20.7, 20.8 and 20.9) near Rosslyn Bay south of Yeppoon is another protected site within the Capricorn volcanics. Walking trails lead to the summit where lookout points offer spectacular 360° views. The major attraction for geotourists is an entablature shaped by cooling columns which radiate from a volcanic conduit, aptly named Fan Rock (Figs. 20.10, 20.11 and 20.12). The irregular fractures of Fan Rock are well defined and display an exceptional entablature typical for the radial cooling of lava and present a rare opportunity to observe the centre of a volcanic vent. To view Fan Rock's entablature without shadows obscuring the columnar joints it is recommended to visit this site at sunrise.

Nearby at the southern end of Kemp Beach, directly opposite the Double Head area another volcanic section of the Capricorn Coast National Park is located known as Bluff Point (Figs. 20.13 and 20.14). Although these small national parks do not have any information centres, they offer some signage at the park entrance with basic information about the geology, ecology and biology of the area (Fig. 20.15).

On the Capricorn Coast it is obvious that tourism is based to a large degree on the rich volcanic heritage combined with a favourable climate. Although not promoted directly as volcanic attractions the islands along the Capricorn coast (Keppel Island Group) share a similar geological heritage as the mainland and remnants of volcanic activity are parts of the protected areas. Instructive signage is increasingly added to areas of interest for tourism and often includes safety advice (Figs. 20.12 and 20.15).

Fig. 20.5 Volcanic plugs of the Hedlow Trachyte Group—from right Iron Pot Mountain, Black Mountain, Rocky Cone Mountain and Mount Hedlow. *Photo* Patricia Erfurt-Cooper

Fig. 20.6 Mount Jim Crow is surrounded by a small national park with direct access from the road between Rockhampton and Yeppoon. *Photo* Patricia Erfurt-Cooper

Table 20.2 Safety advice for visitors of Mt Jim Crow National Park

To enjoy a safe visit to this area, please
Wear suitable clothing and carry plenty of water
Notify someone of your plans and advise them of your safe return
Be prepared to encounter wildlife. Snakes, spiders and stinging insects are common so know how to respond
Do not rely on mobile phones. Coverage is limited in this area
Be aware of the weather and plan for it
Heed safety advice. Signs warn of potential dangers and are there for your safety
Carry a first-aid kit and know how to use it

Source Department of National Parks, Recreation, Sport and Racing (2012)

Fig. 20.7 Double Head Capricorn Coast National Park viewed from Bluff Point. The Fan Rock viewing platform (Fig. 20.12) is located on the lower summit (*arrow*). *Photo* Patricia Erfurt-Cooper

Fig. 20.8 Double Head showing inclined columnar jointing of a cooled rhyolite dome. View from the south. *Photo* Patricia Erfurt-Cooper

Fig. 20.9 Double Head's inclined (*west*) columnar jointing viewed from the north. *Photo* Patricia Erfurt-Cooper

Fig. 20.10 Fan Rock is a unique example of an exposed volcanic conduit. *Photo* Patricia Erfurt-Cooper

Fig. 20.11 The spectacular Fan Rock entablature at the Capricorn Coast shows the inside of a volcanic plug with radiating clusters of cooling columns. *Photo* Patricia Erfurt-Cooper

Fig. 20.12 Safety barriers along the steep cliff opposite the Fan Rock entablature were erected to prevent accidents. *Photo* Patricia Erfurt-Cooper

Fig. 20.13 The headland of Bluff Point National Park viewed from the Fan Rock Lookout. *Photo* Patricia Erfurt-Cooper

Fig. 20.14 Bluff Point National Park, the volcanic headland opposite the Double Head section. *Photo* Patricia Erfurt-Cooper

Fig. 20.15 Information sign at the entrance to the Double Head section of the Capricorn Coast National Park. *Photo* Patricia Erfurt-Cooper

20.5 Central Queensland Inland Region

Moving inland from the Capricorn Coast the Peak Range east of Capella offers panoramic views of a chain of prominent volcanic plugs (Fig. 20.16), which include Browns Peak (807 m), Anvil Peak, Table Mountain, Malvern Hill, Ropers Peak and Scotts Peak. The central vent of the Peak Range volcano has not been determined with accuracy, but it is thought that the individual peaks are representative of a line of vents along the crest of the central

Fig. 20.16 Peak Range scenic mountains—Scotts Peak and Ropers Peak. *Photo* Patricia Erfurt-Cooper

volcano (Willmott 2006). The domes, plugs, pinnacles and dykes of the Peak Range volcanics consist of trachyte, rhyolite and basalt and are surrounded by extensive basalt lava plains (Johnson et al. 1989).

20.6 Mackay Region

Travelling north from the Central Queensland region along the eastern seaboard the dense cluster of seventy four islands of the **Whitsunday Group** are further examples of past volcanism. Combined with a complex ecosystem of coral reefs protected in national parks the Whitsunday Group is an important part of the World Heritage listed Great Barrier Reef Marine Park (GBRMP). Described in local tour guides and brochures as a 'unique geographic location' the Whitsundays attract over 800,000 visitors annually. However, the volcanic landforms are not the primary drawcard for tourists who focus more on rainforest and wildlife adventures or sailing around the islands than on geologic heritage and educational geotourism.

20.6.1 Cape Hillsborough and Other Neighbouring National Parks with Volcanic Heritage

North of Mackay volcanic rock formations are found in several national parks including **Cape Hillsborough**, which is located directly on the coast. Exposed volcanic rocks such as the trachyte core of Pinnacle Rock are small eroded plugs, which represent minor vents and are surrounded by the remains of lava flows and ash layers (Johnson et al. 1989; Willmott 2006). Based on recreational attractions natural and cultural key sites within the protected areas are presented to visitors through a range of interpretations. One of the primary purposes of the national park management is to *'provide safe, sustainable and nature-based recreation opportunities in all parks, particularly at Cape Hillsborough National Park'* as well as informing visitors of the area's natural, cultural and social values (Queensland Parks and Wildlife Services 2011). While the natural attractions of the parks are described by the park management as *'notable for geologically diverse volcanic landscapes with distinct lava seams separated by layers of volcanic ash and pyroclastic material'* this is not used to promote geotourism.

20.6.2 Townsville and Magnetic Island

The **Townsville** area has a distinctive volcanic heritage. On the mainland the pink granite monolith **Castle Hill** (Mt Cutheringa, 286 m) is a major tourist attraction. Opposite Townsville 8 km out to sea lies Magnetic Island, which is also part of the World Heritage listed GBRMP. Magnetic Island is a significant tourist destination with over fifty per cent of the island protected as national park with the granite spine of Mount Cook (497 m) at the centre. Rocky granite headlands around the island are further indicators of a volcanic history. Although these geological features are not directly promoted as volcanic attractions for geotourism they are nonetheless important indicators of a fiery past which adds to the attraction of this geo-destination.

20.7 North Queensland

The **Wet Tropics** of Queensland's north are a large UNESCO World Heritage area which attracts growing visitor numbers with estimated projections of annual figures reaching 4 million by 2016 (Wilson et al. 2004). This makes the Wet Tropics one of Australia's premier tourist destinations which in August 2012 was also recognised as Australia's 14th National Landscape (Australian Government/Environment 2012). The vast area of the Wet Tropics contains several volcanic national parks and protected sites with cinder cones, crater lakes and lava flows as unique tourist attractions. The region's geodiversity offers a great opportunity to take the step from ecotourism to embrace geotourism as an additional segment where tourists can learn more about the geological heritage including the volcanic past.

Apart from the Wet Tropics in geological terms northern Queensland contains several volcanic provinces which reach from the Torres Strait north of Cooktown to the south and southwest of Townsville. Most of the provinces are dominated by small volcanoes and extensive lava flows (Johnson et al. 1989). The provinces reviewed in this chapter as examples of past volcanism in Queensland are the Atherton and McBride Provinces, both extending southwest of Cairns.

20.7.1 Atherton Volcanics

The **Atherton Province** has over fifty eruption centres (Johnson et al. 1989) including the well known Seven Sisters at Yungaburra (scoria cones, Fig. 20.17), Lake Eacham and Lake Barrine (maar craters, Fig. 20.18), Mt Hypipamee crater (diatreme, Fig. 20.22), Bromfield Swamp (maar), Mount Quincan (cinder cone) and Mount Halloran (shield volcano). Lava flows from ancient eruption centres are occasionally exposed on the Tablelands' grazing areas (Fig. 20.19) and waterfalls flowing over layers of basalt have created

Fig. 20.17 Two cinder cones of the group 'Seven Sisters' at Yungaburra near Atherton, North Queensland. *Photo* Patricia Erfurt-Cooper

Fig. 20.18 Lake Barrine, a picturesque maar crater in a rainforest setting with a lake side restaurant and daily boat tours. According to a local Ranger the Crater Lakes National Park receives around 360,000 visitors each year. *Photo* Patricia Erfurt-Cooper

Fig. 20.19 The Atherton Tablelands are covered by extensive basalt layers. Exposed lava flows as seen here are similar to the Stony Rises in Western Victoria. *Photo* Patricia Erfurt-Cooper

Fig. 20.20 During the wet season the Millaa Millaa Falls cover the basalt columns like a curtain. A must-see destination for tourists who are visiting the Atherton Tablelands. *Photo* Patricia Erfurt-Cooper

Fig. 20.21 Sign board at the entrance to the small national park that protects the site of the Hypipamee explosion crater. *Photo* Patricia Erfurt-Cooper

popular volcanic tourist attractions (Figs. 20.20 and 20.25) (Table 20.3).

One of the unique geosites within the Atherton volcanic province is the explosion crater of **Mt Hypipamee**, which is located 110 km south west of Cairns near Herberton in the Hypipamee National Park (Fig. 20.21). This volcanic crater is classed as a diatreme with a diameter of around 20 m. Sheer granite walls (Figs. 20.22, 20.23 and 20.24) are rising up vertically 60 m to a viewing platform, from which visitors can safely look down on the crater lake. Approximately 40,000 people are visiting this extraordinary location every year (pers.com. with Park Ranger 2010).

In the year 2011 a group of divers set out to establish the depth of the Hypipamee diatreme and to investigate a suspected tunnel connection to the nearby Barron River. Their research under water revealed a depth of 75 m, not the originally claimed 81 m. However, the group of divers could not find any caves or tunnels in the crater lake, although the visibility at depth was negotiable (Stickley 2011). The difference in the measurements of the water's depth may be due to fluctuating levels caused by seasonal rainfall.

Tourists who visit the Hypipamee crater can take a short walk to the nearby Dinner Falls, which are fed by the waters of the Barron River. On a tropical summer day tourists enjoy the cooling waters cascading over lava layers (Fig. 20.25), which divide the falls into individual sections. To explore the region in more depth the Tablelands Tourism Network offers adventure tours that include several

Fig. 20.22 Sheer granite walls surround the Hypipamee diatreme. The water surface of the deep crater lake is covered in duckweed which occasionally gets disturbed with rocks and branches thrown down by tourists. Personal observation and photo by Patricia Erfurt-Cooper

Fig. 20.23 Detail of the granite rock walls at the Hypipamee crater. *Photo* Patricia Erfurt-Cooper

waterfalls as well as the craters of Mount Hypipamee and Lake Eacham. Other locations with a volcanic background are the Babinda Boulders south of Cairns, a swimming hole surrounded by huge granite boulders. However, cooling down in the fast flowing waters is not without risk; over the last 50 years 15 people have lost their lives in the deceptive currents (Cairns Attractions 2012).

20.7.2 The North West of Queensland

Further inland to the west of Cairns the **McBride province** is home to shield volcanoes, scoria cones and some of the longest lava flows in the world. Some 164 volcanic eruption centres have been identified (Whitehead 2010) including the Undara volcano (1020 m), the **Kinrara** volcano and the

Fig. 20.24 Visitors on the viewing platform (*insert*) of the Hypipamee crater 60 m above the water level below. *Photo* Patricia Erfurt-Cooper

Fig. 20.25 The cascading Dinner Falls are popular with tourists, especially during summer. *Photo* Patricia Erfurt-Cooper

Kalkani crater. While there is no public access to the Kinrara volcano, the Undara Volcanic National Park (Fig. 20.26) offers numerous unique attractions, although it is mainly known for the famous Undara lava tubes (Figs. 1.12, 1.13 and 20.27). More or less at the centre of the McBride province is the Undara volcano crater with lava plains estimated up to 600 m thick. Some of the lava flows have travelled great distances from the Undara crater (Fig. 20.28) with one of the longest flows measuring 160 km (Johnson et al. 1989). Visitors to Undara can expect a unique geo-experience all year round, although during the wet season guided tours are reduced and access to the Undara Lodge may be restricted. A recommended reading for everybody who is interested in the uniqueness of Undara a

Table 20.3 Examples of Volcanic Heritage in the Wet Tropics

North Queensland Volcanics—part of Wet Tropics World Heritage
Mt Halloran—shield volcano, lookout at summit
Mt Quincan and **Seven Sisters**—cinder (scoria) cones
Undara Volcanic National Park—lava flows, volcanic caves, lava tunnels
Kalkani Crater—sign boards, self guided tours, crater rim walk
Mt Hypipamee National Park—explosion crater (diatreme or volcanic pipe)—crater viewing platform (Fig. 20.24)
Crater Lakes National Park—Lake Barrine and Lake Eacham, maar craters (volcanic explosion craters), information centre, boat tours, restaurant
Millaa Millaa Falls—basalt lava layers, popular photo opportunity (Fig. 20.20)
Millstream Falls National Park—columnar basalt, lava flows
Malanda Falls Conservation Park—lava flows, visitor centre

Fig. 20.26 The entrance to the Undara Volcanic National Park where tourists can either take guided tours or visit some areas like the Kalkani Crater on their own. *Photo* Patricia Erfurt-Cooper

Fig. 20.27 The size of the Undara lava tubes varies with some of them over 20 m wide and 10 m high. *Photo* Patricia Erfurt-Cooper

Fig. 20.28 Basaltic lava flows are piled up into walls and ridges and cover the terrain near the Kalkani Crater. *Photo* Patricia Erfurt-Cooper

book by Anne and Vernon Atkinson (2001) describes in detail the discovery of the Undara lava tubes and draws comparisons to other areas.

Undara is one of the most significant lava flows in the world with outstanding examples of lava caves and flow features (Australian Government 2005).

The 'Undara Experience' offers not just visits to the lava tubes, but includes the whole geodiversity of the area with interesting geological, environmental and historical features explained to visitors in detail during guided tours (Undara Experience 2012).

20.8 South East Queensland

The volcanic provinces of southern Queensland reach from the Hummock basalt outcrops in Bundaberg down to the state border, where the shield of the Tweed Volcano is located with Mt Warning as a central intrusive complex. Within this geographic region the Bunya Mountains south of Kingaroy appear as a northern extension of the Main Range which includes the Toowoomba Basalt region with the Darling Downs and runs parallel to the Great Dividing Range (Johnson et al. 1989).

South east of Bundaberg near Gayndah the **Coalstoun Lakes National Park** includes one of the more recent volcanic formations in Australia. Coalstoun Lakes consists of at least three known volcanoes, from which the lava flows today known as Barambah Basalts were extruded (Johnson et al. 1989). Mount Le Brun (200 m) with two intersecting volcanic craters is one of the three recognised volcanoes east of Gayndah in Queensland (Fig. 20.29). While the crater lakes and parts of the craters are protected in a small national park, most of the crater rims and the exterior of Mount Le Brun are on private land (Geological Society of Australia 2013).

North of Brisbane adjacent to the Bruce Highway, the Glasshouse Mountains, a group of dome shaped hills and sharp peaks (Fig. 20.30) rise abruptly from the coastal plain (Johnson et al. 1989). The remnant volcanic plugs are composed of various forms of trachyte and rhyolite (Table 20.4) and are the most amazing landscape feature of the Sunshine Coast Hinterland. Some of the mountains are part of small individual national parks with car parks, hiking trails and some signage. The majority is either under the control of the State Forestry or surrounded by private land (Ferrett 2005). Tourism in the Glasshouse region is promoted as Hinterland Tourism with a focus on the diversity of the region including activities such as bushwalking all year round. Although geotourism is practiced without mentioning it by embracing the geodiversity of the area, at this point in time the volcanic heritage is only one of the regional attractions and does not appear to receive any more attention than for example the well known Australia Zoo located nearby.

Queensland has an abundance of volcanic national parks and other protected sites and although some well known destinations have been reviewed for this chapter, not all of them can be described in detail. For further information on

Fig. 20.29 The two intersecting craters of Mount Le Brun, known as the Coalstoun Lakes, are located about 25 km south of the town of Biggenden. *Photo* Patricia Erfurt-Cooper

Fig. 20.30 The remarkable Glasshouse Mountains, a cluster of volcanic plugs located in the Sunshine Coast Hinterland can be viewed from a lookout at the Mary Cairncross Scenic Reserve. *Photo* Patricia Erfurt-Cooper

Table 20.4 The Glasshouse Mountains include fourteen peaks of which seven are located in the Glasshouse Mountains National Park and Forest Reserves

Glasshouse Mountains (South East Queensland)
Mt Beerburrum (280 m) *Trachyte*
Mt Beerwah (556 m) *Trachyte*
Mt Cooee (106 m) *Rhyolite*
My Coochin (235 m) *Rhyolite*
Mt Coonoowrin (377 m) *Rhyolite*
Mt Elimbah (109 m) *Rhyolite*
Mt Mellum (406 m) *Volcanic Crater*
Mt Miketeebumulgrai (202 m) *Trachyte*
Mt Ngungun (253 m) *Rhyolite*
Round Mountain (97 m) *Rhyolite*
Mt Tibberoowuccum (220 m) *Rhyolite*
Mt Tibrogargan (364 m) *Rhyolite*
Mt Tunbubudla (338 m) *Rhyolite*
Wild Horse Mountain (123 m) *Rhyolite*

the subject of unique geosites a book series by Warwick Willmott (geologist) covering 'Rocks and Landscapes of the Queensland National Parks' is highly recommended.

20.9 South of the Queensland Border

The Queensland and New South Wales state border region is partly incorporated in the vast area of the World Heritage listed Gondwana Rainforests and includes a number of volcanic national parks (Table 20.5). This serial UNESCO World Heritage property represents outstanding examples of exceptional geodiversity with distinctive geological features including shield volcanic craters and is of international significance for science and conservation. The deeply dissected Tweed Shield erosion caldera is one of the best preserved calderas in the world with the mountain mass of Mt Warning (Woolumbin) at its centre and is displaying a large number of volcanic features (Australian Government 2005; UNESCO World Heritage Centre 2012; Fig. 20.31).

Table 20.5 The listed National Parks are integrated in the World Heritage Area known collectively as the Gondwana Rain Forests. Since 2012 the area is also promoted as the Green Cauldron of Australia

Gondwana Rain Forest World Heritage—Volcanic National Parks
Barrington Tops National Park—carved out of ancient volcanic flows, extinct volcanic peaks, located on a volcanic flow plateau
Main Range National Park—shaped by volcanic activity, eroded lava layers, volcanic peaks
Border Ranges National Park—Mt Warning volcanic rim caldera, several remaining volcanic plugs
Lamington National Park—volcanic formation, result of two large, ancient volcanoes, part of the Tweed Shield Volcano caldera
Springbrook National Park—eroded basalt caves, natural stone bridge
Nightcap National Park—volcanic heritage with massive peaks
Mt Warning National Park—central volcanic remnant (eroded core) of an ancient shield volcano
Mount Barney National Park—rugged volcanic peaks
Washpool National Park—rich volcanic soils
Dorrigo National Park—past volcanic eruptions, eroded volcanic layers
Gibraltar Ranges National Park—landforms shaped by volcanic activity
New England National Park—extensive sheet of tertiary volcanics, volcanic rock layers

One specific example is the cooling columns of the lava flows from the Tweed Volcano on the Fingal Headland. This rock formation bares a remarkable similarity to the Giant's Causeway in Ireland and is known by the same name and features on the cover of a geotourism map for New South Wales.

Mount Warning National Park is a major volcanic tourist destination together with several other national parks in this region and which include:

- Lamington National Park
- Border Ranges National Park
- Nightcap National Park
- Springbrook National Park
- Main Range National Park
- Richmond Range National Park
- Bunya Mountains
- Parts of the Darling Downs

Due to its volcanic origins in 2012 the region was accepted into the National Landscapes Program as Australia's Green Cauldron (Fig. 20.32); a landscape with a distinctive unspoilt natural character as well as cultural identity, which provides authentic experiences for visitors (Corbett and Mackay 2012). Collectively these volcanic environments are major attractions for geotourists as well as for general recreational purposes and contribute significantly to the individual local economies. Visitor safety and risk management is largely in the hands of the relevant state authorities, who are in charge of national parks and other protected sites.

Volcano tourism in Queensland is not yet promoted as an individual tourism sector. Even geotourism, which includes volcano tourism, is only slowly starting to develop, although the vast amount of volcanic heritage in Queensland's national parks and world heritage sites provides additional attractions for domestic and international visitors.

Fig. 20.31 Mount Warning rising out of the Tweed Caldera. *Photo* Patricia Erfurt-Cooper

Fig. 20.32 The remnant volcanoes of the Gold Coast Hinterland viewed from Coolangatta Airport. The volcanic national parks of the Scenic Rim and the Border Ranges are part of the Gondwana Rain Forest World Heritage and the Green Cauldron National Landscape. *Photo* Patricia Erfurt-Cooper

20.10 New South Wales

The volcanic heritage of Australia continues throughout the State of New South Wales where the remnants of the Ebor Volcano (Dorrigo National Park) provide one of the outstanding examples of a shield volcano (UNESCO World Heritage Centre 2012). Several other national parks include volcanic sites of significance and are popular tourist destinations:

- Nandewar Volcano—Mt Kaputar National Park
- Warrumbungle Volcano—Warrumbungle National Park
- Mount Canabolas—State Conservation Area
- Monaro Range (Volcanic Province)—Snowy Mountains Region.

The Dorrigo National Park in northern New South Wales overlies the eroded Ebor shield volcano and is geographically part of the Gondwana Rainforests of Australia World Heritage site. Ebor is a central volcano province with ancient lava flows up to 400 m thick. Due to extensive erosion over time it is thought that only 10 % of the original volume of the volcano remains today. While the main focus of the Dorrigo National Park is the unique rainforest environment and the remarkable wildlife, which is attractive especially for birdwatchers, interesting geological features such as lava flows and columnar jointing covered by waterfalls certainly extend the appeal to a wider audience.

Further inland near Narrabri the Nandewar Range is a reminder of the previously active Nandewar shield volcano, which is thought to have once reached a height of 2,100 m with a diameter of 50 km (Fox 1994). Remnant volcanic landforms located within the Mt. Kaputar National Park give an indication of this violent past. Lava flows include different rock types such as rhyolite, basalt and trachyte. One of the most remarkable features of the park is **Sawn Rocks**, a stunning example of columnar jointing of a trachyte flow that resembles a wall of organ pipes.

Fig. 20.33 The volcanic landscape of the Warrumbungles in a view from the southern entrance to the National Park. In the centre Crater Bluff. *Photo* Patricia Erfurt-Cooper

> Mount Kaputar National Park contains many outstanding examples of landforms associated with volcanism. These include a circular set of dykes and the remains of numerous tiered lava terraces (NSW National Parks and Wildlife Service 2006).

The Warrumbungles are one of the most distinctive volcanic landscapes in Australia and were described by Cochrane and Joyce (1986) as an extensive and spectacular geomorphological site (Fig. 20.33) with a wide selection of outstanding volcanic features including remnant plugs and domes, dykes and sills, lava flows and tuff layers as well as horizontal and vertical columns (Duggan and Knutson 1993; Ferrett 2005; Johnson 2004). The Warrumbungle National Park includes the central region of the original Warrumbungle volcano which is considered as one of the best examples of a central shield volcano at the eastern seaboard of Australia (Yeates 2001). Because of their geological heritage and their unique biodiversity the Warrumbungles have been included as a site of significance in the National Heritage List (Australian Government 2005).

The dormant Mount **Canabolas** (1,395 m) is an eroded shield volcano located in a State Conservation Area south west of the city of Orange. Evidence for the violent past of the Canabolas volcanic complex can be observed throughout the area with basalt cliffs and rocky outcrops, lava flows, vents, dikes, domes and plugs as visible elements of the geological heritage in New South Wales. Many other volcanic sites exist, perhaps less known and not yet discovered for tourism (e.g. Comboyne shield volcano, Belmore Volcanic Province).

> The high number of volcanic features that characterise the Warrumbungles include many plugs, dykes, and flows, which give their skyline a distinctive and unusual jagged silhouette (Australian Government 2005).

20.10.1 Mount Kosciuszko National Park

Although strictly speaking Mount Kosciuszko (2,228 m) is unrelated to recent volcanic activity, it is interesting to note that the National Park includes vast areas of granite and granodiorite, as well as the Gooandra Volcanics and the Snowy River Volcanics. Mount Kosciuszko and the surrounding Snowy Mountains are part of the Great Dividing Range, which stretches more than 3,500 km from Cape York in northern Queensland along the coastline through New South Wales to western Victoria in the south (Australian Government 2013). Perhaps better known for the nearby ski fields than for geotourism the area nevertheless harbours interesting geological features in the form of eroded granite intrusions on Mount Kosciuszko, which stand up as large boulders where the surrounding sedimentary rocks have been eroded away over time (Fig. 20.34).

Fig. 20.34 Mount Kosciuszko's granite intrusions have been exposed by weathering over time. *Photo* Patricia Erfurt-Cooper

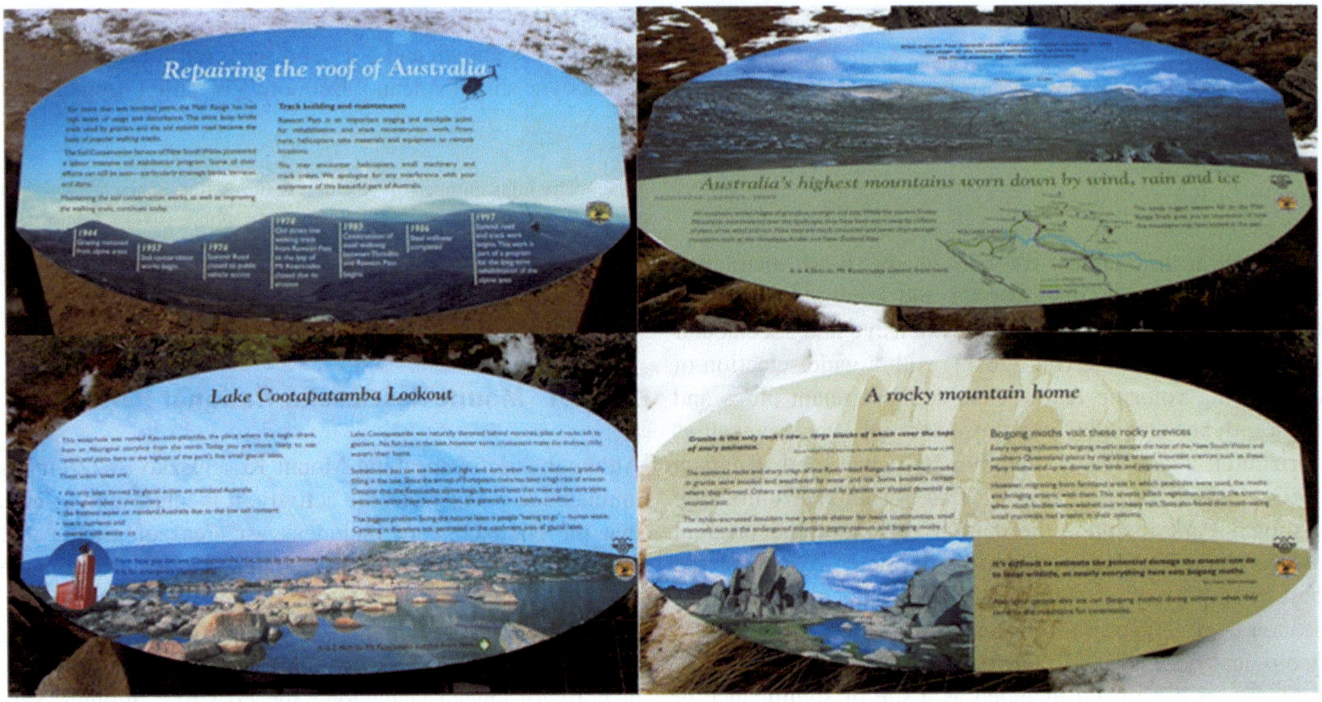

Fig. 20.35 Educational sign boards along the track to the summit of Mount Kosciusko. *Photo* Patricia Erfurt-Cooper

Left behind are glacial moraines and glacier lakes (e.g. Lake Cootapatamba) with prominent rock formations that were exposed by wind, rain and snow and which are protected within the national park of Mount Kosciuszko (Mount Kosciuszko National Park 2013). Raised walking trails to the summit are rather educational with signs posted next to the metal walkway informing about the geology, the vegetation and the local wildlife. Using a camera to capture the abundance of information for later perusal is highly recommended, as there are numerous sign boards (Fig. 20.35) that describe the landforms and the vegetation as well as the measures taken to protect the environment.

20.11 Victoria and South Australia

20.11.1 Newer Volcanic Provinces

The geological heritage of the Newer Volcanic Provinces in the southern States of Victoria and South Australia includes approximately 400 small shield volcanoes and explosive vents (Table 20.6). Countless national parks and the large Kanawinka Geopark are popular attractions between Melbourne and Mount Gambier where tourists can visit many well known volcanic sites. Examples include the Organ Pipes National Park with its basalt columns and Mount Eccles National Park, a volcano complex with lava caves, a crater lake and stony rises. Another spectacular example is the Mount Napier complex, where volcanic processes have been more recent and much of the history of the Newer Volcanic Provinces in Victoria can be interpreted here (Australian Government 2005). Located within the Kanawinka Geopark, not far from Warrnambool, the Tower Hill State Game Reserve lies inside a dormant volcano where natural and cultural resources are protected. Volcanic features of the complex include a crater lake and several scoria cones in the centre of the caldera. Tower Hill is also known as a 'nested maar' and is surrounded by a tuff ring with large sections of exposed ash layers (Fig. 20.36) which represent the record of eruptive events and their sequences (Department of Primary Industries 2013). In many areas it was common to quarry volcanic rock material (e.g. blue stone or basalt and scoria) for buildings and road works. At Tower Hill and the scoria cone Mount Elephant traces of this commercial activity can still be observed (Fig. 20.37).

The Newer Volcanic Provinces and their heritage extend into **South Australia** with the Mount Gambier maar complex and Mount Schank (Fig. 20.38), a remnant volcanic cone located 15 km south of the city of Mount Gambier. The Blue Lake of Mount Gambier is used as the city's water supply (Fig. 20.39) and is famous for its annual change of colour ranging from cobalt-blue between December and March to a steely blue-grey between April and November, which is most likely based on temperature changes. The craters of the Blue

Table 20.6 Examples of popular tourism sites with volcanic heritage

Newer Volcanic Provinces (Victoria, South Australia)
VICTORIA
Newer Volcanic Provinces—approximately 400 small shield volcanoes and explosive vents including
Tower Hill Complex—nested maar, crater lake, scoria cones, tuff ring with exposed ash layers
Mount Napier State Park—scoria cone, lava flows, tumuli cluster, Byaduk Caves
Mount Eccles National Park—scoria cone, crater lake, lava caves, lava flows
Lakes and Craters Region—Leura Maar, Mount Leura, Mount Sugarloaf, scoria cone, last eruption possibly 5,000 years ago
Mount Elephant—scoria cone, last eruption possibly 5,000 years ago
Mount Noorat—scoria cone with crater, last eruption possibly 5,000 years ago
Red Rock Reserve—scoria cone, maar craters
Mount Porndon—composite volcano, stony rises, lava flows, tuff deposits
Lake Gnotuk—maar, volcanic crater lake, tuff ring
Lake Bullenmerri—maar, volcanic crater lake, tuff ring
Lake Purrumbete—maar, volcanic crater lake
SOUTH AUSTRALIA
Mount Gambier Volcanic Complex State Heritage Area—series of volcanic cones, maar craters (Blue Lake, Valley Lake, Leg of Mutton Lake, Brownes Lake), last eruption appr. 4,300 to 4,500 years ago, expected to reactivate at some stage in the future
Mount Schank Volcano State Heritage Area—South Australian Geological Monument, volcanic cinder cone, distinctive cone shape, walking tracks, some interpretive signs, last eruption appr. 4,300 to 4,500 years ago

Fig. 20.36 The crater rim of the Tower Hill volcano reveals an exceptional display of exposed ash layers which represent eruption sequences. *Photo* Patricia Erfurt-Cooper

Lake, the Valley Lake (or Green Lake), the Leg of Mutton Lake and Brownes Lake (both dried up) are surrounded by walking tracks, viewing areas, boardwalks and picnic spots for recreation (Fig. 20.40). Mount Gambier's and Mount Schank's eruption history is considerably recent with an estimate of between 4,300 and 4,500 years.

The Newer Volcanic Provinces and their attractions are too plentiful to explore in detail here. To do them justice it would take a separate book to describe the geologic heritage of Victoria and South Australia and how this heritage is part of numerous tourism destinations including local national parks and protected sites as well as the Kanawinka Geopark.

Fig. 20.37 Mount Elephant, a scoria cone formerly used as a quarry for building material. *Photo* Patricia Erfurt-Cooper

Fig. 20.38 Mount Schank is located close to the city of Mount Gambier (SA). The photo shows the southern maar while the main crater is to the left of the sign. Insert: Mount Schank. *Photos* Patricia Erfurt-Cooper

Fig. 20.39 Mount Gambier's Blue Lake, the city's water reservoir. *Photo* Patricia Erfurt-Cooper

Fig. 20.40 The Valley Lake (Green Lake) is the second largest crater lake of the Mount Gambier volcano complex. *Photo* Patricia Erfurt-Cooper

Fig. 20.41 Red Rock, a scoria outcrop with nearby maar craters and Lake Corangamite in the background. *Photo* Patricia Erfurt-Cooper

Worth mentioning is the Lakes and Crater Precinct, which is included in the Kanawinka Geopark area and contains significant volcanic features like Red Rock (Fig. 20.41), one of Victoria's youngest volcanoes. Red Rock is located near the small town of Alvie between Lake Colac and Lake Corangamite, the largest permanent salt water lake in Australia (Kanawinka Geopark, 2013; Otways Tourism 2013). The ancient landscapes of Victoria and South Australia have long been known for their natural tourist attractions, which are promoted together with the Great Ocean Road.

20.12 Western Australia

20.12.1 West Kimberley Volcanic Zone

The West Kimberley volcanic zone does not appear to be well covered in the literature, but has been described by Ferrett (2005) in a brief overview of the volcanic geoheritage in Western Australia. Geological features include volcanic intrusions, eroded maars and diatremes, remnants of tuff rings, volcanic plugs and columnar joints. Although the volcanic heritage is recognised as outstanding in geological terms, visitors of the Kimberleys are more likely to be fascinated by the unusual rock formations of the World Heritage listed 'tiger striped' sandstone domes of the Bungle Bungles in the Purnululu National Park. So far Western Australia's volcanism has only been of interest for the mining industry with the main focus on the Argyle Diamond mine in the north of the State. Because of the commercial exploitation of this volcanic pipe, the area is out of reach for tourism development.

20.13 Tasmania

The volcanic heritage of Tasmania, although still underreported, is recognised at several geo-sites. One of the more significant sites is 'The Nut', a flat topped volcanic circular headland on the north-west coast of Tasmania (Fig. 20.42). All that remains from erosion and weathering is the basalt core of a solidified former lava lake.

Geological maps (Fig. 20.43) and records show, that Tasmania has a volcanic history, which is exposed in granite intrusions and outcrops of dolerite such as Cradle Mountain (1545 m), a major feature of the Lake St Clair National Park (Fig. 20.44). This national park is part of Tasmania's Wilderness World Heritage Area in the Central Highlands and is located 165 km north-west of Tasmania's capital of Hobart.

With a new focus on geotourism many of Tasmania's unique geological characteristics have been recognised and according to Manchester (2013) a geological trail of 100 sites has been described in recent publications. A number of geo-sites based on volcanic heritage (e.g. basalt columns, granite outcrops, dolerite intrusions, lava flows) are already included in the geo-trails and the existing national parks network.

Fig. 20.42 The Nut (143 m), a circular plug of an ancient volcano, is a landmark on the north-west coast of Tasmania. *Photo* Taspictures, public domain

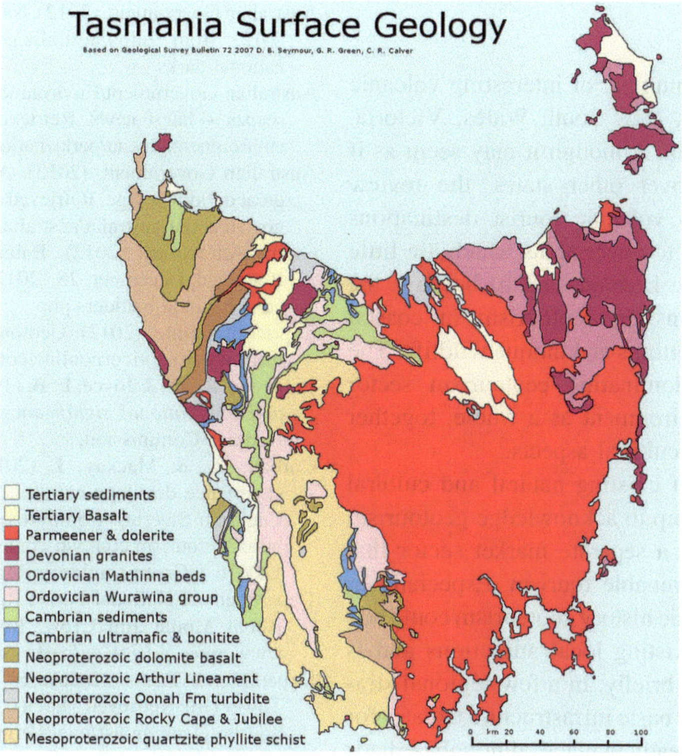

Fig. 20.43 Geological map of Tasmania. *Source* Graeme Bartlett

Fig. 20.44 Tasmania's Cradle Mountain rising behind Dove Lake. *Photo* Bjørn Christian Tørrissen

20.14 Conclusion

This chapter has reviewed a number of interesting volcanic sites throughout Queensland, New South Wales, Victoria, South Australia and Tasmania. Although it may seem as if Queensland was favoured over other states, the review reflects recent research into volcanic tourist destinations with two main objectives: a) to cover areas which are little known yet for their volcanic heritage and b) because the focus in Queensland remains on ecotourism in coastal regions based on rainforest settings and unique wildlife. The major drawcards for the dominating ecotourism sector therefore are the natural environment as a whole, together with local history and socio-cultural aspects.

However, from the use of existing natural and cultural resources it is only a small step to acknowledge geotourism either as an additional or as a separate market sector that shares the principles of sustainable tourism. Especially in areas with a significant volcanic history geotourism could add considerable value to the existing local attractions and is already mentioned, although briefly, in a few regional strategic plans. In many cases the basic infrastructure to cater for this new type of tourism is already in place, although perhaps not yet used to its full potential. More interpretive centres, signage and internet marketing with a focus on the outstanding geological heritage would be of interest for tourists, who want to learn more about the landforms they are visiting. While other states in Australia are slowly starting to consider geotourism, Queensland with its entire geological and in particular volcanic heritage has yet to acknowledge this trend. The only exceptions currently are the Central Coast 'Geotrail' from coastal Yeppoon to Boulia in the Outback and a planned geotrail in north Queensland.

Because the Australian tourism market is not as strong and successful as in the past, all potential attractions should be reviewed for new marketing strategies, whether they receive the attention they deserve and perhaps include volcano tourism as an additional segment. Extensive research of the volcanic geo-destinations in Australia has shown that there is a great potential for further expansion based on these interesting natural resources.

References

Atkinson, A., & Atkinson, V. (2001). *Undara Volcano and its lava tubes. A geological wonder of Australia in Undara Volcanic National Park, North Queensland*. Vernon and Ann Atkinson: Brisbane.

Australian Antarctic Division. (2012). Tourism: frequently asked questions. Retrieved November 20, 2012, from www.antarctica.gov.au/about-antarctica/tourism/frequently-asked-questions

Australian Government. (2005). Australian heritage database—places for decision. Retrieved December 10, 2012, from www.environment.gov.au/heritage/ahc/national-assessmentds/mcg/pubs/mcg.pdf

Australian Government. (2012). National Parks. Retrieved November 20, 2012, from http://australia.gov.au/about-australia/australian-story/national-parks

Australian Government/Environment. (2012). Australia's national landscapes —latest news. Retrieved December 28, 2012, from www.environment.gov.au/parks/national-landscapes/news.html

Australian Government. (2013). Australian rocks and mountains: The great dividing range. Retrieved January 2, 2013, from http://australia.gov.au/about-australia/australian-story/austn-rocks-and-mountains

Cairns Attractions. (2012). Babinda Boulders in Cairns, Australia. Retrieved December 28, 2012, from www.cairnsattractions.com.au/a_babinda-boulders.php

Capricorn Online. (2012). Geotourism. Retrieved November 2, 2012, from www.capricornonline.com.au/pages/geotourism/

Cochrane, R. M., & Joyce, E. B. (1986). *Geological features of national and international significance in Australia*. Canberra: Australian Heritage Commission.

Corbett, R., & Mackay, J. (2012). Australia's Green Cauldron—experience development strategy. Prepared for: Australia's Green Cauldron Steering Committee. Retrieved February 2, 2012, from http://trctourism.com/wp-content/uploads/2012/07/120616-1-Green-Cauldron-Report-EMAIL.pdf

Department of National Parks, Recreation, Sport and Racing. (2012). About Mount Jim Crow. Retrieved November 20, 2012, from www.nprsr.qld.gov.au/parks/mount-jim-crow/about.html

Department of Primary Industries. (2013). Victorian resources online—Tower Hill. Retrieved March 6, 2012, from http://vro.dpi.vic.gov.au/dpi/vro/glenregn.nsf/pages/eruption_points_tower_hill

Destination NSW. (2012). Environmental tours: Mount Gower Trek. Retrieved November 20, 2012, from www.visitnsw.com/things-to-do/tours/environmental-tours

Dig the Tropics. (2013). Dig the Tropic Geo-tourism trail—linking the outback to the reef. Retrieved August 8, 2013, from www.digthetropic.com.au/

Discover Norfolk Island. (2012). Norfolk Island visitor information. Retrieved November 20, 2012, from www.discovernorfolkisland.com/norfolk/

Duggan, M. B., & Knutson, J. (1993). *The Warrumbungle Volcano. A geological guide to the Warrumbungle National Park*. Canberra: Australian Geological Survey Organisation.

Erfurt-Cooper, P., & Cooper, M. (2010). *Volcano and geothermal tourism: Sustainable geo-resources for leisure and recreation*. London: Earthscan.

Ferrett, R. (2005). *Australia's volcanoes*. Sydney: Reed New Holland.

Fox, P. (1994). *Mt Kaputar National Park. Parks of the Western region*. Narrabri, NSW National Parks and Wildlife Service.

Geological Society of Australia. (2013). Coalstoun Lakes Burnett district. Rocks and landscape notes: Brief backgrounds. Retrieved January 3, 2013, from http://qld.gsa.org.au/oalst.pdf

Johnson, W., Knutson, J., & Taylor, S. R. (Eds.). (1989). *Intraplate volcanism in Eastern Australia and New Zealand*. Cambridge, UK: Cambridge University Press.

Johnson, D. (2004). *The geology of Australia*. Cambridge, UK: Cambridge University Press.

Kanawinka Geopark. (2013). Lakes and Craters Precinct: Red Rock. Retrieved March 2, 2013, from www.kanawinkageopark.org.au/wp-content/uploads/2010/03/RedRock.pdf

Lockwood, J. P., & Hazlett, R. W. (2010). *Volcanoes: Global perspectives*. Oxford, UK: Wiley.

Lord Howe Island Tours (2014) Lord Howe Island—Sea to Summit Expeditions. Retrieved 10 April 2014. http://www.lordhoweislandtours.net/

Manchester, P. (2013). Geotourism: energised tourism for Tasmania. Tasmanian geographic. Retrieved November 1, 2012, from www.tasmaniangeographic.com/geotourism-energised-tourism-for-tasmania/

Mount Kosciuszko National Park. (2013). Snowy mountains geology. Retrieved January 2, 2013, from www.gondwananet.com/australia-facts-newsouthwales-mount-kosciuszko-national-park.html

NSW National Parks and Wildlife Service. (2006). *Mount Kaputar National Park plan of management*. Department of Environment and Conservation (NSW).

Otways Tourism. (2013). Red Rock volcanic reserve drive. Retrieved March 3, 2013, from www.colacotway.vic.gov.au/Files/OT_Red_Rock_Volcanic_Reserve_Drive_09.pdf

Park Ranger (2010) Phone conversation with a Park Ranger at Mount Hypipamee National Park.

Queensland Parks and Wildlife Services. (2011). Cape Hillsborough, Pioneer Peaks, Mount Ossa, Mount Martin and Reliance Creek National Parks and adjoining State Waters. Management plan 2011. Queensland Government. Retrieved January 6, 2013, from www.nprsr.qld.gov.au/managing/plan-parl-approved-2011.pdf

Stickley, T. (2011). Mt Hypipamee crater measured and explored by divers who bust its myths. Retrieved April 9, 2014, from http://lizrogersphotography.com/wp-content/uploads/2011/07/Cairns-Post.pdf

Undara Experience. (2012). Australia's accessible outback: Archway explorer. Retrieved December 28, 2012, from http://undara.com.au/tours/archway-explorer/

UNESCO[1]. (2012). Heard and McDonald Islands. Retrieved November 20, 2012, from http://whc.unesco.org/en/list/577

UNESCO[2]. (2012). Macquarie Island. Retrieved November 20, 2012, from http://whc.unesco.org/en/list/629

UNESCO World Heritage Centre. (2012). Gondwana Rainforests of Australia. Retrieved November 23, 2012, from http://whc.unesco.org/en/list/368

Whitehead, P.W. (2010) The Regional Context of the McBride Basalt Province and the Formation of the Undara Lava Flows, Tubes, Rises and Depressions. Proceedings 14th International Symposium on Vulcanospeleology. Retrieved 10 April 2014. http://www.vulcanospeleology.org/sym14/papers/Whitehead-McBride.pdf

Willmott, W. (2006). *Rocks and Landscapes of the National Parks of Central Queensland*. Queensland Division, Brisbane: Geological Society of Australia.

Wilson, R. F., Turton, S. M., Bentrupperbäumer, J. M., & Reser, J. P. (2004). Visitor monitoring systems for the wet tropics world heritage area. Case studies—biophysical assessment. Vol 3. Retrieved February 2, 2013, from www.rainforest-crc.jcu.edu.au/publications/visitor_monitoring.htm

Yeates, A. N. (2001). An assessment of Australian geological sites of possible national or international significance. Volume 1: Rocks and landforms. Report for the Australian Heritage Commission, August 2001.

21 Halema'uma'u's Lava Lake Brightens After Dark: Geotourism in Hawai'i Volcanoes National Park

Lisa M. King and Ted A. Brattstrom

She is Pele-honua-mea, Pele of the sacred land. She is Pele-'ai- honua, Pele the eater of land, when she devours the land with her flames. She rules the volcanoes of Hawai'i, and Mankind has no power to resist her. When Pele is heard from, her word is the final word. — *Herb Kane, 1987*

Fig. 21.1 On a still evening, the bright glow of Halema'uma'u's lava lake reflects high into its gaseous plume as seen for the Jaggar overlook. *Photo* Alan Cressler

21.1 Introduction

Geotourism plays a key economic role throughout the Hawaiian Islands (King 2010a). Perhaps nowhere in the state is this more apparent than on the Island of Hawai'i.

L. M. King (✉)
Curtin Sarawak Research Institute, Curtin University, Miri, Sarawak, Malaysia
e-mail: lisa.m.king@curtin.edu.my

T. A. Brattstrom
Korea International School (KIS), Seoul, Korea
e-mail: ted@hawaii.edu

For volcano collectors, geotourists or those just wanting to watch eruption activity up close in relative safety, few locations in the world match the opportunities presented within Hawai'i Volcanoes National Park (Fig. 21.1). The park, established in 1916, includes the summit and major rift zones of Mauna Loa and Kīlauea, two of the world's most active volcanoes (U.S. National Park Service [NPS] 2008). Hawai'i Volcanoes National Park encompasses 323,431 acres (130,888 ha) (NPS 2008) and showcases a wide array of volcanic features including two calderas, numerous pit craters, pyroclastic cones, steam vents, tree molds, lava tube caves, fissures and volcanic surfaces such as 'a'ā and pahoehoe lava flows. The most popular destination on Hawai'i Island, the park attracted over 1.48 million recreational visitors in 2012 (U.S. National Park

P. Erfurt-Cooper (ed.), *Volcanic Tourist Destinations*, Geoheritage, Geoparks and Geotourism,
DOI: 10.1007/978-3-642-16191-9_21,

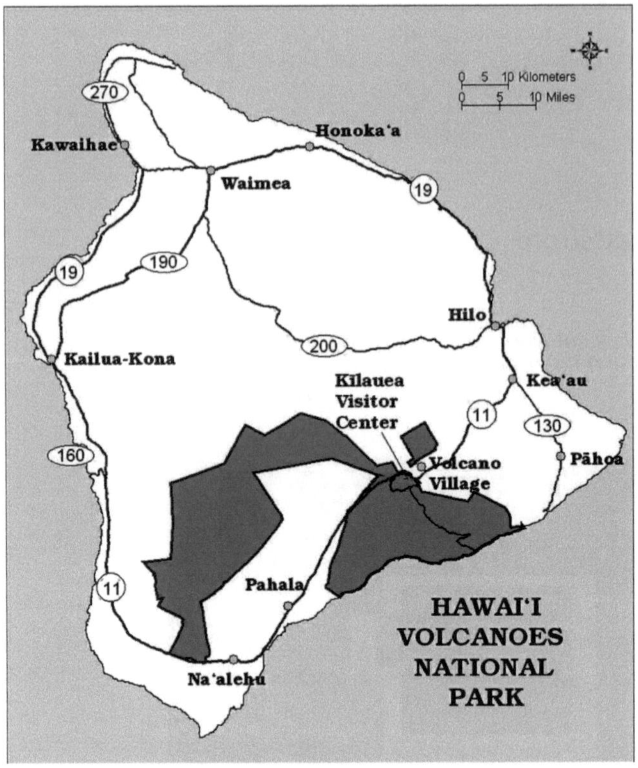

Fig. 21.2 Map of showing location of Hawai'i Volcanoes National Park. Map generated by the U.S. National Park Service and adapted and used with permission by T. Heggie 2005

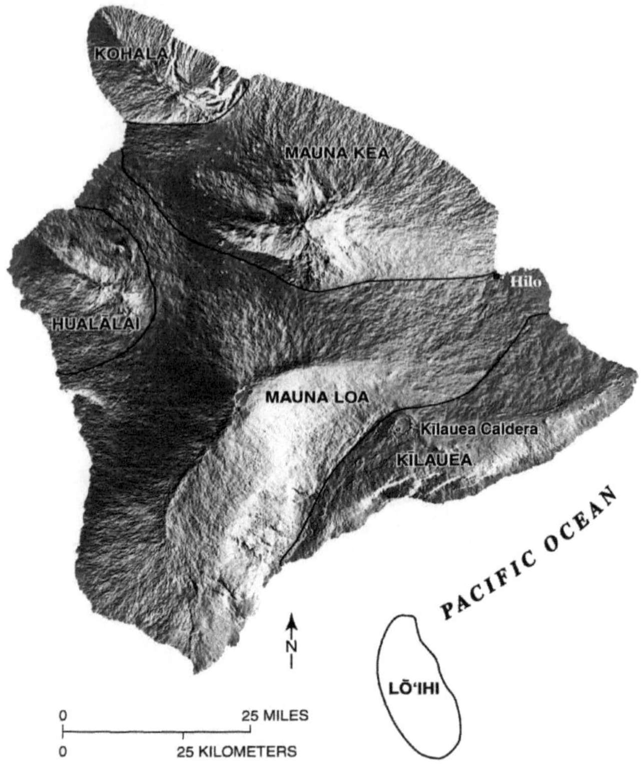

Fig. 21.3 Digital elevation model of Hawai'i Island, showing the surface boundaries between the five volcanoes that make up the island. Model created by Bob Mark, USGS and taken with permission from Duffield 2004

Service 2013). The location of Hawai'i Volcanoes National Park is shown in Fig. 21.2.

This chapter describes the after dark viewing of the Halema'uma'u's summit eruption within Kīlauea caldera inside Hawai'i Volcanoes National Park. Beginning with a brief description of Kīlauea volcano, it's summit caldera and Halema'uma'u pit crater, it then introduces Pele, the quick tempered goddess of Hawaiian volcanoes. The chapter moves on to a discussion of after dark geotourism at the Jaggar Museum overlook and identifies some of the adjustments park management has made to manage visitor risk and accommodate increased evening visitation.

21.2 Kīlauea Volcano

The youngest and most southeastern of the five volcanoes that form Hawai'i Island, Kīlauea volcano rises 4,190 ft (1,277 m) above sea level and covers an area of 552 mi^2 (1,430 km^2). The volcano comprises almost 14 % of the Island of Hawai'i (USGS Hawaiian Volcano Observatory [HVO] 2011a; Fig. 21.3).

Scientists estimate Kīlauea formed sometime during the past 300,000–600,000 years, and only broke the surface of the Pacific Ocean 50,000–100,000 years ago (HVO 2011b). Kīlauea is categorized as a shield volcano, so named because the profile of this type of volcano typically resembles an ancient warrior's shield lying on the ground (Kious and Tilling 1996). Though shield volcanoes may have periods of explosive eruptions (USGS Newsroom 2011c), more often they discharge immense quantities of lavas that flow and fan out over great distances. Over time, thousands of these gently sloping layers cover one another and build the shield volcano upwards and outwards (US Geological Survey 2011a). The slopes of Kīlauea are so gentle and its summit so broad that "visitors accustomed to the steep cones of mainland volcanoes find it difficult to grasp what they are seeing (Hazlett and Hyndman 1996, p. 12)". It is not uncommon for visitors arriving at Hawai'i Volcanoes National Park to walk inside Kīlauea Visitor Center and ask park staff at the interpretive desk, "Excuse me, but where is the volcano?"

Kīlauea is considered to be one of the most active volcanoes in the world (HVO 2011a). Eruptions occur both within the summit area and along its rifts zones. Since

Fig. 21.4 An aerial view of Pu'u 'Ō'ō. *Photo* Lisa M. King

Fig. 21.5 Visitors watching a slow moving pahoehoe lava flow inside Hawai'i Volcanoes National Park. *Photo* John Penisten

January 1983, there has been a near continuous outpouring of lava along Kīlauea's east rift zone. Eruption activity shifts up and down the rift zone over time with a focal point being the Pu'u 'Ō'ō (pronounced poo-ŭ oh-oh) cone (HVO 2011d).

The ongoing eruption of Kīlauea volcano has laid the foundations for the island's helicopter tour industry. While commercial helicopters are not allowed to fly over the Kīlauea summit area, they are allowed to fly over the eastern rift zone and share amazing views of Pu'u 'Ō'ō and any other area eruption activity, including fly-overs of active lava flows and temporary fountaining events (Fig. 21.4).

When active lava flows are accessible to reasonably safe visitation on foot, the park provides designated paths allowing visitors to walk across the lava fields to experience the active flows firsthand. During these periods, hundreds to thousands of visitors walk out to the flows every day (Fig. 21.5).

Fig. 21.6 A view of Kīlauea caldera and Halema'uma'u with newly unfurled golden-red fronds of the 'Ama'u (*Sadleria cyatheoides*) in the foreground. *Photo* Brad Lewis

21.3 Kīlauea Caldera and Halema'uma'u Pit Crater

Both the volcano and the summit caldera are known as Kīlauea (HVO 2011a). A caldera is a huge round or elliptical depression, at least two-thirds of a mile (1.07 km) wide found at the summit of a volcano (Bourseiller and Durieux 2002). Calderas usually form when a shallow underground magma chamber is emptied by a change in eruption activity. The loss of the magma causes the overlying rock roof to collapse, creating a large depression (USGS Volcano Hazards Program 2011b). Calderas typically have flat floors and very steep sides (Hazlett and Hyndman 1996). Kīlauea's caldera is approximately 1.9 miles (3 km) by 3.1 miles (5 km) in size and about 541 ft (165 m) deep (HVO 2011a; Fig. 21.6).

In comparison, a pit crater also has a generally circular shape and steep sides but is much smaller than a caldera (USGS Volcano Hazards Program 2011b). Kīlauea has several pit craters along its rift zones, a common characteristic among shield volcanoes (USGS Volcano Hazards Program 2011c). Halema'uma'u is a pit crater located on the floor of Kīlauea caldera. Though no one has actually observed the formation of a large pit crater, scientists believe the process is similar to that of caldera formation. Additionally, some pit craters may partially fill with lava and develop molten lava lakes (USGS Volcano Hazards Program 2011c).

In March 2008, renewed volcanic activity within Halema'uma'u was signaled by a series of small explosions and the collapse of a portion of the crater floor, forming a new vent. This event was the first activity inside the pit crater since 1924 (Thompson 2008a). Toxic sulfur dioxide emissions, other gases and occasional tephra from this new vent steadily became more intense over the next few weeks. Volcanologists viewing the new vent soon suspected a lava lake had formed as a weak glow was visible at night, but could not verify their suspicions as the constant plume of gases obscured any observation.

Nearly six months later, a helicopter pilot confirmed a lava lake had indeed appeared in the bottom of the crater (Thompson 2008b; Fig. 21.7). Over time, the vent and its lava lake has expanded in size and the fiery orange incandescent glow can be clearly seen at night above the rim of the pit crater.

Fig. 21.7 A view of Halema'uma'u's lava lake with Jaggar Museum and the Hawaiian Volcano Observatory perched on the rim of Kīlauea in the distance. *Photo* USGS, T. R. Orr (March 3, 2011), used with permission

21.4 Pele: The Goddess of Hawaiian Volcanoes

Inside Halema'uma'u lives the fiery and passionate Pele, the goddess of volcanoes in Hawaiian mythology (Fig. 21.8). Hawaiian legends tell of how Pele came to make her permanent home within Halema'uma'u. As Westervelt (1999) relates, Pele decided she wanted to find a new place to live. Thus, she traveled in a large voyaging canoe to the Hawaiian Islands. Pele brought with her a magical digging stick named Pāoa. Pāoa made the great fire pits Pele needed to live in whenever she dug it repeatedly into the ground. Pele worked her way down the Hawaiian island chain, unable to establish a satisfactory home, until at last she found Kīlauea. However, Kīlauea was already occupied by another volcanic god, 'Ai lā'au, the forest eater. 'Ai lā'au had seen Pele when she first landed on Hawai'i Island and he became terribly afraid. As soon as 'Ai lā'au saw Pele walking towards his house, he decided to quickly abandon Kīlauea and move elsewhere. Pele found Kīlauea empty and used Pāoa to make Halema'uma'u her perfect home where she still lives today (Westervelt 1999).

Fig. 21.8 An image of Pele, the quick tempered goddess of Hawaiian volcanoes who makes her home in Halema'uma'u's pit crater. *Image* Herb Kane Estate, used with permission

21.5 Thomas A. Jaggar Museum

One of the best places in the park to view the summit eruption is at the Thomas A. Jaggar Museum, named after one of the first geologists to study Hawaiian volcanoes (Wright, Takahashi and Griggs 1992). The museum, built from rough-hewn basalt blocks, contains videos, displays on Hawaiian volcanoes and presents some of the research conducted by the adjoining Hawaiian Volcano Observatory (Fig. 21.9). Visitors can view real time seismographic information recording earthquakes in a variety of locations within the park along with current measurements of ground tilting as the volcano inflates and deflates. Large, colorful murals present stories about Pele and other Hawaiian gods and goddesses. Displays include examples of lava bombs, Pele's hair, 'a'ā and pahoehoe lava (Fig. 21.10). To allow visitors to view the slowly churning lava lake currently in

Fig. 21.9 Jaggar Museum on the *left* with its famous overlook perched on the rim of Kīlauea looking directly towards Halema'uma'u. The glass tower of the Hawaiian Volcano Observatory tower is seen behind the museum. *Photo* Ted A. Brattstrom

Fig. 21.10 Visitors inside the Jaggar Museum (photo taken before a complete refurbishment a short time later) have the opportunity to learn about different lava types. Displays such as these encourage visitors to touch individual pieces of igneous rocks. *Photo* Ted A. Brattstrom

Halema'uma'u, a live webcam relays images from a camera set up on the crater rim to a computer monitor in the museum. (To view visible and thermal images of the lava lake go to http://hvo.wr.usgs.gov/cams/panorama.php?cam=HMcam and http://hvo.wr.usgs.gov/cams/panorama.php?cam=HTcam). Park rangers are often available to interpret the video images, answer questions and keep an eye on eruption activity.

The Jaggar Museum overlook space is defined by a low, broad rock wall. For most visitors, this is a once in a lifetime opportunity to witness a volcanic eruption firsthand. The gas and ash plume is visible 24 hours a day. However, many visitors forget that sunlight makes the lava lake's incandescence invisible during daylight hours. Thus, one of the most common questions any day ranger stationed at the overlook receives is "Why isn't it glowing?"

21.6 Geotourism After Dark at the Jaggar Overlook

Before dusk, scores of visitors begin gathering at the Jaggar Museum overlook to witness the increasingly visible incandescence of Halema'uma'u's lava lake as night falls. The molten lava lake cannot be viewed directly from the overlook as it is too deep within the pit crater, but its fiery orange-yellow light illuminates the crater walls and causes the gas plume to glow spectacularly as the sky darkens (Fig. 21.11).

On occasion, sections of the vent wall falls into the lava lake, emitting a dense cloud of ash invisible to the after dark geotourist. The ash cloud suddenly and completely obscures the glow, creating the startling illusion that the lava lake has unexpectedly drained and the summit eruption has ceased or moved elsewhere. Minutes later, the cloud's blanketing effect dissipates and the glow once again materializes.

Rock falls and gaseous emissions also create audible rumblings that can be heard by visitors at the overlook on still nights (Okita 2011). When eruption activity along the east rift zone is pronounced, the glow from Pu'u 'Ō'ō may also be spotted from the Jaggar overlook area.

On clear moonless nights, the Milky Way is clearly visible above the Park (Fig. 21.12). In 1919, one eloquent visitor stood in awe during a clear night looking at the crater and wrote "I stood on the brink midst the darkness and the stars, and watched o'er come by the wonder of it all" (Bevens 1992, p. 168).

Not surprisingly, since the beginning of the summit eruption, after dark visitation to the park has risen dramatically. According to the statistics kept by the U.S. National Park Service [NPS], Hawai'i Volcanoes National Park visitation after 5:00 p.m. has skyrocketed since the start of the more active phase of the Halema'uma'u eruption

Fig. 21.11 On a still evening, the bright glow of Halema'uma'u's lava lake reflects high into its gaseous plume as seen from the Jaggar overlook. *Photo* Alan Cressler

Fig. 21.12 Halema'uma'u with the down slope Pu'u 'Ō'ō eruption visible on the *left*. *Photo* Brad Lewis

(NPS staff member, pers. comm., 2011) with the bulk of after-hours visitors arriving is generally between 5:00 and 8:30 p.m. (Fig. 21.13).

Furthermore, local community members are visiting the park more frequently to view the glow. As the quality of the incandescence varies depending on viewing conditions and level of eruption activity, many residents regularly check the webcam image on the Hawaiian Volcano Observatory website to determine if conditions are worthy of a visit. If the webcam indicates Pele is putting on a show, some will choose to make a quick visit to the park to participate in the viewing experience.

Fig. 21.13 Visitors admiring Halema'uma'u's lava lake glow from the Jaggar Museum overlook. *Photo* Ted A. Brattstrom

As Kīlauea's eruption activity and its accessibility plays a significant role in attracting visitors to the park (King 2010a, b), Hawai'i Island tour companies have learned to adapt to the ebb and flow of eruption activity. If lava flows are easily accessible, new tours are developed or variations of previous tours are offered to attract visitors who would like to experience viewing the summit glow or walk out to the lava fields to view active flows. If the lava flows are inaccessible or have briefly paused, these tours are either shelved or redesigned to focus on other park attractions.

The Halema'uma'u eruption provides a unique opportunity for island tour companies. An eruption event of this nature has not occurred for nearly 100 years, making it a 'must-see' part of the national park visit. Hawaii Island tour companies have adjusted their offerings to take advantage of the tourist demand and most offer the opportunity for clients to choose a tour that includes staying at the overlook until dark. Furthermore, the summit eruption is literally 'drive-up' accessible, clearly visible as dusk falls and is very photogenic. This rare event has provided a fresh reason for visitors and Hawai'i residents to travel to the park and stay until after dark. Hawai'i Island tour companies have adjusted their offerings to take advantage of the tourist demand and most offer the opportunity for clients to choose a tour that includes staying at the overlook until dark.

21.7 Managing Safety During Evening Overlook Visits

Keeping visitors safe is Hawai'i Volcanoes National Park's top priority. The park has taken several steps to minimize visitor risk during this eruptive phase.

In the initial stages of the vents opening in 2008, it was seen that trade winds pushed the volcanic plume nearly continuously across the famous Crater Rim Drive encircling Kīlauea caldera. As sulfur dioxide levels within the plume were initially dangerously high and still vary on a daily basis, park management decided to close a section of the popular road in the interest of public safety. At the time of this writing, parts of Crater Rim Drive have been closed for over six years.

Eruption activity in the park is continuously monitored by park staff. During the initial phases of the summit eruption, park rangers were stationed daily at the overlook with a portable air quality monitor to constantly assess air quality and eruption activity. Permanent air quality monitors are now installed in strategic park locations. When Halema'uma'u's toxic plume shifts over high density visitor areas, sections of the park are closed off and evacuated, sometimes for only a few hours, sometimes for an entire day—until the air quality monitors indicate sulfur dioxide levels are low enough that the area can be safely reopened. In rare instances, the entire park has been closed until the air has been declared safe (Associated Press 2008). In the event the plume quickly shifts direction, air filtering systems have been installed in Jaggar Museum and Kīlauea Visitor Center and the exterior doors refitted so the buildings can act as emergency shelters.

Other safety measures have been implemented to insure the public is informed about the park's air quality. U.S. National Park Service and USGS produce a web page informing the public about current SO_2 levels. Additionally, a portable electronic sign board is often used near the park entrance to inform visitors of air quality levels.

21.8 Park Adjusts to New Visitor Demand

With Halema'uma'u's new eruption and visitation increasing after-hours, the park adjusted its resources to meet the increased demand for services in the Jaggar Museum area. The park entrance station hours were expanded to collect visitor entrance fees later in the evening and to monitor visitation levels. The Jaggar Museum overlook viewing area was expanded and improved night-time lighting installed to accommodate the increased number of visitors. The adjacent car park was also redesigned to improve traffic flow. New interpretive signage has been placed along the rock wall rim of the overlook.

To accommodate demand, Jaggar Museum hours were expanded into the evening, staying open as late as 8:30 p.m. Inside the museum, the live lava lake webcam has been enthusiastically received by both visitors and commercial tour operators. New and updated products associated with the summit eruption have been developed by local entrepreneurs for sale inside the non-profit Hawai'i Pacific Parks Association gift shop, such as post cards, calendars, posters and framed images. With the additional evening gift shop purchases, the Association has been able to supply more funds for regional park programs and research activities.

Few interpretive rangers are assigned to the museum area during the day as the Kīlaeua Visitor Center is open to visitors. However, during the early evening hours when the center is closed, interpretive staff are available. This situation is considered a bonus by tour operators as their clients have the opportunity to interact with an expert on the eruption activity who can answer detailed questions.

The closure of Crater Rim Drive denied commercial bus tours a logistically easy tour route around key national park features. The park has compensated for the road closure by upgrading the infrastructure around other parts of the Crater Rim Drive such as steam vent features, rim trails and viewpoints.

During the months October–November 2012 there was not only an increase in the height of the lava lake in Halema'uma'u, but also an increase in the press devoted to the eruption activity. The added media attention resulted in new pressures on the park. As viewing demand increased, additional personnel were added to direct and control parking at the Jaggar Museum area. Upon occasion, the parking lot is completely full and visitors are directed to park at a nearby parking area and walk several hundred meters along a paved trail to the viewing area. Park personnel keep in touch via two-way radio to coordinate parking at both sites to help keep visitors safe.

21.9 Conclusion

Hawai'i's tourism is tightly linked with its outstanding volcanic features. Hawai'i Volcanoes National Park, located in the southeastern part of the island, is one of Hawaii's most iconic geodestinations. The summit eruption is best viewed at the Jaggar Museum overlook. Observing the spectacular orange incandescence of the lava lake inside Pele's home, Halema'uma'u's glow against the night sky is a rare once-in-a-lifetime viewing event for most visitors. Area residents are also visiting the park more often after dark. Park management has adapted operations and installed additional infrastructure to minimize risk and better serve the visitor after dark. The Halema'uma'u eruptions continue to provide an outstanding opportunity for visitors to gain an appreciation of the primal volcanic forces that created the Hawaiian Islands. The exceptional geological heritage of Pele's unique features remain forever protected for future generations within the Hawaii Volcanoes National Park.

Acknowledgments The authors would like to thank David Faber for reviewing and commenting on the draft manuscript. Also a warm mahalo to Alan Cressler, Brad Lewis, and John Penisten for permission to use their images.

References

Associated Press. (2008). *Hawaii Volcanoes National Park closed due to eruption, April 10 2008* [Press release]. Retrieved July 23, 2013, from www.foxnews.com/story/0,2933,349331,00.html

Bevens, D. (1992). *On the rim of Kilauea: Excerpts from the Volcano House Register 1865–1955*. Honolulu: Hawaii Natural History Association.

Bourseiller, P., & Durieux, J. (2002). *Volcanoes*. NY: Harry N. Abrams Inc.

Duffield, W. (2004). *Chasing lava: A geologist's adventures at the Hawaiian Volcano Observatory*. Missoula, MO: Mountain Press Publishing Company.

Hazlett, R., & Hyndman, D. (1996). *Roadside geology of Hawai'i*. Missoula, MO: Mountain Press Publishing Company.

Heggie, T. (2005). *The epidemiology and etiology of visitor injuries in Hawaii Volcanoes National Park*. PhD thesis. College Station: Texas A&M University.

Kane, H. (1987). *Pele: Goddess of Hawai'i's volcanoes*. Captain Cook, HI: Kawainui Press.

King, L. (2010a). Geotourism in the Hawaiian Islands. In D. Newsome & R. Dowling (Eds.), *Geotourism: The tourism of geology and landscape*. Oxford: Goodfellow Publishing Ltd.

King, L. (2010b). The lure of lava tubes: Exploring lava tube tourism on the island of Hawai'i. In P. Erfurt-Cooper & M. Cooper (Eds.), *Volcanic and geothermal tourism: Sustainable georesources for leisure and recreation* (pp. 67–76). London: Earthscan.

Kious, J., & Tilling, R. (1996). *This dynamic Earth: The story of plate tectonics*. USGS Special Interest Publication (On-line version). Retrieved July 23, 2013, from http://pubs.usgs.gov/gip/dynamic/dynamic.html

Okita, T. (2011). *Lava lake at Kilauea has reached one of its highest levels ever*. Hawaii News Now. Retrieved July 23, 2013, from http://www.hawaiinewsnow.com/Global/story.asp?S=14060931

Thompson, R. (2008a). *Halemaumau blast sends boulders aloft*. Star Bulletin. Retrieved July 23, 2013, from http://archives.starbulletin.com/2008/03/20/news/story02.html

Thompson, R. (2008b). *Lava lake seen on Halemaumau floor*. Star Bulletin. Retrieved July 23, 2013, from http://archives.starbulletin.com/2008/09/06/news/story08.html

U.S. Geological Survey. (2011a). *Principal types of volcanoes*. Retrieved July 23, 2013, from http://pubs.usgs.gov/gip/volc/types.html

USGS Volcano Hazards Program. (2011b). *VHP Photo glossary: Caldera*. Retrieved July 23, 2013, from http://volcanoes.usgs.gov/images/pglossary/caldera.php

USGS Volcano Hazards Program. (2011c). *VHP Photo glossary: Pit crater*. Retrieved July 23, 2013, from http://volcanoes.usgs.gov/images/pglossary/PitCrater.php

USGS Hawaiian Volcano Observatory. (2011a). *Kīlauea: Perhaps the world's most active volcano*. Retrieved July 23, 2013, from http://hvo.wr.usgs.gov/kilauea/

USGS Hawaiian Volcano Observatory. (2011b). *Kīlauea—Eruption history*. Retrieved July 23, 2013, from http://hvo.wr.usgs.gov/kilauea/history/main.html

USGS Newsroom. (2011c). *A new view of Kīlauea's explosive past* [Press release]. Retrieved July 23, 2013, from http://www.usgs.gov/newsroom/article.asp?ID=3061#.Ue5NDa1-_b0

USGS Hawaiian Volcano Observatory. (2011d). *Recent Kīlauea status reports, updates, and information releases*. Hawaiian Volcano Observatory Daily Update: Monday, November 14, 2011. Retrieved July 23, 2013, from http://volcanoes.usgs.gov/hvo/activity/kilaueastatus.php

USGS Hawaiian Volcano Observatory. (2011d). *Kīlauea volcano—tour of the summit caldera*. Retrieved July 23, 2013, from http://hvo.wr.usgs.gov/gallery/kilauea/caldera/main.html

U.S. National Park Service. (2008). *Hawai'i Volcanoes National Park Business Plan 2008*. U.S. National Park Service.

U.S. National Park Service Public Use Statistics Office. (2013). *Hawaii Volcanoes NP (HAVO)* reports—annual park recreation visitation (1904-Last calendar year). Retrieved October 27, 2013, from https://irma.nps.gov/Stats/SSRSReports/ParkSpecificReports/AnnualParkRecreationVisitation(1904-LastCalendarYear)?Park=HAVO

Westervelt, W. (1999). *Hawaiian legends of volcanoes*. Honolulu: Mutual Publishing.

Wright, T., Takahashi, T., & Griggs, J. (1992). *Hawai'i volcano watch: A pictorial history, 1779–1991*. Honolulu: University of Hawaii Press and Hawai'i Natural History Association.

Additional Information

For further information related to Hawai'i Volcanoes National Park click on the following websites.

The official Hawai'i Volcanoes National Park website. http://www.nps.gov/havo/index.htm

The U.S. Geological Survey Hawaiian Volcano Observatory website. http://hvo.wr.usgs.gov/

The Hawai'i Pacific Parks Association website. http://www.hawaiipacificparks.org/

The Friends of Hawai'i Volcanoes National Park website. http://www.fhvnp.org/

22 The Volcanic Heritage of Iran

Alireza Amrikazemi

Geographical Location of Iran

Type of Volcanoes:
Stratovolcanoes

Last known eruptions:
Quaternary

Typical eruption style:
Strombolian, Plinian & Vulcanian

Current status:
Dormant

Major attractions:
Hiking, hot mineral springs, skiing, volcanic features (Geotourism), natural landscapes

Status of protection: Registered in the Natural Heritage List of National Sites

Visitors per year:
~ 27,000

22.1 Introduction

Iran is situated in the Alpine-Himalayan orogenic belt. Two mountain ranges, the east-west trending Alborz Mountains and the northwest-southeast trending Zagros mountain ranges dominate the tectonic features of the country (Fig. 22.1).

A. Amrikazemi (✉)
Department of Geoheritage, Geological Survey of Iran, Tehran, Iran
e-mail: aamrikazemi@gmail.com

The lower to middle Eocene Epoch was a period of extensive volcanism in most of Iran, but the present physiography of the country is related to the late Tertiary period, and the volcanism which began in this period continues to the present (Amrikazemi and Mehrpooya 2006). Mountains in fact cover half of Iran's surface area, with 4 mountains of particular interest to volcano tourism. These are *Damavand, Sabalan, Taftan,* and *Sahand.* In addition there are two mud volcano regions in Iran—in Chahbahar (on the Sea of Oman) and Bandar-e-Tourkmen (near the Caspian Sea).

Since 1979 the majority of visitors to Iran have been religious pilgrims and business people (Economist Intelligence Unit 2008). While official figures do not distinguish

P. Erfurt-Cooper (ed.), *Volcanic Tourist Destinations*, Geoheritage, Geoparks and Geotourism,
DOI: 10.1007/978-3-642-16191-9_22,

Fig. 22.1 The Iranian Mountain Ranges and the location of the volcanoes described in this chapter. *Source* Map by Uwe Dedering, modified by Editor

between those travelling to Iran for business and those coming for pleasure, and they also include a large number of diaspora Iranians returning to visit their families in Iran or making pilgrimages to Mashhad and elsewhere, the number of tourists visiting Iran reached 3.2 million in 2011. Areas considered for further expansion in the tourism sector in Iran are ecotourism (geotourism), coastal tourism, restoration of historical relics, handicraft townships, and health tourism (Press TV 2010) with 30,000 people coming to Iran to receive medical treatment in 2012 (Fars News Agency 2012). Iran has in recent years earned about US$ 2 billion a year from tourism and close to 2.5 % of national employment is generated in the tourism sector. However, weak advertising, unstable regional conditions, a poor public image in some parts of the world, and absence of efficient planning schemes in the tourism sector have all hindered the growth of tourism. Nevertheless, Iran's '20-Year Vision' document projects investment of over $ 32 billion in the country's tourism sector and targets 20 million tourists by 2025 (Tehran Times 2012). In order to encourage domestic and foreign direct investment in this sector, the 50 % tax exemption previously granted to tourism enterprises has been extended to include five-star hotels (Austrade 2012). Tariffs for utilities comply with industrial ones.

22.2 Volcano Tourism in Iran

Volcano tourism in Iran is considered a part of ecotourism-type activities and is rarely mentioned as geotourism at this time, although government authorities are promoting the use of the term geotourism to describe the use of such resource (Amrikazemi and Mehrpooya 2006). The first aims of tourists in visiting volcanoes in Iran are: mountaineering as sport (hiking, climbing and skiing), and leisure and health (nature therapy using hot mineral water springs). Recently, after the geotourism and geopark concepts began to be promoted in society, tourists to Iran have begun to notice volcanoes from a new direction, geology. This interest can be seen mostly in the experiences of educational tourists, including geology and geography students, where a certain section of their field excursions are assigned to volcano visits.

In this chapter the Damavand, Sabalan and Sahand volcanoes are introduced. All of these are Quaternary volcanoes and belong to the Alborz Mountain Chain in North and Northwest of Iran. Other resources such as Mt Taftan, Iran's only semi-active volcano, and the mud volcanoes and other volcano related resources will also be briefly described.

22.3 Mount Damavand

Mount Damavand at 5,620 m is the highest peak in Iran and the Middle East and the highest volcano in Asia (Fig. 22.2). It is also considered as one of the highest (2nd) conical volcanoes in the world as well as probably being the world's biggest inter-continental volcano (Amrikazemi 2013). Mount Damavand is very close to Tehran, the capital city of Iran (60 km). The main road from Tehran to the north passes by this volcano, so it is very accessible for both domestic and international tourists. Several spa centres on the eastern side of the volcano host numerous visitors at weekends and holidays. Geo-attractions include basalt columns (Fig. 22.3), a permanent glacier (Fig. 22.4) and a deep valley (Fig. 22.5) where visitors can observe the makeup of the volcano's geology. The most important of the hot springs are located in *Abe Garm Larijan* in the Lar Valley. The water from these springs is useful in the treatment of chronic wounds and skin diseases. Near these springs there are public baths with small pools for public use.

Fig. 22.2 The eastern side of Mt Damavand viewed from Nava Village. *Photo* Alireza Amrikazemi

Fig. 22.3 Mt Damavand—basalt columns as geo-attractions which are easy to access, as they are situated near main roads (Haraz Road). *Photo* Alireza Amrikazemi

Fig. 22.4 Yakhar, the biggest glacier of Mt Damavand on the eastern side. *Photo* Alireza Amrikazemi

Fig. 22.5 Mt Damavand—view from Reyneh into a deep valley. *Photo* Alireza Amrikazemi

Mount Damavand is a stratovolcano which is currently dormant, although there are fumaroles near the summit which emit sulphur gas (Fig. 22.6). While no historic eruptions have been recorded, hot springs at the base and on the flanks as well as fumaroles and solfatara near the summit indicate a hot or cooling magma body still present beneath the volcano. Therefore Damavand is classed as a potentially active volcano with the last major eruption estimated to have occurred around 5350 BC. The main flows from Damavand have included lava flows, lahars, block lava, pyroclastic flows,

Fig. 22.6 Fumaroles near the summit of Mt Damavand emitting sulphur gas. *Photo* Alireza Amrikazemi

Fig. 22.7 The main flows from Mt Damavand include lava flows, lahars, block lava, pyroclastic flows, tuffaceous flows, and ash flows. *Photo* Alireza Amrikazemi

tuffaceous flows, and ash flows (Fig. 22.7). Pumice outflow deposits have spread over an area of 400 km^2. The main petrological components of Damavand include trachyte, trachyandesite, and alkalibasalt (Siebert and Simkins 2002). The volcano is registered in the Iranian Natural Heritage List of National Sites, which places the mountain under national protection. Mount Damavand has a special place in Persian mythology and folklore being mentioned in poems and other pieces of literature written by poets, writers, historians and tourists, old and modern (Yazdani 2010).

22.4 Mount Sabalan

Mount Sabalan (4,811 m) is the 3rd highest peak and the 2nd highest volcano in Iran (Fig. 22.8). Local people call it Savalan in the Azeri language. Sabalan is located at the center of Ardabil Province near Ardabil City and Meshkin Sharh City. As a Quaternary andesite stratovolcano Sabalan produced Strombolian and Plinian style eruptions and main flows from the volcano include lava flows, block lava, pyroclastic flows, epiclastic flows, prismatic flows,

Fig. 22.8 Mt Sabalan has in the past produced Strombolian as well as Plinian style eruptions. *Photo* Alireza Amrikazemi

Fig. 22.9 Mt Sabalan's main flows include lava flows, block lava, pyroclastic flows, epiclastic flows, prismatic flows, lahars and volcanic breccias. *Photo* Alireza Amrikazemi

lahars and volcanic breccias (Fig. 22.9). The crater lake (Fig. 22.10) at the summit is one of the highest fresh water lakes in the world and one the most important attractions of this volcano with an area of approximately 7,000 m^2 and a depth between 25 and 35 m (Amrikazemi 2013).

There are also numerous hot mineral springs around Mount Sabalan. In the town of Sarein on the south side of the volcano are several spa centers, which make this small town an important spa destination in Iran (Sarein Travel 2011). On the northwest side of the volcano are other hot springs as well

Fig. 22.10 Mount Sabalan's Crater Lake is one of the highest fresh water lakes in the world. *Photo* Alireza Amrikazemi

as cold springs which together are traditionally used for health and wellness tourism with purpose built facilities. North of Sabalan Volcano a long deep valley (Shirvan) was carved by erosion into volcanic material. A sharp layering of pyroclastic material can be seen in the walls of the valley (Fig. 22.11). These walls have a rough morphology, and columnar joints that remain after the erosion such as chimney rocks and pillars which can be seen in this valley. The variety of colours in the igneous deposits and abundance of vegetation have created unequaled landscapes in the valley.

Sabalan is a legendary mountain for Iranians, especially for the Azari people. This volcano is almost also registered in the Natural Heritage List of National Sites. Fantastic biodiversity can be observed around Sabalan Volcano and surrounding radial valleys. Added to this is the culture of the famous nomad tribe named Shahsavan who lives in the Sabalan foothills; they produce a famous Persian pile rug also named the Shahsavan Rug. The daily life of this tribe is a cultural attraction for both domestic and international tourists, which complements the volcanic tourism resources in the area (Amrikazemi 2013). The Alvarez Ski fields and cable car are among other sport and leisure facilities in the area with some 50,000 tourists per year now visiting this volcano and its surrounds.

22.5 Mount Taftan

Taftan is a Persian word which means "The Place of Heat". Mount Taftan (4,100 m) is a semi-active stratovolcano and the highest mountain in southeastern Iran located near the city of Khash in the Sistan and Baluchestan Provinces (Siebert and Simkin 2002; Amrikazemi and Mehrpooya 2006). Traces of activity at the peak and in its vicinity include sulphurous fumaroles.

According to the Global Volcanism Program (Siebert and Simkin 2002), Mt Taftan is "Highly active, sulfur-encrusted fumaroles occur at the summit of the southeast cone. The deeply dissected northwest cone is of Pleistocene age. In January 1902 the volcano was reported to be smoking heavily for several days, with occasional strong night-time glow. A lava flow was reported at Taftan in 1993, but may have been a mistaken observation of a molten sulfur flow".

Fig. 22.11 Layering of pyroclastic material can be seen in the walls of Shirvan Valley. *Photo* Alireza Amrikazemi

22.6 Mount Sahand

Mount Sahand is a massive, heavily eroded stratovolcano complex in northwestern Iran (Fig. 22.12). At 3,707 m it is the highest mountain in the Iranian province of East Azarbaijan. A diversity of craters exists together with the peak, one of them containing Lake Uromiyeh. The Sahand Mountains are directly south of the city of Tabriz, with Mount Kamal Daqi as the highest peak. Approximately 17 peaks can be accounted for reaching elevations of over 3,000 m in height. On the northern foothills of the mountain and not far from Tabriz, a winter sports complex has been established. This complex, also known as the Sahand Skiing Stadium, greets visitors with a variety of sports including skiing and snowboarding. The snow statue competition, which runs once a year at mid-winter in the stadium, is a famous and entertaining event, that attracts spectators and competitors from all over the country.

Sahand volcano has created the widest spread of alternation of cinder and lava in Iran (4,500 km^2). This volcano has numerous cinder cones some of which also have crater lakes. Two smaller peaks of Sahand are located in the Bostan Abad region: Chénaq Daqi and Almalogol. Chénaq Daqi (Fig. 22.13) is a volcanic crater with a short shield-

Fig. 22.12 Mount Sahand Volcano and Kamal Daqi Peak. *Photo* Alireza Amrikazemi

Fig. 22.13 Chénaq Daqi Crater, Mount Sahand. *Photo* Alireza Amrikazemi

like cone which is easily accessible by all-terrain vehicles (4WD). The morphology of this cinder cone and its surrounding radial waterways characterize it as one of the most spectacular features of Sahand. The Almalogol cinder cone is in the vicinity of Chénaq Daqi and is mostly characterized by the water being gathered in its crater to form a seasonal lake which has a special beauty itself (Amrikazemi 2013).

22.7 Other Volcanic Tourism Resources

There are other volcano tourism resources in Iran. These range from lava flows, basalt columns (Bates and Jackson 1982), isolated craters and volcanic villages to mineral springs (Amrikazemi and Mehrpooya 2006). In the northwest of Iran a large area is covered by lava flows from Mount Ararat in neighbouring Turkey with local faults and fractures in the base rock evident in the Maku City region. An ancient lava flow can be observed in the Maku area, where the force of water erosion has carved a channel inside the cooled lava resulting in black glossy walls adjacent to the river giving rise to the local name Zangmar or 'black snake' in Persian (Amrikazemi 2010). The basalt columns are also located near Maku as well as occurring in relation to Mount Damavand, and in the east of the country at Birjand City.

Iran also has a number of isolated volcanic craters (Fig. 22.14) in the Ghorveh region, and in the Rayen region (Fig. 22.15) (Amrikazemi and Mehrpooya 2006). In southeastern Iran the village of Meymand and in north-

Fig. 22.14 Quaternary Crater, Qorveh. *Photo* Alireza Amrikazemi

Fig. 22.15 Volcanic craters near the Rayen region. *Source* Google Earth

western Iran, Kandovan comprise dwellings either carved out of volcanic deposits or built with volcanic conglomerates (Figs. 22.16 and 22.17). Moreover, Iran has over 400 hot and cold mineral springs, most of which are found in volcanic areas. Although their mineral compositions can vary, most springs contain Carbonates, Sulphur, Sodium and Iron compounds with their temperatures ranging from cold to over 80 °C. There are also Travertine deposits associated with some of these springs, especially near Mount Sahand.

Fig. 22.16 The unique cave village Meymand in the Iranian province of Kerman has been inhabited continuously for the last 3000 years. *Photo* Alireza Amrikazemi

Fig. 22.17 Kandovan village in the province of East Azerbaijan not far from the city of Tabriz is thought to have existed for at least 800 years and was carved out of layers of volcanic remains. *Photo* Alireza Amrikazemi

22.8 Conclusion

This chapter has described a number of areas in Iran that are suitable for volcano tourism. Nature-based tourism is popular amongst both inbound and domestic tourists, and the attractions in the volcanic areas include developments based on this form of tourism. While these are more related to sports such as skiing and/or hot springs, the opportunities for more active volcano visitation do exist. Considerable work needs to be done on the conservation and interpretation of geological sites such as the volcanoes, as well as enabling the local owners of such resources to participate fully in tourism. Nevertheless, the outlook for volcano tourism in Iran is positive as more awareness is raised about natural tourism resources and their protection for future use.

References

Amrikazemi, A. (2013). *Geoheritage Atlas of Iran*. Tehran: Geological Survey of Iran Publications.

Amrikazemi, A. (2010). *Atlas of Geopark & Geotourism resources of Iran*. Tehran: National Geoscience Database of Iran & Geological Survey of Iran Publications.

Amrikazemi, A., & Mehrpooya, A. (2006). Geotourism resources of Iran. In R. K. Dowling & D. Newsome (Eds.), *Geotourism* (pp. 78–92). Amsterdam: Elsevier Butterworth-Heinemann.

Austrade. (2012). *Doing business in Iran—For Australian exporters*. Austrade. Retrieved December 20, 2012, from www.austrade.gov.au/Doing-business-in-Iran/default.aspx

Bates, R. L., & Jackson, J. A. (1982). *Our modern stone age: Los Altos*. New York: William Kaufmann.

Economist Intelligence Unit. (2008). *Iran Travel and Tourism Forecast*. August 18.

Fars News Agency. (2012). *Ahmadinejad Stresses Iran's Growing Medical Tourism Industry*. 2012-01-17. Retrieved December 20, 2012 from English.farsnews.com

Press TV. (2010). *Breaking News*. Press TV. 2010-12-06. Retrieved December 20, 2012 from http://presstv.com/Program/142594.html

Sarein Travel. (2011). *Sarein Travel, Ardabil, South Azerbaijan*. Retrieved September 20, 2013, from www.youtube.com/watch?v=QatLSQHx8VA

Siebert, L., & Simkin, T. (2002). *Volcanoes of the World: an Illustrated Catalog of Holocene Volcanoes and their Eruptions*. Smithsonian Institution, Global Volcanism Program Digital Information Series, GVP-3.

Tehran Times. (2012). *Iran's foreign tourist arrivals continue to increase*. Retrieved February 16, 2012.

Yazdani, A. (2010). *Lar, the Paradise I knew*. Tabriz: Payam Resan Publishing Institute.

Erta Ale and the Danakil Rift Zone

23

Rosaly M. C. Lopes and Jani Radebaugh

23.1 Introduction and Tectonic Setting

Ethiopia may not be the first country to come to mind when considering where to find active volcanoes to visit. Located on the horn of Africa, at the southeastern margin of the Sahara Desert, Ethiopia is better known for its archeological and historical tourism, as well as its struggles with famine and concerns with neighbors South Sudan, Somalia and Eritrea. However, one of the rare treasures of active volcanoes, a roiling and churning lake of lava (Fig. 23.1), resides within Ethiopia's borders and draws volcanologists and adventurers from around the world to its edge. Aside from the volcano, the Dallol geothermal region is one of the most wonderfully colorful and photogenic volcanic landscapes in the world, still largely unknown, though at risk of having its fragile environment damaged by unprotected tourism.

Ethiopia is the most populous land-locked nation in the world, with over 82 million inhabitants. It is also one of the earliest locations where humans settled, and its archeological sites attract many tourists. The National Museum of Ethiopia in the capital, Addis Ababa, houses "Lucy", the world's oldest and most complete adult specimen of *Australopithecus afarensis*—estimated to be 3.2 million years old. Hominid bones discovered in eastern Ethiopia dating back 4.4 million years make the country one of the earliest known locations of human ancestors. Ethiopia is largely regarded as the region from which *Homo sapiens* first set out for the Middle East and other locations. *Homo sapiens idaltu*, who lived about 1,60,000 years ago, was found at the Middle Awash site, an archeological site along the Awash River in the Afar Depression.

Tourism has increased in Ethiopia in recent years, mostly from Europe, and largely encouraged by the government, which is committed to develop tourism through a number of initiatives. Tourism is a featured component of Ethiopia's Poverty Reduction Strategy Paper that aims to combat poverty and encourage economic development. Several National Parks have been established, which include the Semien Mountains National Park, located in the Gondar Zone, which is home to a number of rare species including the Ethiopian wolf (*Canis simensis*), the Gelada Baboon (*Theropithecus gelada*) and the Walia Ibex (*Capra walie*), the latter found uniquely in Ethiopia. Tourists also come to see numerous historic sites such as the cities of Axum, the original capital of the kingdom of Axum (also known as Aksumite Empire and Abyssinia), and Lalibela, world-famous for its monolithic (rock-hewn) churches.

Ethiopia is a large country, with an area of over 1 million km^2. The Ethiopian plateau, or Roof of Africa, forms the highest continuous landmass on the continent, and ranges from 1,500 to 4,900 m. This plateau was uplifted 75 million years ago, as magma was rising up from the mantle, upwarping the crust, and preparing to split it apart. The precious Nile originates on the Ethiopian plateau as the Blue Nile, which is a point of pride for Ethiopians. The lush, subtropical plateau, producing much of Africa's coffee, honey and the grain teff, belies the occurrences of debilitating drought that mainly happen at the fringes of the country. The capital, Addis Ababa, with about 3 million residents, was established in 1886, fairly recently for this long occupied region, reportedly because the wife of the ruler Menelik III preferred to live in the cool highlands, next to abundant springs. Addis Ababa, at 2,300 m altitude, has a cool and pleasant climate, though some visitors can feel the effects of altitude sickness.

Away from the plateau, Ethiopia drops precipitously to low elevations laid bare by the ravages of a subtropical sun. The north of the country gives way to the Nile River Valley, carving out giant chasms that rival the Grand Canyon in

R. M. C. Lopes (✉)
Jet Propulsion Laboratory, California Institute of Technology, Pasadena, CA, USA
e-mail: rosaly.m.lopes-gautier@jpl.nasa.gov

J. Radebaugh
Brigham Young University, Provo, UT, USA
e-mail: janirad@byu.edu

P. Erfurt-Cooper (ed.), *Volcanic Tourist Destinations*, Geoheritage, Geoparks and Geotourism,
DOI: 10.1007/978-3-642-16191-9_23, © Springer-Verlag Berlin Heidelberg 2014

Fig. 23.1 Erta Ale lava lake in Feb. 2011. View was at night, from about 10 feet from the South Pit crater edge. Lavas filled the crater to about 30 feet below the western rim. *Photo* J. Radebaugh, February 2011

size, depth, and climate. From the northeast of Ethiopia, across the entire eastern border to the south, runs the Afar, or Danakil, Depression and the massive Great Rift Valley that form two arms of the East African Rift Zone (EARZ) triple junction (Fig. 23.2). Here, the African continent is rifting apart from itself, forming long gashes in the Earth that will eventually open up so wide that ocean will invade them. The Red Sea, the major arm of this triple junction

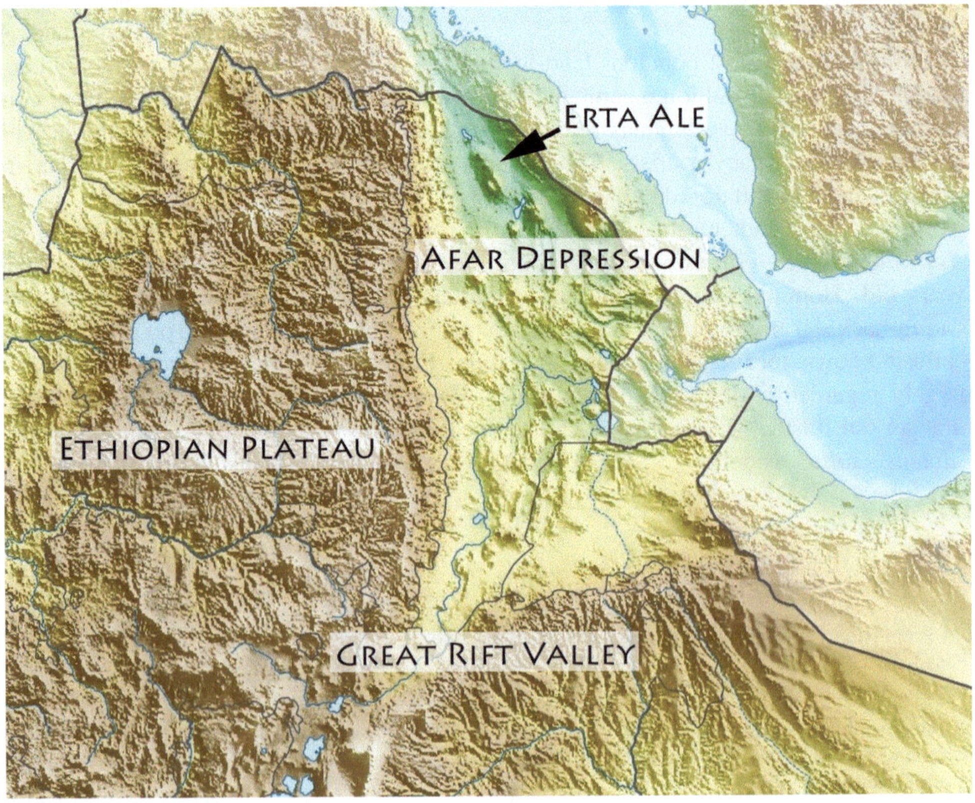

Fig. 23.2 Location of the Afar Depression and Erta Ale Volcano with respect to the Ethiopian Plateau and Great Rift Valley, on a shaded relief map. From SRTM topography

Fig. 23.3 Walls of the Afar Depression are visible rising 7,000 feet above the valley floor. A train of camels carries salt from the Great Salt Lake, past Hamad Ela base camp, up to the market at Mekele. *Photo* J. Radebaugh, February 2011

Fig. 23.4 **a** The Dallol, northern end of the Afar Depression. Spring water has percolated down to magmatic regions, been heated and suffused with minerals that it has brought to the surface and deposited on salt formations. Sulfur and iron make the coloration. *Photo* J. Radebaugh, February 2011. **b** The spectacular lake in Dallol with its vivid green waters, attributed to alophile algae. *Photo* J. Radebaugh, February 2011

parallel to the Afar Valley, is a region where rifting progressed far enough that ocean floor replaced high and dry continent. Currently, rifting in the Afar Depression and Great Rift Valley has produced large fractures in the ground, high cliff walls and low valley floors (Fig. 23.3) and, most telling of the current forces ripping the region apart, active volcanoes. Eruptions have formed low shields, gently sloped and covering broad regions with black basalt lavas. These shields and the lava flows that form them are dotted across the Afar Valley. Water emanating from

Fig. 23.4 continued

springs at the valley floor, in the Dallol geothermal region, interacts with magma, creating boiling mud pots, yellow and red sulfur mounds, and green lakes of sulfur and iron mixtures (Fig. 23.4a, b).

The Afar Depression, on the northeast border of Ethiopia (Fig. 23.2), contains at least eight major volcanic centers and numerous sites of smaller eruptions that have been active over historical times. Many of these are part of the Axial Volcanic Ranges, a zone of volcanic shields, summit calderas and fractures that run NW to SE across the center of the Afar Valley. The volcanoes are thought to be surface expressions of seafloor spreading, in which new "ocean floor" material is being produced at the bottom of the Afar Depression. Just like on the ocean floor, fractures associated with the volcanoes parallel valley walls, lavas are youngest at the center of the Axial Volcanic Ranges and older at the valley margins, and the depression is lower than sea level near the base of the volcanoes. Nearly all of the volcanoes are considered active, and several of the volcanoes have erupted recently or are currently erupting. Dabbahu (in Ethiopia) began recent eruptions in 2005, and Nabro (in Eritrea) erupted in June of 2011, with abundant ash falls and blocks. Erta Ale is the most active volcano in Ethiopia and contains one, sometimes two, lava lakes, making it one of the few places on Earth where one can see this rare volcanic phenomenon.

23.2 The Erta Ale Volcano

Erta Ale is one of the largest volcanoes in the Afar Depression, at 50 km wide. In comparison, Mount St. Helens in the Western USA is a stratovolcano about 8.5 km in diameter and Kilauea in Hawaii, a shield volcano, is about 180 km × 75 km. Erta Ale, like Kilauea, is made of fluid, dark basaltic lava flows. It is a low shield with elevation just 613 m (2,011 feet) (Fig. 23.5). Erta Ale's summit has a 0.8 km × 1.8 km (about 0.5 miles × 1 mile), elliptical caldera that currently contains two smaller, steep-sided pit craters. The summit craters are renowned for one, or sometimes two, long-term lava lakes (Figs. 23.6 and 23.7) inside the two main pit craters. Another caldera, larger at 1.8 × 3.1 km wide and elongated parallel to the trend of the Erta Ale range, is located to the SE of the summit and is likely a previously active summit caldera. Lava flows from within the northern portion of the larger caldera have locally overflowed its rim. The most recent lava flows on Erta Ale have come from the smaller, northern caldera, and have overflowed the SE and NW caldera rim. Additionally, diffuse materials surround the northern caldera and blanket older flows; these could have resulted from explosive eruptions centered in the caldera.

The lava lakes within the pit craters have been active since 1967, likely since at least 1906 (Barberi and Varet 1970; LeGuern et al. 1979). There are no reliable written

Fig. 23.5 Overlapping flows of basaltic lavas make up the slopes of Erta Ale Volcano. These flows are several decades old, as evidenced by their dull texture and vegetation. Another large shield, Ale Bagu, can be seen to the south of Erta Ale, part of the Axial Volcanic Ranges. *Photo* J. Radebaugh, February 2011

Fig. 23.6 South Pit crater, as viewed from the edge of the South Sink, contained within the Erta Ale summit caldera. *Photo* Marco Fulle, Stromboli-on-line, 2002

records for the activity and it is possible that eruptions have gone on for much longer. The Smithsonian Global Volcanism program reports three uncertain eruptions in its website (www.volcano.si.edu) in 1873, 1903, and 1904. These eruptions appear to have been more explosive (with a VEI or Volcanic Explosivity Index of 2) than the more recent lava lake activity (VEI = 0). The lack of historical records is largely due to the Afar region being an isolated, very difficult region to travel in, populated by fierce tribes that did not welcome visitors until recently. Consistent observations of Erta Ale's activity started in 1967 by Le Guern and colleagues (Le Guern et al. 1979). They reported that, between 1967 and 1973, overflows from the 70–200 m diameter lava lake filled the northern part of the North Pit

Fig. 23.7 Erta Ale summit caldera and pit craters. The final stages of our track from Dodom to the camp, the most elevated location in the image, and across the lower caldera floor are shown as *white lines*. The caldera is covered in younger, dark lava flows that have also breached the northern caldera margin. The South Sink contains the smaller South Pit, the location of the current lava lake (shown as a *red circle*). Older lava flows on the volcano summit have been blanketed in *brownish*, diffuse materials erupted explosively in the past. *Source* Image from GeoEye, Google Earth 2011

(Fig. 23.7) summit crater to a depth of about 100 m by January 1973. Continuous overflows beginning in 1971 from a 100 m diameter lava lake in the southern crater, or South Sink, had covered the entire South Sink crater floor by 1973 and, towards the end of that year, flowed across the caldera floor, over the rim and down the south flank of the volcano. Many other expeditions to Erta Ale have documented the activity, particularly in recent years (e.g. Harris et al. 2005). We observed the active lava lake in February 2011, when it was remarkably active, with churning, rifting of the lake crust, and frequent small explosions that sent incandescent fragments into the air.

The activity of Erta Ale has changed since the 1967–1968 observations of Le Guern and colleagues, shifting from the North Pit (now 350 m in diameter) to the smaller South Sink (now 180 m × 140 m in diameter; Fig. 23.7). The level of the lake can vary, although it is typically located in a smaller crater, the South Pit, 100 m below the rim of the South Sink (Oppenheimer and Francis 1998). In 2009, the lava lake surface was at a high level (Edelmann and Roscoe 2010) and the South Sink crater terrace was covered with fresh lava. Reports on the Smithsonian Global Volcanism website indicate that in February 2010 activity at the South Sink was comparable to that noted years earlier in December 2002. The upper platform inside the crater was, on 10 February 2010, covered by a lava flow, and the lava lake surface was ~20 m below the edge of the South Pit, the smaller crater near the W rim of the South Sink. On 9 February 2009 the 65-m-wide active lake in the W half of the crater rose and overflowed the lower ring terraces. In 2010, scientists from the Afar Consortium Project observed the lava lake at Erta Ale during 21–23 November. As reported to the Smithsonian Global Volcanism project, they noted the lava lake had filled the South Pit crater and breached the W rim of the South Sink, spilling two lava flows into the main caldera floor.

During our visit in February 2011, we found that the lake had overflowed and filled up the entire South Pit to within 10 m below the rim and western (most safe location at writing) vantage point, a sight rarely seen in recent years. The lake was thus nearly 50 m wide (about 150 feet), and was actively roiling and disrupting the thin crust, with fountains occurring on some portion of the lake nearly continuously. A similar situation, although with a slightly less active surface, had been observed in the previous month by Marco Fulle and others (www.swisseduc.ch/stromboli/perm/erta/index-en.html). This level of activity (as of February 2011) is the most spectacular that has been seen for many years. More recent reports of the activity are difficult to obtain, since a tragic attack on tourists in January 2012 (discussed below) slowed visits of research groups considerably.

23.3 Tourism to Erta Ale

Erta Ale, the Dallol, and surrounding attractions are spectacular areas to visit, but travel in the region is not easy and, as discussed below, can be extremely dangerous. Tourism in the region was almost non-existent before the early 1990s, when tourist groups and film crews started visiting the area. Tourism increased in the mid-2000s and, by the time our group visited the volcano in 2011, tourists came to the volcano almost daily. However, since the tragic events of January 2012, only just over a month from the time of writing, tourism has greatly decreased. It remains to be seen what steps the Ethiopian government will take to ensure the safety of visitors.

Tours starting in Ethiopia typically will depart from the city of Mekele in northern Ethiopia. Mekele has an airport which connects the city to Addis Ababa and others in the country. The journey from Mekele to Erta Ale typically takes two days using 4WD vehicles. It is possible to drive to within 7 km of the summit of Erta Ale. The hike to the summit is not a difficult one but, due to the very high temperatures in the region (over 100 °F, or 40 °C, in the day in February), it is usually done at night, taking about 3–4 h. Local Afar tribesmen serve as guides and organize the transport of gear on camels. It is possible, by previous arrangement, for visitors to use camels instead of hiking (Fig. 23.5).

Once at the summit, visitors hike down the caldera rim and across the caldera floor to the South Pit crater edge, guided by the local Afar (Fig. 23.7). The hikes up to the summit, and down to the crater floor, are not particularly technical, but they can be strenuous because of the heat, hike duration, nearly incessant and dry winds, and the necessity to hike through the night with little sleep. Tourists typically will spend an hour or more at the edge of the lava lake at night and then climb back to the huts to spend the few remaining hours of night before hiking back down. If possible, volcano viewers should negotiate spending two nights at the summit and climbing down into the crater during the day to view the lava lake both in daytime and nighttime. Be aware that temperatures are very high, easily 40 °C and sometimes reaching 45 °C. Huts provide some shade during the day, but tourists are highly exposed at the rim of the lava lake, where radiant lava and hot gases quickly drain viewers of fluids and energy. Huts for tourists are also available in Dodom, the Afar village at the base of Erta Ale.

Those visiting Erta Ale should make sure to visit other spectacular locations in the area. Dallol is an active volcano under a layer of salt, deposited when a large, former lake basin dried up. The Dallol volcano is marked by a crater formed by a phreatic, or water-rich, explosion and is the top attraction in this region northeast of Erta Ale. Several other explosion craters are present in the region, and are the lowest known subaerial volcanic vents in the world, at over 45 m below sea level. Dallol (Carniel et al. 2010) is sometimes referred to as the "gateway to Hell" though we prefer to think of it as "Io on Earth". The striking colors of the landscape, yellows, oranges, reds and greens (Fig. 23.4) reminded us of images of Jupiter's moon Io. It is truly a landscape out of this world. It last erupted in 1926, when a phreatic eruption formed the 30 m wide crater. The current activity is in the form of hot briny springs. Salts are washed out of the layers over the volcano and are transported to the surface by geothermally heated water, where they rapidly crystallize as the water evaporates. The characteristic white, yellow and red colors are the result of sulfur and potassium salts colored by sulfur and traces of ferric chloride, iron oxide, ferrous chloride and iron hydroxide. Fresh deposits are usually nearly white while older ones effectively rust and become reddish in color due to the gradual formation of iron oxides. There are also small, temporary geysers which form cones of salt. When we visited in 2011, there was a spectacular acid green lake and numerous fumaroles. The lake is rare; its presence was due to the unusually long-lasting rainy season a few months before.

The Dallol settlement, now abandoned and a ghost town, is located 130 m below sea level and holds the record high average temperature for an inhabited region on Earth. Temperatures have reached 63 °C (145 °F) in the sun. Needless to say, the Dallol region is best visited early in the morning. Potash and sylvite were mined here in the first half of the 20th century and a railway, now dismantled, connected Dallol to Mersa Fatma in Eritrea. From Dallol, it is a short drive to other attractions including the Great Salt Lake, where salt is mined, cut into slabs, and transported by camel caravans across the desert. In one day, our group visited Dallol, the mining operation at the Great Salt Lake, salt canyons, and the bubbling "Yellow Lake", located only 8 km from the border with Eritrea. The salt canyons in the SW flank of Dallol mountain are impressive, with salt pillars up to 40 m high. Since tourists had been kidnapped a few years ago in this region, we were escorted by the Ethiopian military. At that time, Erta Ale itself was not considered to be as dangerous a location, being far away from roads which could be used for a quick getaway.

On the human side, a visit to this region offers a rare glimpse at the way of life of the Afar people. Villages such as Dodom, Hamad Ela and even the larger Berhaile (which has electricity and, in comparison, appears to be back in civilization) seem to be lost in time and experiences at these places are unforgettable, such as taking a primitive shower while watching camel caravans go by in Hamad Ela (see Fig. 23.3), feeling welcomed in Dodom even though every teenager and adult male seemed to be carrying an AK-47, and having our first cold drinks in days at the local bar in Berhaile. There is no tourist infrastructure and the region is certainly dangerous, but for the adventurous it is uniquely rewarding.

23.4 Risk Management

The natural risks of Erta Ale and surrounding region are relatively low. The volcano's activity is usually confined to the lava lake and explosions are small though there is some risk of wall collapse at the crater edge (see Fig. 23.1). However, there are presently significant risks of a human nature. Despite visible security by both the military and the local Afar, tourists are still in danger from the Afar Separatists who call themselves the Afar Revolutionary Democratic Front (ARDUF). In 1995, a small group of Italian tourists were kidnapped though released after a few months. The attack resulted in an Ethiopian military campaign against the group, coordinated with the Eritrean government. During 2007 and 2008, ARDUF was responsible for several attacks on tourists, although there were no fatalities. Much worse was the recent attack during the early hours of January 17, 2012. Tourists and some scientists were sleeping in the huts at the summit of Erta Ale when they were attacked by a group of ARDUF militia. Three tourists and two scientists were shot; according to survivors, they were lined up by their attackers and executed. Two German tourists and two Ethiopian guides were abducted, but were later released. Several others were wounded but survived. ARDUF has claimed responsibility for the attack. The tourists were not robbed; the sole purpose of the attack appears to be terrorism. ARDUF does not accept the Ethiopia-Eritrea border and its purpose is to unite the Afar people who currently live on both sides of the border. The latest news is that the Ethiopian government has established a semi-permanent military unit to increase safety for tourists at the volcano, and that tourist groups are starting to return.

The volcano, containing one of the few currently active lava lakes in the world, and the surrounding Afar region are truly spectacular, but the region remains a challenging and dangerous one to venture in. Anyone wishing to go should make sure to hire a knowledgeable guide and security escort, preferably the Ethiopian military, and to work with the local Dodom Afar people, who understand the value of tourism to their village.

Acknowledgements We thank Marco Fulle and Roberto Carniel for reviewing this chapter and answering many of our questions about Erta Ale. We thank our 2011 expedition team colleagues Serina Diniega, James and Irene Dorsey, Ken Freund, and Ralph Lorenz, our guide Getnet Tilahun and 13 Suns Tours for logistical arrangements. Part of this work was carried out at the Jet Propulsion Laboratory, California Institute of Technology, under contract with NASA.

References

Barberi, F., & Varet, J. (1970). The Erta Ale volcanic range (Danakil Depression, Northern Afar, Ethiopia), *Bulletin of Volcanology, 34*(4).

Carniel, R., Jolis, E. M., & Jones, J. (2010). A geophysical multi-parametric analysis of hydrothermal activity at Dallol, Ethiopia. *Journal of African Earth Sciences, 58*(5), 812–819.

Edelmann, J., & Roscoe, R. (2010). Volcano Tourism in Ethiopia and the Danakil Rift Zone. In P. Erfurt-Cooper & M. Cooper (Eds.), *Volcano and Geothermal Tourism: Sustainable geo-resources for leisure and recreation* (pp. 45–53). Washington: Earthscan Publishing.

Harris, A. J. L., Carniel, R., & Jones, J. (2005). Identification of variable convective regimes at Erta Ale Lava Lake. *Journal of Volcanology and Geothermal Research, 142*(3–4), 207–223.

LeGuern, F., Carbonelle, J., & Tazieff, H. (1979). Erta Ale lava lake: Heat and gas transfer to the atmosphere. *Journal of Volcanology and Geothermal Research, 6*, 27–48.

Oppenheimer, C., & Francis, P. (1998). Implications of longeval lava lakes for geomorphological and plutonic processes at Erta 'Ale volcano. *Afar: Journal of Volcanology and Geothermal Research, 80*, 101–111.

24 Fernando de Noronha Archipelago: A Paradise Formed by Volcanism in Brazil

Jasmine Cardozo Moreira and Gilson Burigo Guimarães

Fig. 24.1 Location map of the Fernando de Noronha Archipelago, Brazil. *Source* Mysid—Wikimedia Commons 2011

J. C. Moreira (✉)
Tourism Department, Ponta Grossa State University,
Ponta Grossa, Brazil
e-mail: jasmine@uepg.br

G. B. Guimarães
Geography Department, Ponta Grossa State University,
Ponta Grossa, Brazil
e-mail: gburigo@ig.com.br

24.1 Introduction

This chapter presents aspects of the volcanic Fernando de Noronha Archipelago (Fig. 24.1), one of the leading eco-tourism destinations in South America, which is composed of two protected areas and recognized as World Heritage by UNESCO since 2001. The archipelago is an exclusive destination, which controls the entry and stay of visitors. Concern for the carrying capacity and strict guidelines

P. Erfurt-Cooper (ed.), *Volcanic Tourist Destinations*, Geoheritage, Geoparks and Geotourism,
DOI: 10.1007/978-3-642-16191-9_24, © Springer-Verlag Berlin Heidelberg 2014

prevent negative environmental and social impacts. The area has the most beautiful beaches in Brazil according to the ranking of the main Brazilian tourist guide, the Guia Quatro Rodas (2010). Besides the beaches and an overall natural beauty, Fernando de Noronha also has historical and geological monuments that attract domestic as well as foreign tourists. The Archipelago is a very popular honeymoon destination and is considered the best place in Brazil to go scuba diving, mainly because of the geological features that can be observed under water and which are forming an impressive backdrop. The preparation for this chapter involved the consultation of literature related to tourism in the archipelago based on its geological and geomorphological features, as well as field trips and participant observation. The objective is to present the archipelago as the best tourist attraction featuring Brazils volcanic heritage. To contextualize the archipelago, its main features and the geological and geomorphological features, many of them linked to volcanism, are presented.

24.2 Volcanism in Brazil

The Brazilian territory, in geotectonic terms, is located exclusively in areas within the South American Plate. As about 95 % of the active volcanoes in the world occur along the margins of tectonic plates involving divergent boundaries (plates spreading apart) or convergent boundaries (plates colliding), it is hardly surprising to find an absence of active volcanoes in modern Brazil. However, this does not mean that Brazil lacks typical volcanic terrain in the country. Spectacular examples in scientific terms or economic and scenic terms, which are related to magmatic events at different times in the planet's geologic history, are recorded here. In the last decade a number of studies have been carried out (e.g. Juliani and Fernandes 2010) on the older, still recognizable volcanoes in Brazil, which are associated with mineral deposits (especially gold and copper), and are located in the Amazon region of São Felix do Xingu in Southern Pará. On the other hand, the visual impact of the Iguazu Falls (Natural World Heritage under the auspices of UNESCO), is due to a spectacular sequence of basaltic lava flows of the Early Cretaceous ($\sim$133 million years), related to one of the most voluminous volcanic processes of its kind on Earth (continental flood volcanism) linked to the formation of the Atlantic Ocean and the fragmentation of the Gondwana Continent, culminating in the separation of Africa and South America (Marques and Ernest 2004).

Volcanic manifestations have dotted Brazil over the past 250 million years, mainly linked to the opening of the South Atlantic Ocean, reactivation of oceanic fracture zones and/or displacement of the South American Plate over areas of anomalously warm mantle, or *hot spots* (Mizusaki and Thomaz Filho 2004; Ulbrich et al. 2004; Alves et al. 2002). Fernando de Noronha (12–1.5 million years) and Trindade (3.7 million years to less than 350,000 years) stand out as the most recent volcanic and oceanic islands, consisting of alkaline rocks in the form of portions of large orogenic processes (Ulbrich et al. 2004).

24.3 Fernando de Noronha Archipelago

Islands typically create a positive image for tourists, because in contrast to many continental regions they have a distinct and immediately recognizable identity (Pearce 2003). This is the case with the Fernando de Noronha Archipelago, where the isolation contributes to the sensation of breaking routines, and their geological and geomorphological attributes further strengthen the visitor's vocation for geotourism. Fernando de Noronha is located in northeastern Brazil, with access only by sea or air. The archipelago has a main island of 26 km^2 (9 km^2 of marine platform and 17 km^2 of land area), which is surrounded by 18 smaller islands and dozens of rocks (Fig. 24.1). Its administration is divided into two protected areas: a Marine National Park and an Environmental Protection Area.

The north-western coast is known as the 'Inner Sea' because it is protected from the winds and ocean currents by the topography of the island. There are ten beaches, which receive large waves during the summer, coming from the northern hemisphere. The 'Outer Sea' at the southeastern coast receives wind and waves all year round and there are three sandy beaches protected by islands. The vegetation is mainly shrub and herbaceous. Regarding the fauna this is one of the most important regions for the breeding of marine birds in the Atlantic and a nursery for species such as dolphins and sea turtles (IBAMA et al. 2005). For the UNESCO (2002), the geographical position of the archipelago, the peculiarities of the geomorphology above water as well as the underwater complex formation (with caves, canyons and coral reefs) are some of the factors that contributed to its recognition as World Heritage in 2001.

24.3.1 Aspects of Tourism

Fernando de Noronha has many beaches, forts and historical monuments, natural beauty and geological monuments, where the geological aspects related to volcanism are evident in the landscape. According to Smith (2005), tourism on the island began in 1974. In 1992 there were 10,094 visitors. 10 years later, in 2002, the number increased 520 % to 62,551. This growth does not cause major environmental impacts because these are protected areas and ecotourism is the main source of income for islanders (Silva 2003).

The archipelago has many lodging options (from the most luxurious accommodation to very basic and locally classified as one, two or three dolphins), as well as bars and restaurants (seafood, regional and international cuisine). Among other options there are craft shops, travel agencies, transport (taxis and rental cars), diving operators, boat tours, dolphin watching tours and deep sea fishing. There is also the NAVI project, a Russian ship with a glass bottom (3 m long by 2.5 m wide), through which tourists can observe the sea below. Two visitor centers assist in the interpretation of the environment. In the Tamar Project every night environmentally-themed videos and lectures are held on topics related to the Archipelago. The ICMBio[1] Visitor Center is currently undergoing renovation, but there is also a museum for sharks and a memorial, that represents the historical aspects of the island. Community training is performed continuously. The 'Centro do Golfinho Rotador' (Spinner Dolphin Center) offers free courses such as English, management, scuba diving and dolphin watching guidance, while in 2007 the first course for Geotourism guides in Brazil was offered.

24.3.2 Guidelines and Restrictions

Population growth in Fernando de Noronha is strictly controlled. The entry and stay of visitors is limited and everyone must pay an Environmental Preservation Tax. This fee ensures the maintenance of the environmental conditions of the archipelago, and preserves the safety and comfort suitable for the wellbeing of the community and visitors. The daily rate is R$ 40.40 (approximately US$ 24.00). The more time on the island, the higher the rate, reaching R$ 3,333 (US$ 1,984) for 30 days. This fee must be paid in advance online or on landing. There are several trails and natural viewpoints and guidelines must be followed:

- Restrictions on diving in certain areas are in place;
- Local guides are required on certain trails;
- Hours of visitation are limited at the National Park trails and beaches;
- For some geological monuments visitation is prohibited.

An example is the Atalaia Beach, which has an officially limited carrying capacity of 100 people per day. Twenty people each for thirty minutes can enter the natural swimming pools. This accuracy is due to the fact that natural pools are important environments for life and growth of reef fish. Very few accidents happen in the archipelago, and as volcanism is no longer active, the accidents are mainly related to slippery rocks, loose stones causing falls, drowning due to accidental falls in hazardous areas, plants that burns (e.g. nettles of the genus *Urtica*) as well as bites from Moray eels.

[1] Brazilian agency responsible for national protected areas.

24.3.3 Geological Aspects Related to Volcanism

The archipelago is the submerged part of a large volcanic edifice currently inactive, about four thousand meters deep and about seventy kilometers in diameter. It is part of a chain of underwater volcanic mountains that stretch out from the Atlantic Ridge to the Brazilian continental platform, near the coast of Ceará State (Teixeira et al. 2003; Fig. 24.2).

The edifice consists of a substrate of pyroclastic rocks (incandescent fragments, which originate from the mantle and are ejected from volcanic vents over varying distances) penetrated by igneous intrusions (rising magma cooling towards the surface), which after a long period of erosion was covered by lava flows (Almeida 2002; Fig. 24.3).

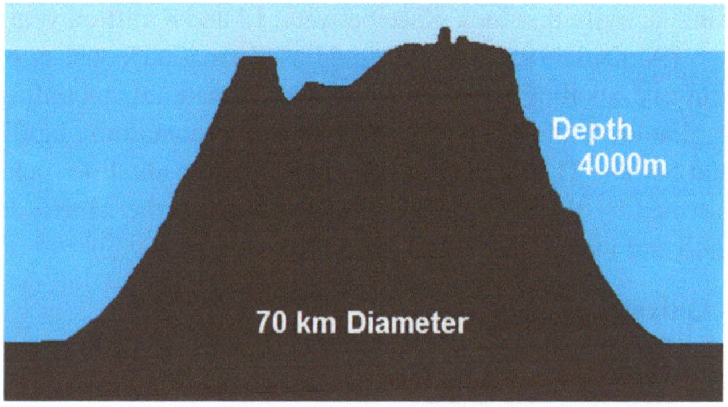

Fig. 24.2 The volcanic edifice in Fernando de Noronha region. *Source* Moreira 2009, based on Teixeira et al. 2003

Fig. 24.3 Simplified geological map, with the three formations. *Source* Moreira 2009

> Fernando de Noronha is the most didactic and only example of the result of observable phenomena related to volcanism, as other parts of Brazil suffered evolution or different occupation. It constitutes the site in a natural laboratory for the study of volcanology, with rich rocks produced by the movement of magma and the Atlantic Ocean (Ibama and Funatura 1990, p. 110).

The eruptions would begin with violent explosions that cleared the throat of the volcano, ejecting huge quantities of pyroclastic rocks into the air (bombs, lapilli and ash), and then once the pressure of the gases had been released, the lava would begin to pour out. These lavas were very fluid, sufficient to form layers less than three inches thick (Almeida 1955). At the island we can find three different formations (Fig. 24.3):

- Remédios Formation

This is the oldest formation and corresponds to the volcanic activity that took place between 12 and 8 million years ago (Almeida 1955). The Remédios Formation consists of eruptive alkaline rocks and pyroclastic material, including agglomerates, tuffs and breccias, as well as containing lapilli and bombs. This formation also left intrusive alkaline rocks (formed by the intrusion of magma), such as the Morro do Pico and most of the dikes of the island.

- Quixaba Formation

After the Remédios Formation, there was a long period of erosion. It was only between 3.3 and 1.5 million years ago that new magmatic activity occurred including ankaratritic lava flows (Cordani et al. 2003), which to the untrained eye resemble basalt, the most common volcanic rock on the planet. Since there was a major period of erosion, the lava flows settled on the uneven surfaces where they filled in the depressions of varying depths—from a few centimeters up to more than forty meters (Ibama 2005). The effusive rocks (magma rising to the surface) are black and present a major textural and mineralogical uniformity with a very fine grain. This Formation is the most abundant in the Archipelago and can be observed at most of its beaches (Fig. 24.4).

- Caracas Formation

The rocks from the Caracas Formation are sedimentary deposits resulting from modern and Pleistocene processes (from 1.8 million years ago until the present day), involving the calcarenite Caracas. These are old sediments, mostly of eolian origin (eoalianites) which formed the ancient shifting formation, which is now petrified through chemical processes (Fig. 24.5).

The Archipelago also has reefs that are inhabited by colonies of vermetid and annelid mollusks and algal limestone. The composition of the sand on the beaches is special, as there is no quartz, only fragments of volcanic rocks and grains of limestone from shells, worm-casts, crustaceans and mollusks. Dunes can be seen at two beaches and at Enseada da Caieira.

Fig. 24.4 Quixaba formation: black rocks with spheroidal decomposition. *Source* Jasmine Moreira

Fig. 24.5 Deposits of calcarenite in the Caracas Formation. *Source* Jasmine Moreira

24.3.4 Main Attractions of the Archipelago Based on Volcanism

- Morro do Pico

At an altitude of 321 m the Morro do Pico is the highest point on the island. The peak is the largest phonolite of the island, consisting of intrusive alkaline rocks, which are the product of magmatic activities from approximately 9 million years ago (Almeida 1955). Morro do Pico has the shape of a *plug* because of the erosive processes that have taken place over several thousand years and which have affected the surface of the rock faults, leading to the fall of massive blocks (Fig. 24.6).

- Ilhas Dois Irmãos

These islands are dark colored volcanic rock formations with large columns formed during the cooling process of the lava demonstrated by the splitting of the rocks (see Ulbrich et al. 2004). The Ilhas Dois Irmãos present a good opportunity to observe columnar jointing, which is common in igneous rocks (Fig. 24.7).

- Baía dos Porcos

The Baía dos Porcos is an enclosed bay with calm waters, good visibility for snorkeling and opportunities to observe volcanic landforms. The transparency of the water together with the yellow sea bed and the reflection of the blue sky give the water an emerald green color (Teixeira et al. 2003; Fig. 24.8).

- Praia do Sancho

This is a well-rounded bay enclosed by a cliff that reaches a height of 40 meters in some areas. Access is via the sea or through a crevice in the rock, which can be climbed up or down using iron steps. At the Praia do Sancho it is possible to observe the layers of several lava flows in sequence, which belong to the Quixaba Formation (Teixeira et al. 2003). This beach is suitable for diving.

- Praia do Leão

Located close to the dunes is the prominent Morro Branco, a small isolated rise, that is an interesting intrusive trachyte dome. On both sides of the beach are calcareous algae reefs. When the tide is coming in, great spouts of water can be seen, caused by the waves that are compressed into these tight spaces (blow holes). This is the island's most sought after beach by the green turtles that come to lay their eggs in the sand.

- Praia da Atalaia

At low tide, the reefs form a natural pool that is perfect for snorkeling, although visiting times are strictly controlled. This beach is geologically interesting because it has three formations, including the Quixaba Formation, a phonolite (Ilha do Frade) and dikes (Fig. 24.9).

Fig. 24.6 Morro do Pico viewed from the Praia da Conceição. *Source* Jasmine Moreira

Fig. 24.7 Ilhas Dois Irmãos, seen from above the Baía dos Porcos. *Source* Jasmine Moreira

Fig. 24.8 The Baía dos Porcos is ideal for snorkeling. Located on the *left side* of the picture is a pool worn into the rock, but where diving is prohibited. *Source* Jasmine Moreira

Fig. 24.9 Atalaia Beach and the Ilha do Frade phonolite. *Source* Jasmine Moreira

- Enseada da Caieira

This bay is of particular educational and scientific value, because it is possible to find in one small area more than 14 different types of eruptive rocks, which is one of the world's best examples of the phenomenon of magmatic differentiation (Santos et al. 1999; Fig. 24.10).

- Buraco da Raquel

The Buraco da Raquel is a geological monument where visible forms of marine erosion are present and can be observed. While visits to this monument are not allowed, it can be viewed from a lookout point close to the Shark Museum (Fig. 24.11).

Fig. 24.10 Enseada da Caieira. *Source* Jasmine Moreira

Fig. 24.11 Buraco da Raquel, a geological monument. *Source* Jasmine Moreira

24.4 Conclusion

All these features related to volcanism make the Archipelago of Fernando de Noronha a remarkable place in terms of geology and geomorphology. This potential can be used for Geotourism and should be included in the interpretive activities offered to visitors and the community (Moreira and Bigarella 2008). According to the Brazilian Geological Survey (CPRM 2011), the area presents important geological, geomorphological and geotourism aspects, highlighting the exceptional beauty of the landscape. Those factors, when combined with other attributes observed in the area, justify the creation of a geopark under the auspices of UNESCO.

References

Almeida, F. F. M. (1955). *Geologia e petrologia do arquipélago de Fernando de Noronha*. Rio de Janeiro: DNPM/DGM. 181.

Almeida, F. F. M. (2002). Arquipélago de Fernando de Noronha. In C. Schobbenhaus, D. A. Campos, E. T. Queiroz, M. Winge, & M. Berbert-Born (Eds.), *Sítios Geológicos e Paleontológicos do Brasil*. Brasília: DNPM, CPRM e SIGEP.

Alves, E. C., Maia, M., Sichel, S. E., & Campos, C. M. P. (2002). Zona de Fratura Vitória-Trindade no Oceano Atlântico sudeste e suas implicações tectônicas. *Revista Brasileira de Geofísica., 24*(1), 117–127.

Cordani, U. G., Ulbrich, M. N., Menor, E. A., & Lopes, R. P. (2003). *Cenozoic alkaline volcanism of Fernando de Noronha Island: South American Symposium on Isotope Geology* (vol. 4, pp. 1–24). Field trip guide. Salvador: CBPM/IRD.

CPRM (2011) Geoparque Fernando de Noronha. www.cprm.gov.br/geoecoturismo/geoparques/noronha/index.php?GEOPARQUE=2.

Guia Quatro Rodas (2010). São Paulo: Abril.

IBAMA, FUNATURA (1990). *Plano de Manejo*. Parque Nacional Marinho de Fernando de Noronha – PE. Brasília.

IBAMA, PNUD, TETRAPLAN (2005). *Plano de Manejo* APA F. de Noronha-Rocas–S.Pedro e S.Paulo. Fernando de Noronha.

Juliani, C., & Fernandes, C. M. D. (2010). Well-preserved late paleoproterozoic volcanic centers in the São Félix do Xingu region, Amazonian Craton, Brazil. *Journal of Volcanology and Geothermal Research, 191*(3–4), 167–179.

Marques, L. S. & Ernesto, M. O. (2004). Magmatismo toleiítico da Bacia do Paraná. In V. Mantesso-Neto, A. Bartorelli, C. D. R. Carneiro & B. B. de Brito Neves (Eds.), *Geologia do Continente Sul-Americano: Evolução da Obra de Fernando Flavio Marques de Almeida* (pp. 245–263). São Paulo: Beca.

Mizusaki, A. M. P. & Thomaz Filho, A. O. (2004). Magmatismo Pós-Paleozoico no Brasil. In V. Mantesso-Neto, A. Bartorelli, C. D. R. Carneiro & B. B. de Brito Neves (Eds.), *Geologia do Continente Sul-Americano: Evolução da Obra de Fernando Flavio Marques de Almeida* (pp. 282–291). São Paulo: Beca.

Moreira, J. C. (2009). *Geological guide—Fernando de Noronha* (vol. 01, p. 42). São Paulo: Nícia Guerriero Edições.

Moreira, J. C. & Bigarella, J. J. (2008). *The geotourism guide training at the Fernando de Noronha Archipelago, Brazil: Proceedings of Inaugural Global Geotourism Conference, 2008*, Fremantle-WA (vol. 01, p. 457).

Pearce, D. (2003). *Geografia do Turismo: Fluxos e regiões no mercado de viagens* (p. 392). Sao Paolo: Aleph.

Santos, A., Jodar, J. M., & Menor, E. A. (1999). The Fernando de Noronha Archipelago: Presentation of the geological heritage. Prospects. In D. Barretino, M. Vallejo & E. Gallego (Eds.), *Towards the balanced management and conservation of the geological heritage in the new millenium* (pp. 214–218). Madrid: Sociedad Geológica de. España.

Silva, J. M. (2003) *Parque Nacional Marinho de Fernando de Noronha: Uso público, importância econômica e proposta de manejo*. Simposio De Areas Protegidas: conservação no âmbito do cone sul. 2, Pelotas-RS. Anais. Universidade Católica de Pelotas, pp. 335–342.

Teixeira, M. G., & Linsker, R. (Eds.). (2003). *Arquipélago Fernando de Noronha o paraíso do vulcão*. São Paulo: Terra Virgem.

Ulbrich, M. N. C; Marques, L. S. & Lopes, R. P. (2004). As ilhas vulcânicas brasileiras: Fernando de Noronha e Trindade. In: V. Mantesso-Neto, A. Bartorelli, C. D. R. Carneiro, B.B. de Brito Neves (Eds.), *Geologia do Continente Sul-Americano: Evolução da Obra de Fernando Flavio Marques de Almeida*. São Paulo:Beca.

UNESCO. (2002). *Patrimônio mundial no Brasil* (2nd ed.). Brasília: Caixa Econômica Federal.

Wikimedia Commons (2011). *Map of Fernando de Noronha*. Disponivel em http://commons.wikimedia.org/wiki/File:Map_of_Fernando_de_Noronha-en.svg. Acesso em 11 Aug 2013.

Reunion Island: A Volcanic Showcase

Ross Dowling and Julie Marguerittte

Reunion Island, France	
	**Geographical location including map co-ordinates:** 800 km east of the coast of Madagascar and 200 km to the south west of Mauritius **Latitude:** 21°14′33″S **Longitude:** 55°42′32″E
	Type of volcano: Basaltic shield volcano
	Summit Elevation: 2631m
	Major attractions: Tropical climate, volcanic landforms, lava flows, collapsed caldera, Piton de la Fournaise, three Cirques, adventure activities
Typical eruption style: Effusive, Hawaiian, fluid basaltic lava flows **Current status:** ongoing activity since 2010	**Status of protection:** Parc National des Hauts de la Réunion (2007) World Heritage Site La Réunion National Park
Visitor numbers: 446,500 in 2012	

R. Dowling (✉) · J. Marguerittte
Edith Cowan University, Joondalup, WA 6027, Australia
e-mail: r.dowling@ecu.edu.au

J. Marguerittte
e-mail: julie_margueritte@hotmail.com

25.1 Introduction

Réunion Island is a French overseas department situated in the Indian Ocean (Map above). It is around 800 km east of the coast of Madagascar and 200 km to the south west of Mauritius, the nearest island (D'Haese et al. 2009). It is

P. Erfurt-Cooper (ed.), *Volcanic Tourist Destinations*, Geoheritage, Geoparks and Geotourism,
DOI: 10.1007/978-3-642-16191-9_25, © Springer-Verlag Berlin Heidelberg 2014

63 km long and 45 km wide. Moreover, with 2,512 sq. km, it is the largest of the Mascarene Islands. Discovered in 1516 by the Portuguese explorer Pedro de Mascarentas and named Mascarin after him, it became French a century later with the neighbouring Islands, 'Les Mascareignes' (The Mascarenes), and renamed after several years 'Bourbon' in honour of the King of France. It then became 'Reunion Island', after the abolition of slavery (Monod 2008). Administratively, Réunion is one of the one of the 27 regions of France and although it is overseas, it is an integral part of the Republic with the same status as those situated on the European mainland (http://en.wikipedia.org).

25.2 The Natural Environment

About 3 million years ago, a volcano rose from the bottom of the Indian Ocean, between Madagascar and Mauritius, and gave birth to Réunion Island. It has grown by the successive eruptions of the volcanoes Piton des Neiges (Snow Peak) and the Piton de la Fournaise (Furnace Peak). The lava has erupted from an oceanic hot spot which is 2,400 km under the earth's surface (Reunion Island 2013). The Island is constantly moving and transforming and it has been shaped latterly by subsidence and erosion creating a remarkable landscape which includes three cirques (Burson 2008).

25.3 The Volcanoes

The ***Piton de la Fournaise*** is a basaltic shield volcano forming the south eastern end of Réunion Island (Fig. 25.1). It is created by a volcanic hotspot which also created the Piton des Neiges and the islands of Mauritius and Rodrigues. The volcano rises more than 2631 m above sea level and is a twin shield volcano known to volcanologists as the Hawaiian type, because of the similarity of climate and volcanic nature. Much of its more than 530,000 year history overlapped with eruptions of the deeply dissected Piton des Neiges shield volcano to the north west of the island. Three calderas formed at about 250000, 65000 and 5000 years ago by progressive eastward slumping of the volcano. The volcano is surrounded by a horseshoe-shaped lava cliff enclosure (Enclos Fouqué; Fig. 25.2), which is 10 km (6 mi) in diameter and encloses two craters (Insight Guides 2010).

Numerous pyroclastic cones dot the floor of the calderas and their outer flanks (www.reunionisland.fr/volcano). The volcano is one of the world's most active oceanic volcanoes and it has erupted more than 100 times since 1640 and is under constant monitoring. Since 1998 it has erupted between two and four eruptions annually, most recently on 2 January 2010. During April 2007 a lava flow occurred which was estimated at 3 million cubic metres per day.

The ***Piton des Neiges*** is an extinct volcano forming the highest point on the island at 3070 m above sea level (Fig. 25.3). Collapsed calderas and canyons are south west of the mountain (Oehler et al. 2007). Despite its name, snow (French: *neige*) practically never falls on the summit (http://en.wikipedia.org). Réunion also has three calderas—the Cirque de Salazie, the Cirque de Cilaos and the Cirque de Mafate. The last is accessible only on foot or by helicopter.

The slopes of both volcanoes are heavily forested. Cultivated land and cities like the capital city of Saint-Denis are concentrated on the surrounding coastal lowlands. Offshore, part of the west coast is characterised by a coral reef system. The climate of the Island is tropical and humid. Summer, from December to March, is hot and humid with an average temperature of 26 °C on the coast, and winter, from April to November, is cool and dry with an average temperature of 20 °C. There is also a cyclonic season from December to February which is hot and rainy, but cyclones touching the Island are not frequent (Burson 2008).

Reunion Island is known for its remarkable landscapes and biodiversity with many rare flora and fauna and numerous endemic species including more than 500 flowering plant species (Knehti 2009). It has been classified by the International Union for Conservation of Nature as one of 25 natural areas around the world to be protected as a priority. In 2007 the Parc National des Hauts de la Réunion was established resulting in half of the total area of the island now being under protection. The park is 105,000 acres in area and is located at altitudes between 1,300 and 1,700 m. In 2010 the 'Pitons, cirques and remparts of Réunion Island' were ascribed as a World Heritage Site coinciding with the core zone of La Réunion National Park (whc.unesco.org). The area covers more than 100,000 ha or 40 % of Réunion and is dominated by the two towering volcanic peaks, massive walls and three cliff-rimmed cirques. It 'includes a great variety of rugged terrain and impressive escarpments, forested gorges and basins creating a visually striking landscape. It is the natural habitat for a wide diversity of plants, presenting a high level of endemism. There are subtropical rainforests, cloud forests and heaths creating a remarkable and visually appealing mosaic of ecosystems and landscape features' (whc.unesco.org/en/list/1317).

About 800,000 people live on Réunion which has a mix of ethnic groups from Mozambique, India, China, Madagascar, Comores and France. The principal language is

Fig. 25.1 Piton de la Fournaise (Furnace Peak). *Photo* Shane Evans

Fig. 25.2 The Enclos Fouqué lava cliff enclosure. Photo: Ross Dowling

French but Creole is also spoken. The majority of the people live near the coast but in the three cirques descendants of fugitive slaves who escaped from the coast, live in small hamlets. These people often live without water or electricity, however, in recent years the installation of solar panels has brought about some elements of modernisation. The population of the island is growing rapidly and it is estimated that it will reach one million by 2020.

Fig. 25.3 The Piton des Neiges (Snow Peak) with one of the hamlets of the Cirque de Mafate in the foreground. *Photo* Julie Marguerritte

25.4 Tourism

Together with the agricultural sector, tourism is the major contributor to the economy of Réunion Island (Ghosh et al. 2003). Its tourist appeal includes its natural and geologic features. Most of the island's visitors are seeking relaxing holidays or adventure holidays attracted by its impressive natural resources including its volcanoes, mountains, cirques and forests. In 2001 the number of tourists in the Island was 424,000 (Sautron 2009) but by 2010 the number of tourists had increased to 800,000. They spent more than €149 million which was a major contribution to the island's economy (Tourmag.com 2010).

Tourists travel to Réunion Island from all parts of the world, but the majority (81 %) come from France. Because of the remoteness of the island, it is not overrun by mass tourism. This is because it lacks high quality beaches and so most visitors to this part of the world who are seeking sun, sand and sea vacation, prefer going to nearby Mauritius. The island's tourist infrastructure is not as developed as that of its neighbour nor are they as attractive as those in Mauritius. Another reason why tourism is not as developed is that the island's economy is in Euros making it relatively more expensive for tourists. In addition those working in the travel and hospitality industry often have little training and they do not always speak a second language. Despite this situation the island is investing in tourism most notably in the construction of new hotels.

Réunion's tourist attractions include its favourable climate and impressive scenery (http://en.wikipedia.org). Réunion's capital Saint-Denis and the lagoons of St-Gilles-les Bains also attract tourists. The Domaine du Grand Hazier, an 18th century home of a sugar planter and official French historical monument, has a large garden with fruit trees and tropical flowers. Festivals take place year round throughout the island. The island has many natural tourist attractions, including Piton de la Fournaise, Piton des Neiges, lava cliffs and deep canyons. The collapsed calderas of Cirque de Cilaos, Cirque de Mafate, and Cirque de Salazie also draw tourists. The Maïdo viewpoint is located at an altitude of 2205 m and has excellent views of the coast.

The main activities are its adventure activities focussed on the volcanic landforms. According to the Insight Guide (2010: 183) 'the spectacular landscape of Reunion is punctuated by trails, craters, peaks, rivers, forests and waterfalls, tailor-made for a vast array of outdoor pursuits'. The Lonely Planet Guide states that 'Reunion is a mini-Hawaii, with astounding geographical diversity which…. provides an incredible stage for the stimulus-needy' (Carillet and Presser 2010: 175). Adventure activities include hiking, the main activity, as well as canyoning, paragliding, rafting, horseback riding and climbing.

25.5 Reunion's Volcanic Tourism Attractions

Natural landscapes, geology, rocks, are the main features that attract geotourists to a place. Geotourism (essentially geological tourism) is a form of nature-based tourism that based on the geosystem that specifically focuses on geology and landscape (Dowling and Newsome 2010). It promotes tourism to geosites and the conservation of geodiversity and an understanding of earth sciences through appreciation and learning. This is achieved through visits to geological features, use of geo-trails and viewpoints, guided tours, geo-activities and patronage of geosite visitor centres (Dowling and Newsome 2006; Dowling and Newsome 2010). Geotourists can comprise both independent travellers and group tourists, and they may visit natural areas or urban/built areas wherever there is a geological attraction (Newsome et al. 2012). A subset of geotourism is volcano tourism, where the focus of the tourist attraction and activity is volcanoes and volcanic landscapes (Erfurt-Cooper and Cooper 2010).

In Réunion the primary geological attraction is the volcanic landscape which provides the setting for a range of, mainly adventure oriented, activities. It is the volcanoes which tourists come to see. Reunion is a volcanic Island and it has excellent natural resources including the Piton de la Fournaise which stands within a lunar landscape, as well as the three Cirques. The Island is well served by trekking paths, and trekker huts have been built near geosites such as the volcanoes and cirques so that tourists can spend a night there before starting their trek early in the morning. Réunion also has relatively easy access to the volcanoes and many buses are available to drive tourists right to the starting place of the trek. The Island also has many guided tours and educational programs available for adults and children. They are designed for a better understanding of what and how the Island has been formed.

The volcanic activity of this tropical island has formed a mountainous relief unique in the world and has made it one of the best destinations for those who love nature and outdoor activities (Villon 2007). The Island is well known for its mountains, cirque and volcanoes as well as its national park and botanic gardens. The two volcanoes, the Piton des Neiges (Snow Peak) and Piton de la Fournaise (The Furnace), are the main tourist attractions. When tourists arrive at the ocher Sandy plain (Plaine des Sables; Figs. 25.4 and 25.5) that leads to the Piton de la Fournaise, they are arriving at a lunar landscape, they will never forget. This sandy plain is covered by volcanic slags belched out of Piton Chisny while in eruption. The volcano has an effusive eruption making it not as dangerous as those with more explosive eruptions (Monod 2008). It sometimes erupts several times a year and these frequent eruptions have made the reputation of the island as a highly spectacular place to witness volcanism in action. 'When eruptions happen, volcano-watchers flock to this corner of the south-east coast to watch the glowing rivers of

Fig. 25.4 The Plaine des Sables (Sandy Plain). *Photo* Ross Dowling

Fig. 25.5 Interpretive Sign. *Photo* Ross Dowling

Fig. 25.6 Trekkers climbing Formica Léo. *Photo* Julie Margueritte

lava roll quickly and quietly down the mountainside before sizzling into the sea' (Richardson 2010: 116).

Each year approximately 200,000 visitors climb the volcano (Figs. 25.6 and 25.7). The starting point for the climb is the Pas de Bellecombe (2311 m) where there is a carpark, toilets, and an unmanned visitor centre providing information on the geology and other features of the natural environment (Fig. 25.8). The display includes a relief model of the volcano which shows the routes and distances of the various hiking trails. A popular hike is the 13 km (8 mi) circuit of the volcano which takes in the scoria cone and active vent, Formica Léo (Figs. 25.9 and 25.10). Other volcanic attractions include the lava lunar landscape and sulphur emissions, combined with the ever-present possibility of an eruption occurring.

Even though it is called 'Snow Peak', the snow at the top of the volcano is relatively rare, and the last time it snowed there was on 1st August 2003 but the snow disappeared few days after. The volcano is situated in the north central part of the Island and it attracts tourists from all around the world (Oehler et al. 2007).

In order to learn more about the two volcanoes, tourists can visit the Maison du Volcan (Volcano Museum) located in the village of Bourg-Murat (Fig. 25.11). It includes

Fig. 25.7 At the top of Piton de la Fournaise. *Photo* Robert Margueritte

Fig. 25.8 Display at the Piton de la Fournaise visitor centre. *Photo* Ross Dowling

Fig. 25.9 Formica Léo scoria cone and active vent, Piton de la Fournaise. *Photo* Ross Dowling

Fig. 25.10 Tourists exploring Formica Léo. *Photo* Julie Margueritte

Fig. 25.11 Maison du Volcan (Volcano Museum). *Source* Maison du Volcan

interactive displays and videos of the eruptions including a 40 min film on the 2007 eruption and a shorter one on the 2010 eruption. The museum was created by Maurice and Katia Krafft, two well-known vulcanologists who died during the Mount Unzen, eruption in Shimabara, Japan, in 1991. The centre has a display of their photos, shows their films and has a lot of information about volcanic activity on the island.

25.6 The Cirques

Other sites visited by tourists are the three cirques named Cilaos, Mafate and Salazie which have been named because of their circular shapes. They are huge spectacular caldera-like valleys which radiate out from the Piton des Neiges in a clover leaf pattern. They have been formed by the collapse of the walls of the Snow Peak. Each is delimited by high cliffs which form impressive gorges. The town of Cilaos (popn. 6000, altitude, 1220 m; Fig. 25.12), in the Cirque de Cilaos, is the starting point for an ascent of the Piton des Neiges. It is similar in many ways to the French alpine resort of Chamonix, its twin town (Insight Guide 2010). The town developed as a spa resort at the end of the nineteenth century, after thermal springs (thermes) were discovered and its natural healing cures were sought after. It is also renowned for its quality wine (vin de Cilaos) and water.

Mafate is the least populated and least accessible of the cirques but is popular with tourists because of its numerous trails. It is considered to be a paradise for hikers because it provides picturesque treks amongst spectacular scenery (Fig. 25.13). Salazie is considered the most beautiful and the largest of the three cirques. It includes stunning waterfalls including le Voile de la Mariée (The Bride's Veil). It is the 'greenest' of the three cirques and has plantations of vanilla and coffee beans. The village of Hell-Bourg situated on the edge of Salazie has been given the award of the most beautiful French village due to its stunning landscape (beachcomber.com.au). According to Lonely Planet, 'you can't but be bedazzled by the fabulous backdrop—the majestic mountain walls that encase Hell-Bourg like a grandiose amphitheatre' (Carillet and Presser 2010: 210). With its magnificent viewpoint on the Mafate cirque, Piton Maïdo is a favourite tourism destination easily accessible by car. While trekking there tourists can visit the Geranium distillery which is part of the heritage of Reunion, and they can discover too the very varied vegetation all along the way.

Réunion's scenery is awe-inspiring and the best way to see it is by a helicopter tour. The helicopters seat up to eight passengers and 'these tours may be expensive but they are worth every euro'. Pilots fly in and out of deep ravines and hover precariously over isolated hamlets and the cirques, allowing plenty of time for photographs, before heading east to the volcano (Insight Guide 2010: 220).

Fig. 25.12 The town of Silaos. *Photo* Julie Margueritte

Fig. 25.13 The spectacular scenery in Mafate Cirque. *Photo* Julie Margueritte

25.7 Conclusion

Réunion Island is a relatively remote island in the Indian Ocean whose main attractions are its volcanic landforms, providing the basis of a well-developed adventure tourism industry, combined with the opportunity to see one of the world's most active volcanoes. As the island does not boast spectacular white sand tropical beaches, Reunion has escaped invasion form international tourists (Insight Guide 2010). However, it is well known to volcano tourists who visit the island to see and experience The Cirques and the Piton de la Fournaise volcano, both listed in the book *1001 Natural Wonders You Must See Before You Die* (Bright 2007). The cirques are deep bowl shaped hollows that superficially resemble volcanic craters but are actually caused by erosion which surround the highest mountain on the island, Piton des Neiges. These spectacular landforms offer superb mountain scenery combined with excellent outdoor adventure activity infrastructure. Since a large tract of the cirques is inaccessible, it is still largely undisturbed providing visitors with an awe-inspiring landscape of mountain ridges, forested valleys and spectacular waterfalls.

In contrast the Piton de la Fournaise is 'the magnum opus of Mother Nature in Réunion' (Carillet and Presser 2010: 217). It is an active geological wonder that erupts with great regularity. Like the volcanoes of Hawaii, the volcano 'is a shield volcano which produces spectacular fire fountains and churns out rivers of basalt lava, which flow over great distances' (Insight Guide 2010: 189). The surrounding lava flows include examples of both *pahoehoe* (relatively smooth) lava and 'a'ā (irregular angular) lava. Between the volcano and the coast is a lava lunar landscape. The Plaine des Sables is a plain of outstanding raw beauty with the reddish brown landscape comprised of ash, dotted with bizarre rock formations. On the coast where the lava flows meet the sea, the waves crashing against the land create spectacular water jets. Overall this volcanic hotspot in the Indian Ocean is a volcano tourist's paradise, and thus it is well worth making the trek to this remote location as the rewards are manifold.

References

Bright, M. (2007). *1001 natural wonders you must see before you die*. London: Quintet Publishing.

Burson, A. (2008). About Reunion Island. Retrieved 24 July 2013 www.reunionisland.fr/geo.html

Carillet, J.-B., & Presser, B. (2010). *Mauritius, Réunion and Seychelles* (7th ed.). Victoria: Lonely Planet Publications.

D'Haese, M., Speelman, S., Alary, V., Tillard, E., & D'Haese, L. (2009). Efficiency in milk production on Reunion Island. *Journal of Dairy Science, 92*, p14.

Dowling, R. K., & Newsome, D. (Eds.). (2006). *Geotourism*. Oxford: Elsevier.

Dowling, R. K., & Newsome, D. (Eds.). (2010). *Global geotourism perspectives*. Oxford: Goodfellow Publishers Limited.

Erfurt-Cooper, P., & Cooper, M. (Eds.). (2010). *Volcano and Geothermal Tourism: Sustainable geo-resources for leisure and recreation*. London: Earthscan.

Ghosh, R. N., Siddique, M. A. B., & Gabbay, R. (2003). *Tourism and economic development: Case studies from the Indian Ocean*. Aldershot: Ashgate Publishing Ltd.

Guide, Insight. (2010). *Mauritius, Réunion and Seychelles insight guides*. Singapore: APA Publications.

Knehti, I. (2009). La Reunion—the World on an Island. Retrieved July 24 2013 from www.buzzle.com/articles/la-reunion-the-world-on-an-island.html

Monod, D. (2008). *En Parcourant la Reunion*. Sainte Clotilde: Clin d'oeil Publisher.

Newsome, D., & Dowling, R. K. (Eds.). (2010). *Geotourism: The tourism of geology and landscape*. Oxford: Goodfellow Publishers Limited.

Newsome, D., Dowling, R. K., & Leung, Y.-F. (2012). The nature and management of geotourism: A case study of two established iconic geotourism destinations. *Tourism Management Perspectives, 2–3*, 19–27.

Oehler, J., Lénat, J., & Labazuy, P. (2007). Growth and collapse of the Reunion Island volcanoes. *Research Article, 70*, 117–142.

Reunion Island. (2013). About Reunion Island geography. Retrieved July 24, 2013, from http://www.reunionisland.fr/geo.html

Richardson, P. (2010). Grande passion. Condé Nast Traveller, December, pp.107-118.

Sautron, J. (2009). Why tourism in reunion island is not successful anymore? Retrieved July 24, 2013, from http://www.free-press-release.com/news-why-tourism-in-reunion-island-is-not-successful-anymore-1252149936.html

Tourmag.com (2010). Réunion: nombre de touristes en hausse au 1er semestre 2009 [in French](Meeting: number of tourists up in 2009). Retrieved July 24, 2013, from http://www.tourmag.com/Reunion-nombre-de-touristes-en-hausse-au-1er-semestre-2009_a35090.html

Villon, J. (2007). *Travel around a tropical paradise*. Mauritius: Jade Reunion Island editions.

Volcano Tourism: The Effect of Eruptions and Disasters

26

Henry Gaudru

26.1 Introduction

There are many factors which affect tourism globally; this chapter aims to discuss the impact natural disasters can have on the tourism industry, looking in particular at certain events such as the eruption of the Soufrière Hills volcano in Montserrat, the unexpected Chaitén volcano eruption in Chile in 2008, the December 2009 eruption of Mayon volcano in the Philippines and last but not least the eruption of the Icelandic volcano Eyjafjallajökull in 2010.

A volcanic disaster can affect tourism for obvious reasons because there is a very close link between tourism and natural disasters. In the event of a natural disaster tourism is significantly disrupted and in many cases the industry is destroyed, requiring redevelopment and adjustment. Tourism is an important part of the economy in many countries around the world. The massive affect a natural disaster can have on tourism is related to the negative multiplier effect, which involves the impact of the event spreading through the economy. There are several case studies which show the varying effect a natural disaster has on tourism with the following studies providing specific information regarding volcanic areas and tourism during eruptive events. However, in some cases disasters based on volcanic eruptions can actually increase tourism. Some places affected by past volcanic disasters have become sought after tourist destinations (dark tourism; disaster tourism) which often become places of educational and historical interest. Many people visit these locations as full holidays, or as a part of another holiday in a proximate area (i.e. Vesuvius, Italy; Unzen, Japan; Mt St.Helens, USA).

Four recent spectacular eruptions, Chaitén (Chile), Mayon (Philippines), Soufrière Hills (Montserrat) and in March 2010 the Icelandic eruption of the Fimmvörðuháls (Eyjafjallajökull) have focused attention on the problems for tourism management and safety during volcanic activity. This chapter looks at each event and its lessons for risk management in volcano tourism.

26.2 Soufrière Hills: Montserrat

The eruption sequence of the Soufrière Hills volcano on the island of Montserrat is an example of how natural disasters can affect tourism, and illustrates the changes in the industry which can take place due to natural disasters. Montserrat is an island of the Lesser Antilles in the Caribbean Sea (Fig. 26.1 and 26.2). It is a volcanic island, mostly mountainous, subjected to severe hurricanes (June-November) and volcanic eruptions.

The Soufrière Hills volcano erupted between 1995 and 2010, but in 1997 it buried streets and buildings around the island and wiped out the capital city of Plymouth (Fig. 26.3a, b). Also destroyed by the eruption were the island's airport and the ferry terminal. This had an obvious impact on the once thriving tourism of Montserrat. Prior to the eruption Montserrat's tourism relied mainly on villa rentals and tourists returning year after year. The tourism sector used to contribute around 20 % of the GDP and was a major foreign exchange earner. Before the eruption, during the year 1994, more than 36.000 tourists visited the island. After the volcanic eruptions the sector virtually disappeared, mainly due to the destruction of Plymouth. As a result, despite continuing volcanic activity reconstruction of infrastructure and marketing of the island's attractions was an early priority.

The effect of the eruption on the island's tourism industry is varied. Naturally there was an initial decrease of visitor numbers during the years after the first eruption, but in 2001 the number of tourists once again increased, with 14,000 visitors arriving. Ten years after the initial volcanic

H. Gaudru (✉)
SVE (Société Volcanologique Européenne), Volcanic Risk Mitigation UNISDR (United Nations International Strategy for Disaster Reduction), Geneva, Switzerland
e-mail: hgaudru@sveurop.org

P. Erfurt-Cooper (ed.), *Volcanic Tourist Destinations*, Geoheritage, Geoparks and Geotourism,
DOI: 10.1007/978-3-642-16191-9_26, © Springer-Verlag Berlin Heidelberg 2014

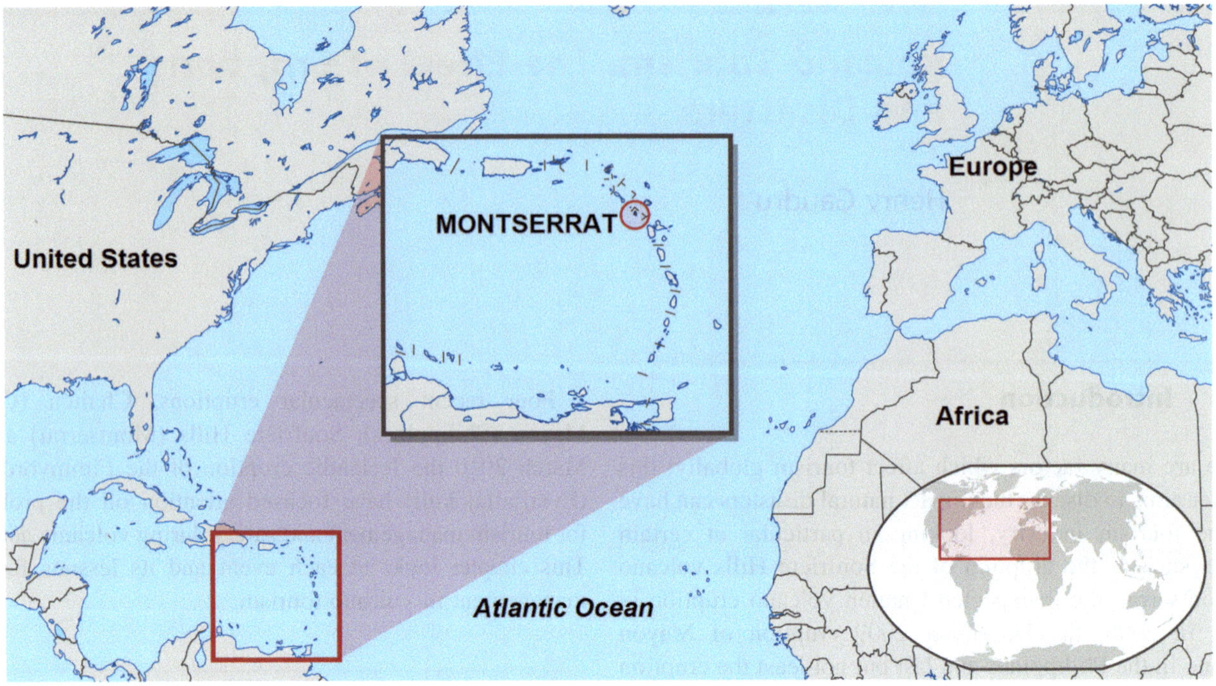

Fig. 26.1 Location of Montserrat in relation to the continents bordering the Atlantic Ocean. *Source* TUBS

eruption the island of Montserrat had a new £ 10 million airport (opened in 2005), and is determined to further recover from the disaster. Instead of ignoring the events that took place Montserrat's tourist board decided to embrace its history and as tourism officials said God '*gave us a volcano, so we're selling volcano tourism*' (quote taken from Times article: Jewel Rises from Ashes, Dawe 2005). The island offers visitors the chance to walk around in the streets of Plymouth with a police officer to explore the ruins of the once busy capital of the island. Also available is a new volcano observatory to provide visitors with a different view of the Caribbean island and emphasise the island's natural assets (there were almost 500 visitors to the MVO in the first months of 2009). For volcano chasers and general tourists Montserrat offers a very rare opportunity to witness an active volcano from a safe distance. While the volcanic 'Exclusion Zone' covers the entire south-eastern half of the island (Fig. 26.2), as well as extending 4 km off-shore along the eastern coast in what is known as the Maritime Exclusion Zone, there are various vantage points from where it is possible to view the volcano and the destruction it has wrought.

In the north east of the island there are tourist facilities at *Jack Boy Hill*, which also provides an excellent vantage point for volcano viewing. This observation point overlooks the destroyed W. H. Bramble Airport, the old estate house, the sites of destroyed eastern villages, now covered by volcanic pyroclastic flows, and of course offers views of the volcano (Figs. 26.4 and 26.5). The facility includes

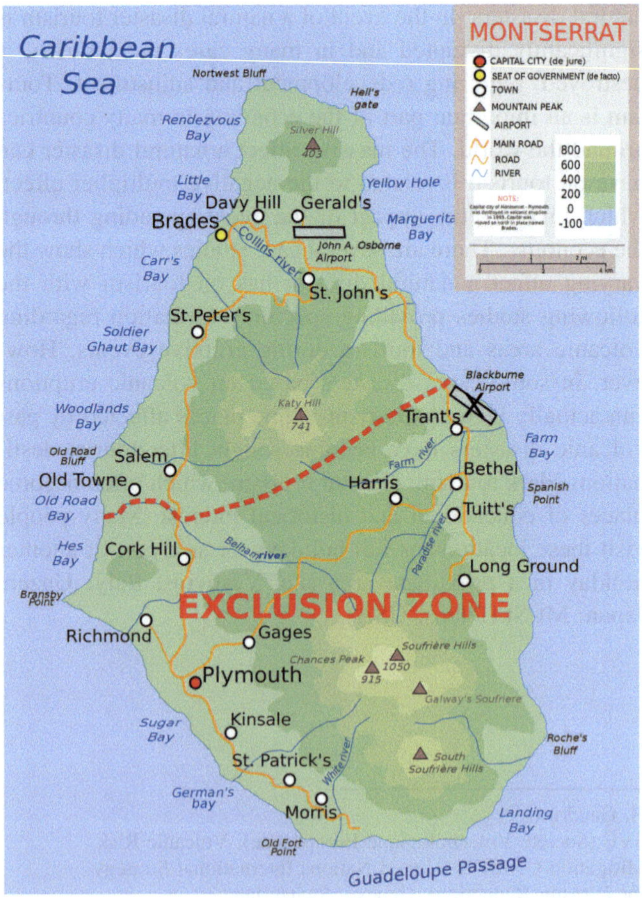

Fig. 26.2 Montserrat and its designated exclusion zone. *Source* Ivan25

Fig. 26.3 **a** Aerial view of Plymouth, the capital of Montserrat buried in layers of volcanic ash. View across the city towards the erupting Soufrière Hills volcano. *Photo* R. P. Hoblitt (USGS). **b** Plymouth in October 1997 viewed from the sea. *Photo* Henry Gaudru (SVE)

a viewing platform, picnic areas, a viewing telescope, a mini trail, and landscaped grounds. Another facility at *Garibaldi Hill* provides a contrasting view of the lush greenery of Salem, Old Town and surrounding areas, and a bird's eye view of the buried capital and other areas in the South (Visit Montserrat 2013). During periods of heavy rainfall or high volcanic activity visitors are advised not to cross the Belham Valley that leads to Garibaldi Hill (which is a natural conduit for floods and pyroclastic flows). Nevertheless, tours into the exclusion zone may be possible, depending on the volcano risk level (Fig. 26.2). Growing numbers of tourists are attracted to Montserrat by the active volcano, with visitors from the English-speaking Caribbean making up around 40 % of stay-over tourist arrivals.

Montserrat provides an interesting case study which illustrates the adaptability of the tourism industry following a large scale natural disaster. While the tourism industry may not be as flourishing as before the volcanic eruptions, visitor numbers have been increasing gradually despite the catastrophe Montserrat has been subjected to. The devastation caused by the eruptive events of Soufrière Hills shows how catastrophic events have the effect of initially decreasing tourism. Affected destinations however try to adjust to the point where the tourism industry is able to integrate the side effects of the disaster as an attraction of the region.

Fig. 26.4 Volcanic plume rose above the Soufrière Hills at sunset, October 1997. *Photo* H. Gaudru (SVE)

Fig. 26.5 Soufrière Hills volcano in 2007 on the Caribbean island of Montserrat. *Photo* Paddy 1983

26.3 Chaitén Volcano: When Danger Becomes an Attraction

Until 2008, the Chaitén volcano, Chile, consisted of a 2.5 × 4 km diameter caldera largely filled by a rhyolitic obsidian lava dome. No historical activity had being recorded until the volcano suddenly burst into life. Following a short seismic crisis, the Chaitén volcano started erupting on Friday May 2 in 2008 for the first time since about 7400 BCE. The Chaitén volcano ejected gas and ash on Thursday night (1st to 2nd of May 2008), causing more than 60 small tremors in the Los Lagos region 750 miles (1,207 km) south of Chile's capital Santiago (Figs. 26.6 and 26.7).

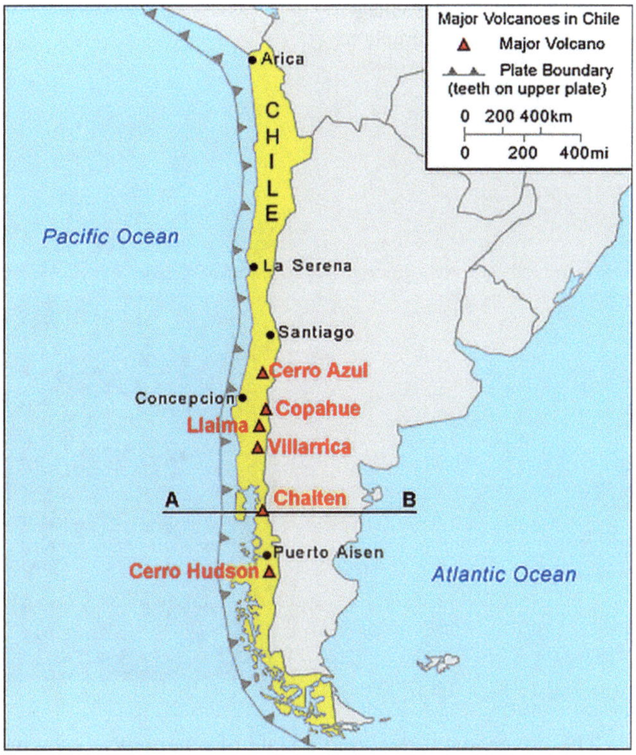

Fig. 26.6 Map showing the location of Chaitén Volcano near the west coast of South America. *Source* Map by Geology.com and MapResources

Fig. 26.7 The Chaitén volcano in eruption, October 2008. Viewed from a commercial flight between Puerto Montt and Balmaceda. *Photo* Gerard Prins

Fig. 26.8 A destroyed village, in the background the eruption column of the Chaitén volcano. *Photo* Javier Rubilar

The government declared a state of emergency, evacuating as many as 1,500 people from the town of Chaitén and nearby villages, just over 6 miles (9.65 km) away when the volcano, which was considered dormant for thousands of years, erupted unexpectedly. The town of Chaitén was not significantly affected by the eruption until heavy rainfall swept ash deposited in the crater and on the flanks of the volcano into drainage channels feeding the Chaitén River (Rio Blanco). Large Lahars resulted, which gradually filled the original river-bed. These lahars reached the city and were the cause of a series of consecutive floods of the Blanco River over the city during May, June and July 2008 (Fig. 26.8).

Chaitén town is strategically located at the entrance to the Patagonia Region of Chile from the north and is a point that previously provided access for ecotourism (Fig. 26.6). The eruptions of the Chaitén Volcano and their consequences destroyed almost the entire town, damaged the aquaculture industry, drastically affected ecotourism and closed nearby national parks and natural reserves. Moreover, the volcanic ash contaminated the area with toxic material and polluted the waterways (Santiago Times 2010). The volcano, which has been erupting since May 2008, has gradually settled down to slow, but constant dome growth, implying the continuation of activity for the foreseeable future. In response to the disaster the Chilean government told the residents of Chaitén that they must be evacuated from the town after the powerful outburst from the volcano only 10 km away, but a few dozen people have decided that they will not leave (Santiago Times 2009). In fact, there are many people who are still residing in the surrounding area of Chaitén. Therefore some assistance is made available to people living in the town such as the supply of emergency water, provision of street lighting and clearing of ash from the streets. The volcano has been at a high alert status since it started erupting, although the ensuing relative calm has caused hundreds of tourists to decide it is safe to approach the volcano (Arenas 2010). This has resulted in the re-issuing of a red alert as a response by the authorities.

In 2009 Chaitén town remained permanently on red alert and was still lacking safe drinking water in 2012 because of the volcanic ash. Nevertheless, many German and Argentinian tourists have selected the town as a vacation destination with more than 500 tourists arriving each month (Santiago Times 2010). The main reason for its popularity is scientific curiosity; many people are coming to see what happened during the eruption and they want to see this with their own eyes. It is probably very exciting for these tourists to visit a town buried in volcanic ash, but it is also quite dangerous as volcanoes are known to erupt anytime with little or no prior warning. As a result of this realisation, at the end of October 2009 a tourism meeting was held in Futaleufu near Chaitén town, where the current state of the volcano was discussed. Following the meeting the local and central government authorities told tourists to stay away from Chaitén as the volcano is still active and can potentially produce ash falls, pyroclastic flows or lahars. But despite these considerations the local newspaper (La Tercera 2012a, b) reported that visitors to Chaitén increased by up to 50 % in 2012 compared to 2011. Much of the destruction remains and while essentials such as electricity and water have only been partially restored, this does not prevent the growing numbers of tourists from visiting the area.

In response to the eruptions of Chaitén the Chilean tourism authorities have introduced security measures to ensure the safety of tourists visiting the area. Newsletters, brochures and signage ensure that visitors use safe routes and avoid delays and potential dangers. Police control and barriers were installed on the road to Chaitén in recognition that a long-term on-going eruption such as Chaitén with a tendency towards unpredictability is neither an ideal nor a safe tourist attraction. But despite the localized devastation the Chaitén eruption has brought, much of southern Chile is either completely unaffected or has recovered from the volcano's activity. Therefore the message from the Chilean tourist industry is that this region, which is one of the most beautiful in the world, is very much open for business. In fact, only 5 months after the evacuation of urban Chaitén, the other three regions of the province—Hualaihué, Palena and Futaleufú, were ready to show the world that their mountain valleys, rivers and lakes were not buried, but were more attractive than ever. In this situation it is difficult to see what more effective control measures might be introduced—visitors and locals want tourism to continue to grow despite the threat of disaster.

Fig. 26.9 Physical location map of the Philippines. *Source* Hellerick

26.4 Mayon: Volcanic Eruptions as an Attraction

Mayon volcano, an active volcano in the Bicol region in the Philippines (Fig. 26.9), has attracted tourists, both of local and foreign origin following its repeated eruptive activity.

Mayon is considered one of the seven wonders of the natural world due to its near-perfect cone (Fig. 26.10). However, in the province of Albay where Mayon is located, the last days of December 2009 were concentrated on the evacuation of the residents in and near the 7 km danger zone around the volcano. While the province exceeded the target of 47,285 residents for evacuation to a safer distance from Mayon, some foreign tourists and even some reporters and Manila-based cameramen were causing headaches for the Provincial Disaster Coordinating Council. Mayon volcano's activity is attracting increased numbers of tourists during days of activity, despite the warnings of the Philippine Vulcanology and Seismology (PHILVOLCS) organisation. In fact, Albay Province recorded an influx of some 76,000 tourists, foreign and local alike, during Mayon's eruption with displays of glowing lava flows (especially at night) from the crater down the mountain slopes (Figs. 26.11 and 26.12).

During this eruptive phase officials in the Philippines expressed their amazement at the stupidity of tourists who were flocking in their thousands to the fields around a dangerous volcano so they could photograph its spectacular lava and pyroclastic flows.

Fig. 26.10 Mayon volcano in Albay, Philippines. *Photos* Tomas Tam (a) and Tryfon Topalidis (b)

Fig. 26.11 Spectacular pyroclastic flows at Mayon Volcano. *Photo* C. G. Newhall

Fig. 26.12 Lava flows on the flanks of Mayon Volcano at night. *Photo* Tryfon Topalidis

The 'Joint Task Force Mayon', which is composed of the Philippine Army and the National Police, is concerned about the possible breach of the zero-casualty goal of the Albay province because of foreigners, who ignore the warnings related to the potential dangers of the Mayon volcano (Nasol 2009). Leading authorities are worried about casualties of tourists and have introduced specific security measures to mitigate risks from eruptions. During the 2009 eruption troops have been ordered to stop foreigners from getting close to the volatile volcano. After two tourists ventured inside the 'Danger Zone' and other tourists and their local guides were seen riding all-terrain vehicles near its foothills, extra troops were deployed at checkpoints to block access to Mayon. Even so, tourists manage to slip past checkpoints around the danger zone by passing through side roads with the help of local 'tourist guides'. Roadblocks have been placed on some side roads to deter such tourists from sneaking in. The local safety and emergency management office has designated strategic viewing places in the areas surrounding the volcano. In addition, Air Force and disaster officials declared the area around Mount Mayon as a no-fly zone after they were flooded with requests from local and foreign tourists for aerial sorties to take a look and document the rumbling volcano from above. This was because they would be affected by the shockwaves that definitely occur in major eruptions and would certainly impact on any aircraft flying within the vicinity of the Mayon crater.

In May 2012 another eruption of Mayon made headlines. Although officially classed as a 'small phreatic event' this eruption caused the death of several (4) German tourists and their tour guides, as well as injuring at least seven (7) other tourists, all members of a tour group that was climbing Mount Mayon (BBC News 2013). The volcanic eruptions in the Philippine Albay Province are thus both a boon and bane at the same time. The Department of Tourism (DOT) in Bicol noted that the Mayon eruption has greatly contributed to the tourism sector with a hotel occupancy rate of 80 % in recent years as compared to the 46 % rate in December 2008. Albay has 1,700 available rooms, adding that tourist arrivals since the start of the Mayon eruption event have been almost 2,500 daily compared to less than a thousand daily in December 2008. However, authorities said the eruption event, considered as disaster tourism, is not encouraged, promoted or sponsored by the Provincial Government (Arquelles 2011). The local Governor said '*Disaster tourism is certainly not within the contemplation of the provincial development strategy even if disaster risk reduction and tourism are integral components*' (Albay Mabuhay Express 2009; PhilSTAR 2009). The unexpected surge of tourists was an unplanned effect of the volcanic eruption, with the revenue from disaster tourism not matching the amounts that had to be spent on evacuation (GMA News 2009); e.g. PHP1.5 million a day for 9,946 families housed in 29 evacuation camps. The incremental output of 2,500 daily tourists mostly staying for one day added PHP2.4 million additional income at a gross value added factor of 41 % or PHP880,000 (assumed PHP1000 daily spending). Thus, while there may be an incremental benefit from disaster or dark tourism, this is most likely to be outweighed by the cost of the actual destruction of livelihoods, including agriculture and properties (Arquelles 2011).

26.5 Iceland: Eyjaföll Volcano—Fimmvörðuháls Eruption (March–April 2010)

The volcano Eyjafjallajökull (1,666 m) is located to the west of the Katla volcano in southern of Iceland (Fig. 26.13). It is an elongated basaltic andesite stratovolcano with a 2.5 km wide summit caldera trending east-west featuring lava flows from fissures on the flanks of the volcano. Eruptions have taken place during historical times, although Eyjafjallajökull has been less active than other volcanoes of Iceland's eastern volcanic zone, with relatively few Holocene lava flows known. A historical eruption took place from December 1821 to January 1823 and more recently in 1999 an intrusion at Eyjafjallajökull appears to have been followed by a small subglacial eruption in the Katla caldera (Global Volcanism Program 2013).

This volcano near the Eyjafjallajökull glacier however began to erupt again on the 20 March 2010 shortly before midnight, leading to road closures in the area. Several hundred people were evacuated from their homes. The new activity was preceded by intense seismicity and high rates of deformation in the weeks before the eruption, in association with magma recharging the volcano. The eruption broke out with fire fountains and Hawaiian eruptive style lava flows on an approximately 500 m long north east to south west oriented eruptive fissure on the northeast shoulder of the volcano at an elevation of about 1,000 m (Fig. 26.14). It was observed from the air from 4 to 7 am on 21 March. Lava flowed a short distance from the eruptive site, and minor eruption plumes at an elevation of less than 1 km were deflected by wind to the west (Fig. 26.15). The volcanic explosive index (VEI) was 1 or less. Tephra falls were minor or insignificant (Institute of Earth Sciences 2010).

Several days after the eruptive activity started, the eruption site was opened for tourists, although the Civil Protection Department released a statement for the attention of travelers keen on observing the eruption as follows: '*Volcanic eruptions are dangerous by definition and must be treated with respect. People who enter areas that have been declared danger zones are there at their own risk. The condition of roads must be taken into consideration because the environment is particularly sensitive at this time of year, when the ground is beginning to thaw, and therefore there is reason to emphasize that off-road driving is prohibited*' (Iceland Review 2010a). A popular hiking trail that draws thousands of people each year up to the Fimmvörðuháls volcano was re-opened, the Department stressing however that people choosing to take the path were doing so at their own risk (Fig. 26.16). Nevertheless, Reykjavik agencies offered daily bus tours to the volcano in the evening.

The eruption quickly turned into a boon for the island nation's tourism industry, as visitors flocked to catch a glimpse of the eruption (The Sydney Morning Herald 2010). During the first week some 25,000 people visited the area in spite of various dangers. Sudden steam explosions, caused by boiling of melt-water from trapped snow, can throw rock and lava high up in the air and quite far away from the source (Iceland Met Office 2010). As of the 25 March 2010, due to escalating of the eruption a warning was issued about toxic gases in the vicinity of the eruption: sulphur, fluorine, carbon dioxide and carbon monoxide were

Fig. 26.13 Location of the major volcanoes of Iceland including Eyjafjallajökull. *Source* National Land Survey of Iceland

released by the activity. Numerous rescue personnel assisted and guided the visitors and also restricted tourist access especially to the north and west of the eruption. On Saturday 27 March thousands of people walked from Skógar to the eruption site on Fimmvörduháls either as part of organized groups or on their own. The car parks at Skógar were full and as early as 5.00 am people started the hike towards the eruption which for the majority took between 5 and 6 h each way. It is estimated that between 3,000 and 5,000 people took the hike. During the first two weeks about 25,000 people came to the eruption site. Not all went to the crater on foot. Many visitors booked seats on planes or helicopters to get a bird's eye view of the phenomenon (Iceland Review 2010b).

After a new crater formed on Fimmvörðuháls during the night of Wednesday 31 March, the site was closed to all traffic. About 50 people were saved by helicopters, one was supplied by the Icelandic Coast Guard and two from Norðurflug, a private company. Tourists were turned back to Skógar and to Thórsmörk while snowmobiles at Mýrdalsjökull glacier were also turned back (Global Disaster Watch 2010). The atmosphere was described like '*we were in the Vietnam War up on the mountains… suddenly four helicopters appeared*'. Some tourists had planned to drive from Thórsmörk, but were ordered to stay in the hut at Básar. More than 100 people were at Básar when police decided to transport all who wished to leave out of the area. Rivers had started to grow and travel to the area was not deemed safe. However, people did not seem to realize that this could be dangerous in relation to the initially very small lava stream from the ground, but one that grew rapidly (Iceland Review 2010c). The authorities urged people to be careful, stressing that they were there on their own and the responsibility was all theirs.

As of 2 April 2010, the eruption site was re-opened and two hills designated as safe places for observation of the two craters. This added security came after a second fissure opened. The lava stream increased considerably after the other crater opened up. Police and rescue squads stressed the importance of following rules at the site. There was still

Fig. 26.14 The first fissure that opened on Fimmvörðuháls viewed from Austurgígar. *Photo* David Karnå

significant traffic to the location and sometimes drivers had not been considerate of those who were standing at the viewing areas. This created a substantial danger for the people who were on foot. Most visitors generally left the area between 8.00 pm and 10.00 pm in the evening, but some stayed the whole night, including some of the members of rescue crews, who were always on hand. The Iceland Review reported that some days there were close to 1,500 people who made it to the eruption sites, even though the area was closed until noon each day. Many tourists came in SUVs and the place where they parked looked like a crowded parking lot at a supermarket. Sometimes, the traffic was so intense that policemen had to direct cars. The traffic was particularly heavy during the Easter weekend—an indicator west of Hvolsvöllur showed 25,979 trips from Maundy Thursday on 1 April 2010 to the second day of Easter on 5 April, compared to 11,999 trips during Easter 2009. This was an increase of 143.4 % over the previous year (Iceland Review 2010d).

The most significant increase in traffic occurred during the first four days of Easter. On Maundy Thursday 2009, 2631 cars drove past the indicator on the highway west of Hvolsvöllur, while the indicator showed 6,242 vehicles passing by on Maundy Thursday, an increase of 137 % (Iceland Review 2010e). As of 7 April, a 55-year-old man, who was reported missing after traveling to see the eruption, was found dead. The body of a 43-year-old woman, one of his traveling companions, had already been found. A third missing tourist, a 33-year old woman, was found alive. She was cold and exhausted but otherwise in good health. There is no telephone network in the area, and these people were not equipped for walking outside in winter highland conditions, as well as having hardly any supplies. As a result the Chairman of the Association of Icelandic Search and Rescue (ICE-SAR) Sigurgeir Guðmundsson said that the government should consider covering part of the cost for guarding the safety of tourists traveling to see the eruption (Iceland Review 2010f). As of 12 April 2010, one of Iceland's most

Fig. 26.15 Aerial view of the Volcanic plume during 2010 eruption. *Photo* S. Hreinsdóttir (University of Iceland—SVE)

recent volcanic eruptions ended—and the unexpected tourist boom lifted this recession-weary country's financial gloom. The impact on a country's fortunes when an erupting volcano occurs was greeted as good news in this case!

26.6 Conclusion

26.6.1 Tourism during Volcanic Eruptions: A Challenge for Risk Management and Economic Strategies

The recent spectacular eruptions of the Chaitén (Chile), Mayon (Philippines), Soufrière Hills (Montserrat) and Eyjafjallajökull (Iceland) volcanoes have focused the attention on problems for tourism management during volcanic activity. These problems arise because more and more people are considering volcanic eruptions as an attraction and for some of them the danger also becomes an attraction, although most people usually try to get as far away as possible when a volcano erupts. The situation where tourists are visiting a volcano such as Mayon in the Philippines has added revenue to the local economy. Due to the growing numbers of tourists who are attracted by volcanic eruptions, health and safety issues have become an increasing problem for the relevant authorities as well as for the stakeholders of the tourism industry.

It is therefore advisable to design special international guidelines and recommendations for visitors, tourism organisations and tour operators in volcanic environments. Tour guides should have geological knowledge and should have emergency training. Tourists are often more vulnerable than locals during volcanic crisis because they are less familiar with local hazards and therefore might be more exposed than anyone else. Also, tourists are more difficult to coordinate and manage during an eruptive crisis, not least because most are unaware of the potential volcanic risk (Erfurt-Cooper 2010). Despite this potential risk, few tourism organizations and/or national tourism offices have developed such information as an integral part of their plans for this aspect of volcano tourism. Those countries with volcanic destinations must consider aspects of tourist behaviour and should include them within local disaster management plans in order to avoid or at least diminish the risk for visitors during a volcanic eruption or disasters conditions.

Houghton et al. (2006) recently wrote that the growth of tourism on active volcanoes is a new large-scale phenomenon. Indeed, some countries such as USA, Italy, New Zealand, and several Central American nations all have significant tourism ventures built around the concept of safe

Fig. 26.16 Overview of the 2nd fissure on Fimmvörðuháls, close to Eyjafjallajökull, as the lava flows down towards the north, turning snow into steam. *Photo* Henrik Thorburn

exposure of visitors to the spectacle of an erupting volcano. However, these operations have the unintentional effect of increasing drastically the size of the at-risk population, often in settings where the precise intensity of eruptive activity is impossible to forecast on short time scales. Balancing the dependence of local communities on tourist revenues with issues of public safety has been a feature of several 21st century eruptions.

Despite a high level of risk perception, tourism employees lack knowledge about early warning systems and emergency response procedures in most locations. To improve the tourism sector's collective capacity to positively respond during future eruptions, authorities should ensure adequate dissemination of hazard, risk and emergency response information. Most importantly education campaigns should focus on: (a) increasing tourist's knowledge of volcanic hazards and (b) increasing tourist and employee awareness of the early warning and information systems and appropriate behavioural responses when a warning is issued. Further, tourism employees should be required to participate in emergency training and evacuation exercises annually.

References

Albay Mabuhay Express. (2009). *Disaster Tourism not encouraged, promoted nor sponsored.* Official Website of the Province of Albay. Retrieved October 6, 2013, from www.albay.gov.ph/mabuhayxpress/GovDesk/DisasterTourism.html

Arenas, P. (2010). *Chaitén recibe cientos de turistas.* Retrieved October 6, 2013, from http://www.publimetro.cl/nota/noticias/chaiten-recibe-cientos-de-turistas/xIQjbq!pR9Ph496cArI/

Arguelles, M. S. (2011). *Mayon eruption brings biggest surge of tourists to Albay.* Philippine Information Agency. Retrieved October 6, 2013, from http://archives.pia.gov.ph/?m=12&fi=p100111.htm&no=49

BBC News. (2013). *Philippine volcano Mayon in deadly eruption.* 7 May 2013. Retrieved October 6, 2013, from www.bbc.co.uk/news/world-asia-22430378

Dawe, T. (2005). *Jewel rises from the ashes.* The Times. Retrieved October 3, 2013, from www.thetimes.co.uk/tto/travel/destinations/caribbean/article1737058.ece

Erfurt-Cooper, P., & Cooper, M. (2010). *Volcano and Geothermal Tourism: Sustainable Geo-Resources for Leisure and Recreation.* London: Earthscan.

Global Disaster Watch. (2010). *Volcaones.* Retrieved October 8, 2013, from http://globaldisasterwatch.blogspot.com.au/2010/04/Thursday-April-1-2010.html

Global Volcanism Program. (2013). *Eyjafjallajökull*. Retrieved October 6, 2013, from http://www.volcano.si.edu/volcano.cfm?vnum=1702-02=

GMA News. (2009). *Restive Mayon good for tourism, bad for farmers*. Retrieved October 6, 2013, from www.gmanetwork.com/news/story/180575/news/regions/restive-mayon-good-for-tourism-bad-for-farmers

Houghton, B. F., Gregg, C. E., & Klawonn, M. (2006). Volcanic hazards in the 21st century, an Asian-Pacific perspective. *Liaison, 111*, 17–21.

Iceland Met Office. (2010). *Ninth day of the eruption—Fimmvörðuháls: eruption in decline*? Retrieved October 6, 2013, from http://en.vedur.is/about-imo/news/nr/1858

Iceland Review. (2010a). *Iceland eruption open for tourists*. Retrieved October 6, 2013, from www.icelandreview.com/icelandreview/daily_news/?cat_id=28304&ew_0_a_id=359848

Iceland Review. (2010b). *Thousands flock to eruption site*. Retrieved October 6, 2013, from www.icelandreview.com/icelandreview/daily_news/?cat_id=28304&ew_0_a_id=360058

Iceland Review. (2010c). *Iceland eruption: About 25 stayed in Thórsmörk after rescue operation*. Retrieved October 6, 2013, from http://icelandreview.com/news/2010/04/01/iceland-eruption-about-25-stayed-thorsmork-afterrescue-operation

Iceland Review. (2010d). *Volcano enthusiasts crowd south Iceland town*. Retrieved October 6, 2013, from http://icelandreview.com/icelandreview/daily_news/?cat_id=16539&ew_0_a_id=360488

Iceland Review. (2010e). *Many foreign tourists come to see the Iceland eruption*. Retrieved October 6, 2013, from www.icelandreview.com/icelandreview/search/news/Default.asp?ew_0_a_id=360295

Iceland Review. (2010f). *People missing at Iceland eruption zone*. Retrieved October 6, 2013, from http://icelandreview.com/icelandreview/daily_news/?cat_id=16539&ew_0_a_id=360349

Institute of Earth Sciences. (2010). *Eruption in Eyjafjallajökull* 2010. University of Iceland. Retrieved October 6, 2013, from http://earthice.hi.is/eruption_eyjafjallajokull_2010

La Tercera. (2012a). *Minvu entrega informe sobre Chaitén y asegura que "sigue expuesta a riesgos"*. Retrieved October 6, 2013, from www.latercera.com/noticia/nacional/2012/06/680-465379-9-minvu-entrega-informe-sobre-chaiten-y-asegura-que-sigue-expuesta-a-riesgos.shtml

La Tercera. (2012b). *Turistas y residentes suben al Volcán Chaitén para ver su actividad*. Retrieved October 6, 2013, from www.latercera.com/noticia/nacional/2012/02/680-432627-9-turistas-y-residentes-suben-al-volcan-chaiten-para-ver-su-actividad.shtml

Nasol, R. M. (2009). *Foreign tourists ignore Mayon warnings*. Retrieved October 6, 2013, from www.inquirer.net/specialreports/volcanomonitor/view.php?db=1&article=20091228-244356

PhilSTAR. (2009). *Albay executive hits 'disaster tourism'*. Retrieved October 6, 2013, from www.philstar.com/headlines/536143/albay-executive-hits-disaster-tourism

Santiago Times. (2009). *Chile: "Rebel" residents ignore Volcano danger, stay put*. Retrieved October 6, 2013, from http://santiagotimes.cl/content/chile-rebel-residents-ignore-volcano-danger-stay-put/

Santiago Times. (2010). *Tourism up in Chile's Chaitén, home to volcanic eruption in* 2008. Retrieved October 6, 2013, from www.santiagotimes.cl/business/tourism/18216-tourism-up-in-chiles-chaiten-home-to-volcanic-eruption-in-2008

The Sydney Morning Herald. (2010). *Tourists flock to see volcano erupt at glacier*. Retrieved October 6, 2013, from www.smh.com.au/travel/travel-news/tourists-flock-to-see-volcano-erupt-at-glacier-20100329-r6m9.html

Visit Montserrat. (2013). *Places of interest*. Retrieved October 8, 2013, from www.visitmontserrat.com/downloads/Montserrat_Map.pdf

27 A Dilemma for Tourists and Land Managers Alike: What Risks to Take?

Chris Newhall

27.1 Introduction

Volcanoes and tourists are like magnets—they either attract or repel each other, with a force strongly dependent on proximity. Some tourists (let's call them Group A), when nearing a volcano, are drawn ever more strongly to its summit or crater. Other tourists (Group B) are repulsed, stepping back from the volcano. Why the difference? One possibility is that Group A simply perceives a much lower risk than Group B. Think of teenagers. With little life experience, many simply don't see all the risks around them. At the other end of the spectrum, think of people who are frightened of their own shadow, perhaps because of some earlier traumatic experience, so they overestimate the likely risks involved with any given activity. Where differences of risk perception exist, let's assume that the true risk to both Groups is the same. In principle then, education of *both* Group A and B through signage, educational programs, and the like would bring everyone's perception of risk closer to the real risk. Perceptions of risk would converge. However, realistically, differences in life experiences will always cause some differences in perception to remain.

A different explanation for behavior is that Group A has a stronger desire to reach the crater or edge ever closer to an eruption than Group B. Here, the difference is in *tolerance* of risk, i.e., in different judgments about what risks are worthwhile and acceptable. Even if their perceptions of risk are identical, they differ in how they value the risky experience. Think of adventure tourists and those who engage in extreme sports. Generally, they know the risks and they choose to take them. Those in high-risk occupations do the same, though more for the sake of their work than for thrills.

C. Newhall (✉)
Volcanologist, Earth Observatory of Singapore and Formerly with US Geological Survey, University of Washington, Seattle, USA
e-mail: cnewhall@ntu.edu.sg

Analogous situations arise when evacuations are recommended in the face of an impending eruption. Some will evacuate readily and others resist, and the explanation for this may be found in the fact that some differences are in perception of risk, but others lie in the value that individuals give to being allowed to stay in their own homes and work through the crisis.

27.2 How Can Risk be Managed?

In an ideal world we might (a) provide everyone with as much information about true risk as is known, including information about comparable risks, gambling odds, and other easy-to-understand terms; (b) ask them to repeat and certify what they understand; (c) tell them that no one will come to rescue them, as it would unfairly jeopardize the safety of the rescuers, and ask them to certify that they understand that, too; and (d) let tourists make their own considered decisions about what risks to take or not to take. But in reality, tourists who take risks and later need rescue will still expect and demand help, and if help is not forthcoming, they or their surviving family will sue the very people who tried to keep them safe. In some countries including the U.S., even waivers that acknowledge and personally accept risk in such situations do not hold up in court. Under U.S. law, an individual can never waive his rights to rescue, no matter how foolish his action and a public safety officer is generally obligated to attempt rescue.

Those who manage tourist attractions, e.g., the management of a National Park, are constantly torn between providing access and an interesting experience versus keeping everyone safe. The simplest way to keep people safe from volcanoes is to forbid access entirely, but this eliminates the interesting experience and—from a business perspective—will severely limit the number of tourists who come. So land managers in most volcanic areas try to balance access and safety. Safe, well-marked and fenced trails through areas of hot springs are a good example. Safe trails up volcanic peaks

P. Erfurt-Cooper (ed.), *Volcanic Tourist Destinations*, Geoheritage, Geoparks and Geotourism,
DOI: 10.1007/978-3-642-16191-9_27, © Springer-Verlag Berlin Heidelberg 2014

are the same. Sometimes hiker registration, guides, and communication devices are also required, so that any signs of danger can be relayed to tourists immediately. Some useful principles for balanced risk management have been developed in many jurisdictions and a representative example may be found at http://www.hse.gov.uk/risk/principlespoints.htm, which can be applied to tourism as well as to the workplace or living environment.

27.3 How Can an Individual Tourist Decide What Levels of Risk are Tolerable?

One simple but effective tool for individuals is a table of more familiar risks that are comparable to, or slightly higher or lower than that which is entailed in the proposed activity. Since tourism is always voluntary, comparisons should also be voluntary, e.g., driving, skydiving, playing various sports, and other lifestyle risks. Newhall and Hoblitt (2002, Table 5.5) list familiar risks for comparison to risks from a restless volcano. That list could be expanded and a chart of familiar risks that could be used in many different settings could be prepared, with the immediate risk indicated on a graphic that shows many risks. Simple? Yes. But duration of exposure must also be normalized, as it would not make sense to compare the risk that a tourist takes for just one day to another risk they take every day for years. And anyone choosing to accept a risk comparable to one that they already take should understand that they are doubling their risk.

Can the voluntary risks encountered in tourism be compared to involuntary risks (occupational risks or risks posed by someone else's actions)? Strictly speaking, no, but in practice I have noticed that people find such comparisons meaningful and are able to make mental corrections between what they will accept voluntarily and involuntarily. In practice, order-of-magnitude differences are meaningful and smaller differences might better be ignored and those risks considered more or less equal. Nevertheless, no estimation of risk is precise. If we knew the future perfectly, there wouldn't be any risk. Uncertainty in risk estimates around volcanoes can easily reach one or even two orders of magnitude, especially during times of volcanic unrest. In formal decision theory, to make the right risk assessment the decision maker needs to know the level of uncertainty as well as the estimated risk. Cautious individuals might judge acceptability after adding uncertainty to the risk; others might ignore uncertainty and just consider the mean likelihood. As a rule of thumb though, I have always considered uncertainty to be plus or minus at least one order of magnitude. Nature often reminds us that we know less than we think!

27.4 How Can a Tourist Manager or Public Safety Official Decide What Levels of Risk are Tolerable?

The level of risk tolerance an individual has is strongly influenced by cultural norms, including religious faith, and also by pragmatic influences like the drawbacks in alternative actions, threat of litigation, etc. A common approach for judging collective or societal risk is to use pre-defined "F–N" plots of probabilities of failure to protect versus the consequences of failure (deaths) (Whitman 2000; Ale 2005; HSE 2009). Various terms are used to describe the axes of these plots, but typically the Y-axis is frequency (F = annual probability) of N or more deaths, and the X axis is simply the number of potential fatalities (N). Governments and other standards bodies set bounds on such plots for acceptable versus unacceptable risks, sometimes with an intermediate ALARP (*as low as reasonably practical*) zone in between. For tourism, limits on the number of tourists allowed into an area could vary according to the probability of eruption, as long as the plotted values of probability of eruption and number of deaths in the event of an eruption fall within the "acceptable" zone.

An alternative is to be more explicit about the tradeoffs between benefits and costs. Two recent papers by Marzocchi and others (2007) and Woo (2008), citing Katz and Murphy (1997), propose a conceptual framework in which public officials can assess the likely costs of evacuations compared with the likely losses in the event of failure to evacuate before an eruption. A decision to evacuate can be based on the ratio of the costs of evacuation and the likelihood of losses if no evacuation is made. The same principle can be used to decide whether to allow tourist access to dangerous volcanic areas.

Here is how it might be done. Let $\boldsymbol{B}$ be the per capita value of allowing tourist access. It could include both economic benefit to tourist businesses and intangible pleasure (hedonic) value to the tourists. Since a business owner is generally not at risk in the event of an accident, and might even see increased business if access to the volcano is restricted, we consider only the benefit to the tourists themselves. A tourist who insists on approaching a volcano might be willing to pay a fee for being allowed to take such a risk (i.e., for hedonic value). We could use this fee as $\boldsymbol{B}$. In most cases this will not be very high because tourists almost always have backup or alternative destinations that would be almost as desirable. The net value of allowing tourist access will be $\boldsymbol{B}*\boldsymbol{N}$, in which $\boldsymbol{N}$ is the number of tourists.

Then, let $\boldsymbol{L}$ be the per capita loss in case tourists are killed or injured. $\boldsymbol{L}$ can be estimated from the value of a statistical life (VSL), much debated and highly variable from one country to another. For the purpose of this discussion, let's just assume lower and upper limits of USD100,000 and USD1,000,000. A tourist can choose which one to use, or we might use the lower value for injury and the higher value for death. The net loss will be $L*E*N$, in which E is the proportion of N tourists who are killed or injured. Once these parameters have been established, a rational decision about whether to allow tourist access might be to allow access under the following formula. If the probability of an accident within the coming week, leading to deaths or injuries, $p_{(1\ week)}$, is $< B*N*7/L*E*N)$, then access might reasonably be allowed. If $p_{(1\ week)} > B*N*7/L*E*N$, access might reasonably be denied. Estimates of probability should be for the same time window as that for which N is counted.

Consider a hypothetical example, Scenic Volcano, where tourists typically spend one day to enjoy hot springs and a short climbing experience. If Scenic Volcano becomes restless, when might tourist access rationally be stopped? Suppose that each tourist would be willing to pay $50, despite risk, to experience the hot springs and climb. If there are 100 tourists per day, $B*N*7 = \$35{,}000/\text{week}$. Now, suppose that Scenic Volcano might have a small, surprise eruption that kills 5 tourists. If L is $1,000,000 per person, and $E*N = 5$, $L*E*N = \$5{,}000{,}000$. In this case, for the next week, $B*N/L*E*N \equiv \$35{,}000/5{,}000{,}000 = 0.007$. So if $p_{(1\ week)}$ is less than 0.007 (i.e., 0.7 %), then this straight "break-even" cost-benefit analysis would allow tourist access during that week. Clearly, the benefit-loss ratio for an individual tourist will depend strongly on the value he or she places on access, the valuation of that person's life (e.g., VSL), and the probability and severity of an accident.

The ratio for a society might also count business losses, though safety of individuals will usually be weighted more strongly than business losses. Indeed, tasked with public safety, some land managers will choose to discount business costs and use an even lower, more conservative probability threshold than the straight "break-even" cost-benefit analysis might suggest.

27.5 Summary and Conclusions

So which basis should tourists and tourism or land managers use for deciding on acceptable risks? Probably, no one method fits all, so I suggest that they consider all of these approaches and choose the one that makes the most sense for their own needs and responsibilities. Sometimes, land managers may need to make unpopular decisions in the name of safety for all; at other times, individual tourists will need to set their own limits. In between, there are options for managing risk by limiting either tourist numbers or the duration of time spent in zones of relatively higher risk, and requiring other measures such as hiker education and registration, and safety gear including two-way communications.

The dilemma is real, for tourists and land managers alike. It helps for each to try wearing the other's shoes.

References

Ale, B. J. M. (2005). Tolerable or acceptable: Comparison of risk regulation in the United Kingdom and in the Netherlands. *Risk Analysis, 25*, 231–241.

Health and Safety Executive (HSE) (2009). Societal risk: Initial briefing to Societal Risk Technical Advisory Group. Technical report of the Health and Safety Executive Retrieved from http://www.hse.gov.uk/research/rrpdf/rr703.pdf.

Katz, R. W., & Murphy, A. H. (1997). Forecast value: Prototype decision-making models. In R. W. Katz & A. H. Murphy (Eds.), *Economic value of weather and climate forecasts*. Cambridge: Cambridge University Press.

Marzocchi, W., & Woo, G. (2007). Probabilistic eruption forecasting and the call for an evacuation. *Geophysical Research Letters, 34*, L22310.

Newhall, C. G., & Hoblitt, R. P. (2002). Constructing event trees for volcanic crises. *Bulletin of Volcanology, 64*, 3–20.

Whitman, R. V. (2000). Organizing and evaluating uncertainty in geotechnical engineering. *Journal Geotechnical and Geoenvironmental Engineering, 126*, 583–592.

Woo, G. (2008). Probabilistic criteria for volcano evacuation decisions. *Natural Hazards, 45*, 87–97.

Part V

Conclusion

28 Conclusion

Patricia Erfurt-Cooper

28.1 Volcanoes for Everybody

The diversity of their landforms and geothermal features has made volcanic regions extremely attractive to a broad range of visitors. In addition, from thousands of volcanic destinations worldwide many are located in protected sites such as national parks, geoparks, and world heritage areas, adding a further dimension to their attraction as tourist destinations. Access to these environments is in demand, especially when active volcanic and geothermal features can be observed. The growing significance of volcanic tourist destinations becomes even more evident when we see them portrayed in special editions of magazines like the National Geographic (2013) magazine, where a selection of 100 destinations describes *'stunning landscapes, breathtaking vistas and awe-inspiring getaways'*. Nearly twenty per cent of these destinations are based on volcanic geology, either active, dormant or remnant. Included in the list of such sites are the Galápagos Islands (Ecuador), Landmannalaugar (Iceland), Jeju Island (Korea), Hawai'i (USA), Santorini (Greece), Crater Lake (Oregon, USA), Tongariro's Emerald Lakes (New Zealand), Yellowstone (Wyoming, USA) and Pamukkale (Turkey).

In a similar fashion *The Age Traveller* (30 November 2013) published a list of the 'world's hottest destinations 2014', which includes Iceland (*visiting gnarled magma fields, vents belching thermal steam and convulsive mudpots around Lake Myvatn*), Vanuatu (*getting up close to one of the world's most accessible active volcanoes, Yasur on Tanna Island*) and Colombia (*the steaming volcanoes of Los Nevados National Park*) (Richardson 2013).

Moreover, during the final stages of research and preparation for this book the inclusion of several additional volcanic areas as World Heritage sites was announced by UNESCO. In June 2013 the long awaited inclusion of Mount Fuji in the list as an area that reflects the *'essentials of Fujisan's sacred landscape'* was confirmed, along with a number of equally significant volcanic sites. As one of the most visited destinations on the Japanese island of Honshu, Mount Fuji and the surrounding areas are expected to gain even more in popularity after official recognition as a World Heritage site based on their cultural significance. As a currently dormant volcano, which is closely monitored for signs of unrest indicating future eruptive activity, Mount Fuji attracts growing numbers of visitors, especially during the official annual climbing season in July and August. During the 2012 climbing season 318,565 visitors made their way to the top of Fuji, a challenge that has increased in popularity according to the visitor counts started in 2005 by the Environment Ministry (Table 28.1), and carried out every year since to keep track of the visitor pressure on the environment as well as on existing facilities (Kameda 2013).

Official websites for Mount Fuji have been established in Japanese and English by the Environment Ministry and the Prefectures of Shizuoka and Yamanashi. Information includes tips for the long hike to the top and provides updates on the congestion level of the hiking trails. The main emphasis of the web based information however is on safety for the climbers. Apart from disseminating essential information in Japanese and English, other languages such as Korean, Chinese and French are used in brochures and on signage. It is also common that Japanese staff members in tourist environments speak very good English and frequently German.

28.2 Risk Factors in Volcanic Areas

Climbing active volcanoes always comes with a certain risk. A tragic accident occurred in October 2013, when two people were caught out by sudden weather changes at Mount Taranaki, an active volcano on New Zealand's north island. Experts from the local alpine rescue team say that in

P. Erfurt-Cooper (✉)
James Cook University, Cairns, Australia
e-mail: patricia.erfurt@my.jcu.edu.au

P. Erfurt-Cooper (ed.), *Volcanic Tourist Destinations*, Geoheritage, Geoparks and Geotourism,
DOI: 10.1007/978-3-642-16191-9_28, © Springer-Verlag Berlin Heidelberg 2014

Table 28.1 Visitor statistics of recent years show an increase in the number of climbers who hike to the summit of Mount Fuji during the July/August season

Mount Fuji visitors during July/August season	
2005	200,000+
2008	300,000+
2010	320,975
2012	318,565

Source Kameda (2013)

mountain environments the weather can change within minutes, which can make it impossible for rescuers to reach missing people in time. The two climbers lost their lives after sheltering in a shallow snow cave for two nights without adequate protection in conditions described by the rescue personnel as the worst possible. With virtually no visibility due to low cloud cover, snow blowing from the summit in 100 km wind gusts and the mountain covered in ice layers, what was intended as a day trip turned into a disaster. At the same time two other climbers also faced problems on the same mountain, but were able to return although they had to be treated for hypothermia and minor injuries (Leask et al. 2013). This is a sad reminder that dangers for health and safety are not necessarily based on volcanic activity, but can be caused by unexpected weather events which are equally beyond human control.

For example, in Japan a high number of climbing accidents was recorded for the month of July 2013. According to the Japan Times: *'Accidents soared to a new record on Mount Fuji in the first month of the climbing season, leaving one person dead and four seriously injured just over a month after the peak was put on the World Heritage List'* (Kyodo 2013). Large numbers of people are using the short climbing season of July and August to reach the summit and there could be in excess of 5000 people per day on the mountain. An additional concern is the potential for a future eruption of Mount Fuji which would affect not only tourists climbing the mountain, but also between 250,000 and 300,000 tourists visiting other destinations in the vicinity of the popular volcano every day (annual visitor numbers 103–105 million), not counting the actual resident population (Table 1.10). While disaster readiness in Japan is part of everyday life as volcanic activity can and has affected many regions, visitors will be at a disadvantage in case of Mount Fuji erupting unexpectedly, although given the intensive surveillance it is hoped that enough time will be available to issue warnings.

Problems with disaster readiness of temporary visitors are not limited to Mount Fuji; the additional concern for unaware international tourists in volcanic regions puts extra pressure on emergency services in other volcanic regions as well, especially if communication barriers (e.g. language) are likely to hinder rescue operations. Many active volcanic regions receive vast numbers of visitors every year (Table 1.9). Should for example a disaster happen at Mount Vesuvius, which is touted by the media as a *'ticking time bomb'* (BBC One 2004), not just millions of local residents would be affected, but also large numbers of tourists, who may find themselves facing the problem of evacuation at short notice.

In more remote volcanic regions the risk may be even greater, although only a comparatively small number of people might be affected. A 'real adventure' destination that is becoming more popular with volcano tourists is the Russian Kamchatka, a large peninsula known for its remoteness and lack of easy access. Without a convenient road network in most areas visitors are mainly dependent on helicopters to drop them off at their chosen destinations. In return the volcanic wilderness of the Kamchatka offers unique attractions including 'bear viewing' as well as a multitude of volcanic hot springs and geysers. Several websites are promoting the World Heritage listed volcanoes of this far eastern region of Russia and invite tourists to embark on the adventure of a lifetime (GoRussia 2013; Volcano Discovery 2013). With around 30 volcanoes currently displaying various levels of activity, recent media reports state that *'erupting volcanoes of the Kamchatka are attracting crowds of adventurous geotourists'* and have *'become sightseeing hotspots for crowds of thrill-seeking tourists eager to see flows of lava and clouds of ash'*. Also reported was the ignorance of tourists, who pay no attention to signs warning of dangerous lava flows, glowing pyroclastic debris and hot ash clouds (RT News 2012), although this is not an uncommon mind-set among some volcano tourists. To be as close as possible to the action is part of the adventure and people are prepared to accept the risks of encountering lava flows and other potentially dangerous volcanic activity. But if somebody just wants to get a *virtual* taste of the fiery action, the internet based YouTube hosts countless user generated videos featuring active volcanoes worldwide.

While volcanic landscapes are attractive destinations, it is the eruptive activity that seems to draw people even closer to observe nature in action. In Mexico the volcano Popocatépetl has shown prolonged signs of unrest, with growing concerns for the safety of both tourists and locals following the temporary closure of local airspace with the suspension flights to and from Mexico City in July 2013. Incandescent tephra, gas, steam and ash plumes led to warnings to the public not to approach the crater closer than at least 12 km as a safety precaution (Smithsonian Institution 2013; SpiegelOnline 2013). However, despite the ongoing activity people are still interested in climbing this volcano, but have to settle for the nearby Ixtaccíhuatl instead, which offers views of Popocatépetl from a safer distance. Up-to-date safety reports are released by the

National Centre for Disaster Prevention in Mexico to the public, informing about the current alert levels based on the traffic light alert system (CENAPRED 2013). Activity updates are also posted online using social media sites.

28.3 The Protection of Volcanic Heritage

The official recognition of volcanic heritage areas as protected sites has been extended to many sites around the world. One such area is Mount Etna on the Italian island of Sicily, which was also placed on the UNESCO World Heritage list in June 2013 as a *'prime destination for research and education'*. Several other regions, both dormant and active, have been World Heritage listed as well for their volcanic landscapes consisting of lava flows, craters and granite massifs of *'great scientific interest'* (UNESCO 2013). An analysis of the additions to the official UNESCO World Heritage list in 2013 has revealed that twenty per cent of the newly listed sites are based on volcanic heritage. Mount Fuji (Japan), Mount Etna (Italy), Mount Kenya (Africa) and El Pinacate (Mexico) are tourist destinations with major attractions consisting either of active volcanic and geothermal features, volcanic rock formations in currently dormant areas, or landscapes of volcanic origin.

Also of interest for geotourists are igneous rock formations including large granite domes such as Half Dome in the Yosemite National Park (USA) or the granite coastline of the Tasmanian 'Bay of Fires' (Australia). Areas in close proximity of active and dormant volcanic environments generally offer a whole range of geodiversity including cultural features, which are frequently part of trip agendas offered by tour operators. For example World Heritage sites such as the Borobudur Temple and Mount Merapi on the Indonesian island of Java can be visited together as part of a guided day tour. In Bali Mount Batur is regularly included in organised tours to temples and rice fields while in Japan volcano tourists also visit shrines and temples and most importantly the local volcanic hot springs. Examples of regions and countries and their number of active volcanoes are shown in Fig. 28.1 to reemphasise the worldwide distribution of active and currently dormant volcanoes.

28.4 The Need for Research and Interpretation in Volcano Tourism

Volcano tourism will continue to develop, especially at new and emerging destinations. This has an important corollary in areas where monitoring of volcanic activity is not carried out due to economic constraints, where public consultation

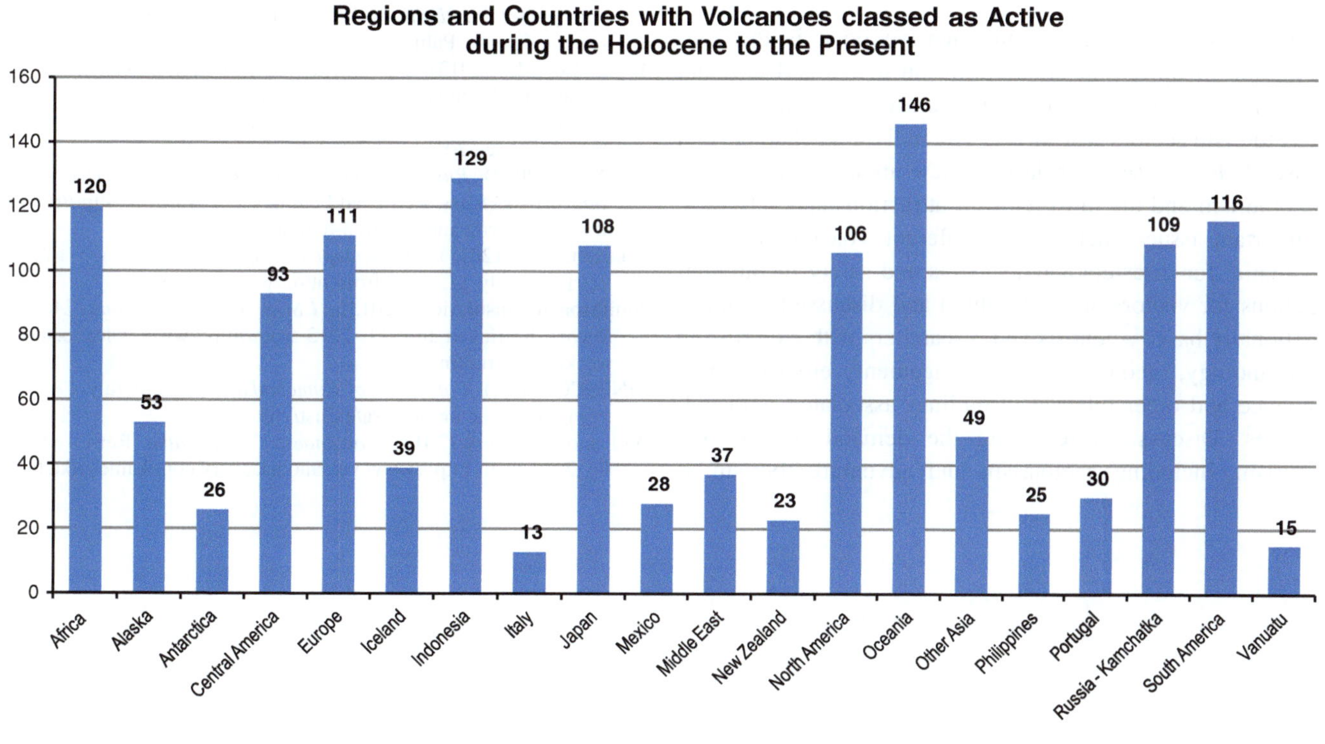

Fig. 28.1 Accurate accounts of volcanoes which are classed as active are difficult to determine as sources can vary considerably. In this table Italy, Iceland and Portugal are displayed as individual countries due to their number of currently active volcanoes. For the same reason North America does not include Mexico and Alaska and Asia and Oceania do not include Japan, New Zealand, Vanuatu and the Philippines. *Source* compiled by Author from various sources

is lacking and where reliable statistics on visitation are not available, indicating that further studies are required to improve management strategies and to provide sufficient information about visitor safety. Often these areas are of great interest for volcano tourists, but are lacking in essential guidelines, warning signs and interpretation which can be understood and recognised by visitors of domestic as well as of international origin. Because visiting potentially dangerous environments is generally done on a voluntary basis, people are frequently left to their own devices while hiking in unfamiliar terrain in a foreign country. Although a certain risk may be accepted by individual tourists, the likely absence of experience in how to cope with an unexpected situation in an unknown environment could increase the risk of getting into difficulties to unacceptable or even dangerous levels. It cannot be stressed enough that tourists in active volcanic and geothermal environments need to obtain up-to-date information about the current state of the area they are visiting. To provide adequate protection for visitors in all volcanic regions more research is needed to be able to draw conclusions between the ratio of visitor numbers and cases of injuries and fatalities. Suggestions for improving visitor safety in active volcanic and geothermal areas are included as Appendix A.

28.5 Final Comments

The stated aim of this comprehensive book is the contribution of up-to-date information on active and dormant volcanic tourist destinations and the role they play in various tourism sectors worldwide. The individual chapters and case studies present a balanced view about volcano-based geotourism and the most relevant attractions and activities. Important issues such as the different volcanic hazards, potential for disasters and accidents and safety recommendations for visitors are highlighted and discussed by internationally based scientists and researchers with expertise in volcanology, geology, risk management, environmental science and other relevant disciplines associated with volcanoes. Discussions regarding the demand for volcano tourism, including geotourism and adventure tourism as well as some historical facts related to volcanoes and case studies of interesting socio-cultural settings are included to increase the knowledge about the geodiversity of volcanic and geothermal environments and to raise awareness about the sustainability of these particular natural settings.

References

BBC One (2004). *Time Machine: The World shaped by Time. Vesuvius: Ticking time bomb*. Video. Retrieved November 2, 2013 from http://www.bbc.co.uk/science/earth/collections/mount_vesuvius#p00h1dhw

CENAPRED (2013). *Semáforo de Alerta Volcánica—Volcán Popocatépetl*. Retrieved November 2, 2013 from https://docs.google.com/file/d/0B38pFJSsUiV4azZsbVB0aXZYMU0/edit?pli=1

GoRussia (2013). *The Best of Kamchatka—All in One Trip—Bears and Volcanoes*. Retrieved February 20, 2013 from http://www.justgorussia.co.uk/en/display_tour/the_best_of_kamchatka_all_in_one_trip_bears_volganoes.html

Kameda, M. (2013). *As new UNESCO site, Fuji set to beckon to masses*. The Japan Times News. Retrieved July 9, 2013 from http://www.japantimes.co.jp/news/2013/07/08/reference/as-new-unesco-site-fuji-set-to-beckon-to-masses/#.Udvnv-p-_VQ

Kyodo, J. (2013). *Record-high number of alpine accidents on Mount Fuji in July*. Retrieved December 2, 2013 from http://www.japantimes.co.jp/news/2013/08/03/national/record-high-number-of-alpine-accidents-on-mount-fuji-in-july/#.Uqxte61-_b0

Leask, A., Thrunissen, M., & Backhouse, M. (2013). *Trapped climbers die on Mt Taranaki*. Retrieved November 2, 2013 from http://www.nzherald.co.nz/taranaki/news/article.cfm?l_id=146&objectid=11147301

National Geographic. (2013). *The world's most beautiful places—100 unforgettable destinations*. Washington D.C: National Geographic Society, Special Publication.

Richardson, L. (2013). *The world's hottest destinations for 2014*. Retrieved December 8, 2013 from http://www.theage.com.au/travel/the-worlds-hottest-destinations-for-2014-countries-20131127-2y9p8.html

RT News (2012). *Russian volcano eruption attracts tourists, sparks 'apocalypse' fears*. Retrieved February 2, 2013 from http://rt.com/news/russia-volcano-eruption-tourists-210/

SpiegelOnline (2013). *Popocatepetl: Angst vor Explosion wächst*. Retrieved July 11, 2013 from http://www.spiegel.de/

Smithsonian Institution (2013). *USGS Weekly Volcanic Activity Report*. Retrieved July 11, 2013 from http://www.volcano.si.edu/weekly_report.cfm

UNESCO (2013). *Volcanoes of Kamchatka*. Retrieved July 17, 2013 from http://whc.unesco.org/en/list/765

Volcano Discovery (2013). *Volcanoes of Kamchatka*. Retrieved July 17, 2013 from http://www.volcanodiscovery.com/kamchatka.html

Appendix A
Safety Recommendations

A.1 Safety Recommendations for Volcano Tourists

Safe conduct in active volcanic and geothermal areas is essential and tourists need to be aware that safety recommendations should be followed for their own benefit. In areas where no visitor safety guidelines are available as yet the following recommendations could contribute significantly to the safety of volcano tourists. Ongoing research related to visitor safety in volcanic areas is central to the development of the correct management strategies in the event of an eruption, whether this is expected or unexpected.

Research articles in science journals analyse volcanic disasters, their health risks and assess mitigation plans, but this information is not generally accessible for the public. Books covering the topic of volcanism sometimes include advice on safety around active volcanoes, but again not everybody is accessing this type of literature. Protected sites such as national parks or geoparks usually provide their own site specific safety rules, although some visitors may not use advice relevant for their personal protection.

The following revised list is a renewed call (based on earlier publications) for safety guidelines specifically developed for volcano tourists visiting active environments worldwide:

1. In case of dangerous environments precise information is essential—logistics with no room for error are needed. Volcano tourists need to be informed about **every potential danger** related to the volcanic environment they are visiting so they are aware of the particular hazards they may encounter.
2. All visitors of active **volcanic areas** should be encouraged to obtain **essential** information from relevant authorities **in advance**. Governments, local governing bodies, geological surveys, tourism organisations, tour organisers and operators should provide **updated** information online and on-site.
3. **Updated advice** could also be provided at hotel receptions, vehicle hire companies and tour booking agencies etc. at the destination in addition to being available through appropriate media outlets.
4. Visitors must be fully aware that it is their **personal choice** between an acceptable and an unacceptable risk.
5. Tour operators and tour guides, travel agents, tourist organisations and all information centres need to communicate very clearly any **possible risk factors** involved in the destination they are promoting.
6. Tour guides need to be specially **trained** for emergencies and should have sufficient geological **knowledge** to be able to **assess** situations of imminent danger.
7. Guidelines and safety instructions must be available in **all majorlanguages** consistent with known visitor patterns at a destination.
8. In many regions **additional signage** is still needed in more languages and with supplementary images or pictograms for quick visual recognition.
9. Suitable **fact sheets** with symbolic signage or pictograms should be handed out to all visitors of volcanic environments, active as well as dormant (see Chap. 1, Fig. 1.23). Fact sheets can be straightforward following a generic template that could be applied to more than one area.

P. Erfurt-Cooper (ed.), *Volcanic Tourist Destinations*, Geoheritage, Geoparks and Geotourism,
DOI: 10.1007/978-3-642-16191-9, © Springer-Verlag Berlin Heidelberg 2014

10. **Essential information** should include emergency phone numbers, a hazard map with danger zones, recommended escape routes, shelter locations and emergency meeting points.
11. Fact sheets should be **short and precise**—no information overload.
12. Precise location maps—easy to interpret and to follow—handed out at visitor centres or at the entrance to volcanic national parks or other protected sites. **Multilingual** editions should include interpretations of any signage within the area.
13. **Escape routes** must be clearly marked—on printed maps, fact sheets and on signage located in the volcanic area.
14. Sign posts or guide posts could be **colour coded** to indicate whether the tourist is in a safe (green) zone or in a dangerous (yellow, orange, red) zone.
15. Rescue efforts also would be made easier if these guide posts were **numbered** (colour and numbers both have to be **reflective** to be useful after dark) and therefore could be used as **markers to identify** individual zones. Tourists need to be **aware** of the zone they are in at all times.
16. More **emergency shelters** are needed in many volcanic areas. They should all contain posters with visual interpretation of signs for imminent danger situations.
17. In particularly dangerous areas **emergencyphones** and possibly either **webcams** or **CCTV** surveillance should be installed at the emergency shelters and connected to a local monitoring station or observatory.
18. **First aid** kits should be available at emergency shelters as well.
19. **Communication services** such as local telecom companies need to supply access (mobile phone reception) to all remote areas that are frequented by large numbers of tourists, as many rely on their mobile phones to contact emergency services.
20. **Emergencyphones** should be installed at certain distance intervals—again with visual explanation of how to use them in a crisis situation. Arrows on colour coded guide posts should indicate in which direction the closest phone is located.
21. Helicopter **rescueservices** should be available in every country for every volcanic tourism destination. The cost factor should be of little relevance as most countries maintain extensive defence forces who are frequently called upon to assist in rescue situations. Defence forces are also able to assist with the **logistics** of communication and **transport** in an emergency. As part of disaster management and prevention it is important to enlist the help of all organisations that are trained for challenging scenarios.
22. Access to active volcanic environments should be **graded** according to potential **risk factors** and based on up-to-date advice from **qualified** scientists (e.g. volcano observatories, geological surveys) who are monitoring the area.
23. Expedition participants should undergo appropriate **training** for emergencies to be prepared for unexpected situations **before** embarking on excursions involving challenging volcanic environments. Relying solely on the leading tour guide is **not sufficient**, especially in the case of an emergency.
24. Insurance companies should provide updated information for travellers, whether visiting **risk prone areas** such as active volcanic environments, that are part of a **general sightseeing itinerary**, is covered in the event of injury or death. Policy holders should be encouraged to seek detailed information about any potential risk beforehand and whether they require **extra cover** in the event of a volcanic accident or disaster, especially if they incur **costly rescue** missions.
25. While there is an obligation for authorities to provide up-to-date information, it is up to the individual traveller to **follow advice**. To encourage this, any rescue service expenses should be recovered from the individuals whose **irresponsible and reckless action** causes rescue personnel to risk their lives to save them. This may encourage responsible behaviour.
26. To **minimize** the overall risk only vehicles suitable for access to volcanic environments should be permitted for use. They need to contain first aid kits and necessary communication equipment in case of emergency. Strict policing is required.
27. In dangerous environments visitors need to wear **protective** clothing and especially safe footwear (closed sturdy shoes at least), hard hats (helmets) and carry a gas mask.
28. Tourist organisations of countries with volcanic environments should insist on **international** safety guidelines to be able to pass these on to their clients.
29. **Smart phone apps** are already in use at many tourist destinations. In the case of volcanic attractions phone apps could include real time status updates for volcano tourists. Special volcano apps can be developed to provide all essential information for a specific volcanic region including emergency advice.
30. With the growing interest in **geotourism** volcanoes are even more in the spotlight as interesting and adventurous destinations. To make volcano tourism **as safe as possible** is the number one the priority to which the above proposed recommendations could contribute.

A.2 Additional Information Examples about Safety on Volcanoes

Websites

www.volcanolive.com/safety.html
Volcano Safety—John Seach
http://volcanoes.usgs.gov/publications/factsheets.php
Volcano Hazards Program
www.volcanodiscovery.com/volcanic_risk_zones.html
Safety on volcanoes: volcanic risk zones
www.seisvol.kishou.go.jp/tokyo/STOCK/kaisetsu/English/level.html
Volcanic Warnings and Alert Levels
www.nps.gov/havo/planyourvisit/yoursafety.htm
Hawai'i Volcanoes National Park
www.tongarirocrossing.org.nz/webapps/i/96914/293446/tongariro-alpine-crossing-must-haves.html
Tongariro Alpine Crossing—New Zealand
http://costaricaexpeditions.com/resourcelibrary/arenal-volcano-safe-visitation-guidelines.php?mnu=2
Arenal Volcano

Books

Lopes, R. (2005) *The Volcano Adventure Guide*. Cambridge: Cambridge University Press.

Appendix B
Volcano Glossary-Terms and Denitions

The following glossary provides a number of volcanological terms which may not be familiar to all readers. Definitions were selected to present a quick reference as well as to clarify similar sounding terminology. For further information the recommended readings (see list in Chap. 1) and the referenced literature throughout this book offer more in-depth technical details.

'A'ā lava Hawaiian term for basaltic lava flows composed of loose and rough angular clinker blocks with a jagged surface. The average temperature of an active 'a'ā lava flow lies between 1,000 and 1,100 °C (compare pāhoehoe lava).

Active volcano (a) currently erupting, (b) eruptive activity during recorded history and/or (c) likely to erupt again in the future. The classification 'active' includes volcanoes with a history that might reach back as far as the beginning of the current geological epoch (Holocene).

Agglomerate coarse, pyroclastic deposit composed of scoria close to an eruptive vent.

Andesite dark, fine-grained igneous rock containing over 50 % silica—the most common rock in volcanic arc regions.

Basalt dark, fine-grained igneous rock of low silica content (between 40 and 50 %)—the most common rock in the earth's crust including sea floors—occurs in lava flows.

Base surge the result of a directed volcanic blast. Base surges move over the ground at high velocity and layers are composed of cross-bedded rock debris (see pyroclastic surge).

Black smoker see hydrothermal vent.

Bombs, blocks rocks and lava lumps ejected during eruptive activity—can cause injuries from impact and burns.

Caldera Spanish for cauldron—large depression caused by the roof collapse of an empty magma chamber after the eruption of large volumes of magma. Examples include: Aso and Aira Caldera (Japan), Ngorongoro Crater (Tanzania), Blake River Mega Caldera (Canada), Vilama Caldera (Argentina), Caldera de las Cañadas (Tenerife/Spain), Santorini (Greece), Crater Lake (Oregon, USA).

Cinder pyroclastic material—largely consisting of vesicular rock fragments of cooled and solidified gas filled lava that has formed during Strombolian and Hawaiian eruptions.

Cinder cone/scoria cone cone-shaped hills built from loose volcanic fragments or air fall deposits including ash sized particles, pumice, tephra and other pyroclastic material from congealed lava. Examples include: Pu'u O'o (Hawai'i), Paricutin (Mexico), Eldfell (Iceland), Red Rock (Australia), Wizard Island (Crater Lake/Oregon), Cerro Negro (Nicaragua), Kometsuka/Aso Caldera (Japan).

Columnar jointing cooling fractures of volcanic rock which frequently display a well defined pentagonal or hexagonal pattern typical for the thermal contraction of hot lava layers—columns are generally oriented perpendicular to their surface layer and are common in all types of lava compositions.

Composite volcano see strato volcano.

P. Erfurt-Cooper (ed.), *Volcanic Tourist Destinations*, Geoheritage, Geoparks and Geotourism,
DOI: 10.1007/978-3-642-16191-9, © Springer-Verlag Berlin Heidelberg 2014

Continental crust outer layer of rock (lithosphere) that forms the continents. The average thickness of the continental crust is estimated to be 35–40 km and is of lesser density than the oceanic crust.

Convergent plate boundary collision area of two plates that can involve the convergence of two continental plates, the meeting of an oceanic plate with a continental plate resulting in subduction of the denser plate, or the collision of two oceanic plates.

Crater funnel or bowl-shaped depression created by eruptive activity at the summit of a volcano or at the centre of a volcanic vent (e.g. summit craters, pit craters, maar craters and diatremes).

Crater rows individual cinder or spatter cones formed close to a volcanic centre or along fissure vents. Examples: Seven Sisters (Queensland, Australia), Sunset Crater (Arizona, USA), Chaîne des Puys (Auvergne, France), Leirhnjúkur and Laki (Iceland), Waimangu Volcanic Valley (New Zealand).

Dacite igneous rock (felsic, extrusive) typically of light grey colour containing between 60 and 70 % silica.

Decade volcanoes a group of 16 active volcanoes selected in 1989 for more comprehensive study of their potential hazards and to raise awareness about volcanoes in general. Included in the list are Avachinsky-Koryaksky (Kamchatka), Colima (Mexico), Etna and Vesuvius (Italy), Galeras (Colombia), Mauna Loa (Hawai'i), Merapi (Indonesia), Nyiragongo (DR Congo), Mt Rainier (Washington State, USA), Sakurajima and Unzen (Japan), Santa Maria/Santiaguito (Guatemala), Santorini (Greece), Taal (Philippines), Teide (Canary Islands, Spain), Ulawun (Papua New Guinea).

Diatreme vertical volcanic conduit within a solid rock environment that can reach great depths. Diatremes can be funnel-shaped or pipe-like and are frequently filled with pyroclastic debris from explosive activity caused by gas-charged magmas interacting with groundwater.

Dike originating from rising magma in a volcanic conduit dikes are sheet-like igneous intrusions that cut into the layering and fractures of pre-existing adjacent rocks.

Divergent plate boundary tectonic plates that spread or move apart caused by rising magma from underneath thereby extending the existing plates. Examples of diverging plate boundaries include the Mid Atlantic Ridge (MAR), which rises to the surface in Iceland, the Great Rift Valley in Africa, the Galápagos Rise, the East Pacific Rise, the Pacific-Antarctic Ridge (PAR) among others (see also seafloor spreading).

Dome see lava dome.

Dormant volcano currently inactive or 'sleeping'. A term used to describe a volcano with the potential to erupt again. Most major volcanoes are now considered to be dormant rather than extinct (see active volcano).

Ejecta pyroclastic material ejected from a volcanic vent during an eruption including tephra, lapilli, blocks and lava bombs.

Eruption ejection of volcanic material from a volcanic vent. Eruptions can range from quiet lava flows to ultra-violent explosions (see VEI).

Eruption cloud formed downwind from an erupting volcano—contains tephra and gas and can spread over vast areas dispersed by wind. Eruption clouds can travel long distances—even encircling the earth in a short time and disrupt air traffic (e.g. eruption of Eyjafjallajökull in 2010).

Eruption column hot gas and tephra rising from a volcanic vent. Depending on size an eruption column can extend many kilometres into the atmosphere, where its content is dispersed by wind forming an eruption cloud.

Eruptive vent opening at the mouth of a volcano from which volcanic material such as lava, gas and tephra is emitted.

Extinct volcano volcano that is unlikely to erupt again in the future.

Extrusion/extrusive the process of emission of magma from a volcanic vent (e.g. lava flows, volcanic dome growth). Extrusive material includes rapidly cooled and solidified igneous rock such as andesite, basalt, obsidian and rhyolite.

Fissure eruption, fissure volcano elongated fracture connected to a volcanic vent that produces lava of low viscosity. Fiery lava fountains or fast flowing lava streams along the fissure or rift are common results of fissure eruptions. Examples include: Eyjafjallajökull (Iceland), Piton de la Fournaise (Reunion Island), Kilauea (Hawai'i), Mt Etna (Italy) (see also volcanic fissure).

Fumarole volcanic vent that emits steam (H_2O) and/or gas such as hydrogen sulphide (H_2S), sulphur dioxide (SO_2) and carbon dioxide (CO_2). Fumaroles can be present at active, dormant and extinct volcanoes. They are also known as Solfatara (from Latin sulfur) based on their emission of sulphurous gases.

Geyser extreme hot spring that intermittently discharges columns of hot water and steam into the air creating spectacular sights. Some well-known examples include: Valley of Geysers (Kamchatka, Russia), Old Faithful—Yellowstone (Wyoming, USA), Strokkur and the original Geysir at Haukadalur in Iceland, Lady Knox Geyser in New Zealand and the El Tatio geyser field in Chile.

Granite most common type of intrusive (plutonic) igneous rock with a high silica content (at least 65 %), quartz (at least 20 %), feldspar (alkali), mica (biotite/muscovite) and other minerals. The granular texture derives from crystallisation during the slow cooling process and makes this rock a popular building material. Granite colours vary from light grey to light brown, pink and red depending on the feldspar content.

Hazard map indication of areas with potential volcanic hazards such as lava flows, pyroclastic flows and lahars. Hazard maps are an important method to reduce the risks from volcanic activity for local residents and temporary visitors including volcano tourists.

Holocene youngest geological epoch following the Pleistocene—to present.

Hot spot persistent volcanic centre that is thought to be fed by a rising plume of hot mantle material. Hot spot activity is independent from plate boundaries and is linked to intraplate volcanism (e.g. Hawai'I, Australia, Cameroon, Antarctica, Yellowstone and Galápagos).

Hot spring (volcanic) continuous and natural flow of warm to hot groundwater following faults and cracks underground. Heated by a subterranean magma body and depending on the volume the water rises to the surface under pressure from unconfined aquifers. Natural mineral springs are generally classed as hot springs if the water temperature is higher than that of the human body (37 °C or 98 °F), although classifications can vary from one country to another. Hot springs occur worldwide including those that are not directly related to volcanic activity.

Hydrothermal relating to hot water—in the context of volcanism associated with water heated in close proximity of a magma body.

Hydrothermal vent/black smoker subaqueous high temperature [hydrothermal] spring continuously spouting water at temperatures of up to 300–350 °C enriched with minerals, which settle around the vent. The particles suspended in the hydrothermal fluid solidify after mixing with cold seawater, creating chimney like structures around individual vents. Hydrothermal vents are commonly known as black smokers although depending on their mineral content 'white smokers' also exist. First discovered in 1977 near the Galápagos Islands they are commonly found along mid-ocean ridges.

Hydrovolcanic [eruption] interaction of rising magma with either ground water or surface water. Hydrovolcanic events such as phreatic and phreatomagmatic eruptions generate pyroclastic flows and base surges. Non-explosive subaqueous eruptions cause the formation of pillow lava. Hydrovolcanic landforms include tuff rings, maars, diatremes and pseudo craters.

Igneous rock cooled and solidified (crystallised) rock formed from magma (plutonic, intrusive, e.g. granite) and lava (volcanic, extrusive, e.g. basalt).

Igneous province/flood basalt large regions where volcanic eruptions have deposited thick layers of basalt, covering vast surface areas. Examples include the Siberian Traps, the Ethiopian Highlands, the Deccan Traps in India, the Brazilian Highlands, the Newer Volcanic Provinces in Australia, the North Atlantic Igneous Province and the Columbia River Plateau.

Ignimbrite formed by the consolidation of pyroclastic flows (pyroclastic surge, pyroclastic density current or Nuée Ardente) with large pumice content resulting in lithified layers of 'welded tuff'. Ignimbrite sheets or layers are generally poorly sorted and can vary in thickness from less than 1 m to hundreds of metres. Loosely or partly consolidated layers are described as non-welded ignimbrite or 'un-welded' tuff.

Intrusion—intrusive rock igneous body of rock (magma) that is emplaced into adjacent rock and can take the shape of dikes, sills or batholiths. Igneous rock solidifies and crystallises while cooling at depth and forms a rock body known as Pluton.

Island arc volcanoes chain of volcanic islands or archipelago located alongside subduction zones (see also volcanic arc).

Jökulhlaup Icelandic term for 'glacial outburst flood'—discharge of melt water from a glacier or icecap that is due to sub glacial volcanic activity. Lakes developing from melting ice underneath a glacier can burst and release enormous volumes of melt water (possibly 100,000 m^3/s) at high speed (7–8 m/s) resulting in catastrophic floods.

Lahar gravity-controlled mudflow carrying volcanic debris and ash saturated with water rapidly downhill. Lahars are dangerous due to their great speed while moving. Embedded in the lahar mass can be large boulders that are able to destroy the foundations of bridges and buildings. Torrential rainfall can cause a lahar by mobilising ash layers and other volcanic debris a long time after a volcanic eruption.

Lava term that describes magma which has reached the surface in a volcanic eruption. Streams of liquid lava can flow from a crater or fissure and cool and solidify into extrusive igneous rock.

Lava cave, lava tube or pyroduct cave which remains after lava flows stop and leave behind an empty chamber in the shape of a tunnel which can be very short and small (Chap. 1, Fig. 1.11) or rather voluminous and up to several kilometres long (see Chap. 1, Table 1.11, Figs. 1.12, 1.13).

Lava dome mass of highly viscous lava extruding slowly (effusive) from a volcanic vent. While a lava dome continues to grow, parts of the extruded material can become unstable and break off (dome collapse), forming pyroclastic flows (e.g. Mt Pelee, Martinique Souffrière Hills, Montserrat), or until the volcano erupts violently due to internal gas build-up and so removes the 'blockage' from its conduit. Lava domes can take different shapes from rounded coulées to uplifted plugs or Peléan domes with spiny ridges. Examples of volcanoes with lava domes include El Chichón (Mexico), Unzen, Tsurumi, Yufudake and Showashinzan (Japan), Novarupta, Redoubt and Augustine (Alaska), Bezymianny (Kamchatka, Russia), Mt St Helens (Washington, USA), Santa Maria (Guatemala), Tarawera (New Zealand).

Lava flow stream of molten rock erupting from a volcanic vent or fissure. Hawai'ian style lava flows are considered the least hazardous type of eruption. Lava flows also refer to the solidified sheets of molten rock.

Lava lake molten lava contained in a volcanic crater. Permanent active lava lakes are rare with known examples of long-term and short-term lava lakes including Ambrym (Vanuatu), Erta Ale (Ethiopia), Pu'u 'Ō'ō (Hawai'i), Erebus (Antarctica), Nyiragingo (DR Congo) and Villarica (Chile).

Lava plateau basalt sheets formed by lava flows pouring from vents and fissures thereby creating broad plateaus or flood basalts (compare igneous province).

Lava Tube solidified tunnel that contained liquid lava which has drained away (see lava cave).

Maar monogenetic volcanic crater formed as the result of a violent phreatomagmatic or phreatic eruption, often filled with water and surrounded by a hydro volcanic landform known as tuff ring. The depth of a maar can vary between 10 and 500 m while their diameter can vary from a few 100 m to several kilometres. Maars are common worldwide. E.g. the Mount Gambier maar complex in South Australia includes two dry maars as well as two permanent lakes, one of which is the water reservoir for the city of Mount Gambier.

Magma subterranean molten rock which can be completely liquid or mixed with rock and dissolved gas. When magma reaches the surface it is called lava.

Magma chamber subterranean reservoir of molten rock (magma) that may have conduits to volcanic vents to reach the surface. Cooled and solidified magma bodies are called Pluton.

Mantle the zone between the overlying crust (oceanic and continental), including the Mohorovičić discontinuity, and the earth's interior core. The mantle is divided into three sections: the upper mantle, the transition zone and the lower mantle.

Mid-ocean ridge see divergent plate boundary.

Moberg or tuya see sub-glacial volcano.

Monogenetic volcano single eruption volcano frequently located within a volcanic field.

Mud volcano similar to magma volcanoes they can erupt by channelling boiling water, steam or pressurized gas to the surface. Mud volcanoes can be related to oil and gas reserves and sometimes erupt in flames. They are found in Iran, Indonesia, Japan, New Zealand, Russia, Italy, China, Pakistan, Romania, Canada, North America, Colombia and Venezuela.

Nuée Ardente French term for high temperature and gas charged 'glowing' or 'fiery' pyroclastic clouds or flows that contain a heated mass of ash and other volcanic debris. Also sometimes described as a 'glowing avalanche' the Nuée Ardente moves at high speed down the sides of a volcano (compare pyroclastic flow).

Obsidian extrusive igneous rock formed from rapidly cooling silica-rich rhyolitic lava without crystallisation. This results in a dark-coloured or black volcanic glass also known as Iceland Agate and which can be used to make tools, jewellery, statues and weapons. Other types of obsidian include rainbow obsidian, snowflake obsidian and mahogany obsidian.

Oceaniccrust layer of rock (lithosphere) that underlies the ocean basins worldwide. The oceanic crust is on average 5–10 km thick and has a much higher density than the continental crust.

Oceanic ridge any submarine mountain range caused by tectonic movement, but not necessarily associated with volcanic activity (compare divergent plate boundary).

Pāhoehoe Hawaiian term for basaltic lava flows with a smooth, rope like or braided surface. Pāhoehoe flows are characterised by low effusion at high temperature (1,000–1,200 °C) and do not undergo the same heat loss process as 'a'ā lava flows.

Peléan eruption eruption type characterised by pyroclastic flows containing hot volcanic gas, ash, and glowing lava fragments. Peléan eruptions frequently occur as a result of the explosive collapse of a lava dome (e.g. Mt. Pelée, Martinique 1902).

Phreatic eruption explosive steam driven eruption caused by the interaction of sub-surface water and magma resulting in violent explosions of steam and pulverised rock (tephra) while magmatic material is not ejected as in phreatomagmatic eruptions.

Phreatomagmatic explosive eruption resulting from the interaction of surface or subsurface water with hot magma ejecting significant amounts of magmatic material.

Pillow lava elongated submarine lava flows which resemble stacks of rounded pillows—frequently hundreds of metres in thickness. Individual pillows are generally not more than 1 m in diameter and are connected to a feeder tube.

Pit crater circular-shaped volcanic craters with steep walls formed by subsiding or collapsing ground (e.g. Erta Ale, Kilauea, Nyiragongo and Galápagos).

Plate boundaries the earth's surface is made up of several large plates which move in certain ways: (a) 'Convergent Plate Boundaries' (plates are moving together, colliding), (b) 'Divergent Plate Boundaries' (plates are moving apart, e.g. seafloor spreading) and (c) 'Transform Boundaries' (two plates are moving in different directions, sliding against each other) (see also subduction).

Plate tectonics theory proposing that the crust of the earth is separated into about a dozen individual plates that move and interact at their boundaries (continental drift), thereby forming new crustal regions, moving continents and causing volcanic activity and earthquakes.

Plinian eruption characterised by tall eruption columns resulting from gas charged explosions which contain large volumes of tephra and fragmented magma and are released from a volcanic vent at high velocity. This type of eruption is named after Pliny the Younger, who observed the eruption of Mt Vesuvius in 79AD and recorded the events for posterity.

Pluton large body of an igneous intrusion formed by consolidating (cooling and crystallisation) deep within the crust.

Polygenetic volcano erupting repeatedly over time with more than one period of eruptive activity interspersed by phases of dormancy. Eruptive events of polygenetic volcanoes can be separated by very short or extremely long periods of quiescence.

Pseudo crater also known as rootless cones formed by exploding steam from interaction with surface water when lava flows over wetlands. No actual vent is present (e.g. Skútustaðagígar, Iceland).

Pumice light-coloured, highly vesicular glassy volcanic rock with a frothy texture. Pumice contains >60 % silica and is formed by rapidly expanding gas in cooling lava. Solidified fragments of pumice can float on water due to their low density. Pumice is also used as construction material because of its insulating properties.

Pyroclastic, pyroclast igneous rock material or solid volcanic fragments formed in an explosive eruption including tephra, bombs and blocks. Also describes the process of explosive fragmentation of magma or lava as well as the resulting deposits of pyroclastic debris.

Pyroclastic flow or Pyroclastic Density Current (PDC) direct result of a volcanic eruption that contains hot gas and particles. Pyroclastic flows and surges can travel along the ground surface at speeds of up to 500 km/h or 300 m/s and their movement is generally determined by gravity. Temperatures inside these flows, also known as Nuée Ardente or glowing clouds, can reach over 800–1000 °C and cause instant death, the destruction of buildings as well as impact damage and burial. They are commonly caused by either a dome collapse, the collapse of an eruption column or by a lateral blast and are considered the most deadly of all volcanic phenomena.

Pyroclastic surge similar to a pyroclastic flow but with the ability to jump over ridges surging uphill, not just following slopes, valleys and water courses. Two types of surges have been observed: 'dry surges' with a low water content and driven by magmatic gases at high temperatures (>100 °C) and 'wet surges' containing steam combined with gas and pyroclasts at lower temperatures (<100 °C).

Repose the time interval between eruptions of a volcano. It has been suggested that the length of the repose period may be directly related to the level of explosivity at the next eruption. The analysis of repose phases between recorded eruptions could indicate patterns that may be used to predict future eruptions of some volcanoes.

Rhyolite fine-grained and light coloured volcanic rock equivalent to granite which contains >68 % silica and is rich in potassium and sodium. Colours range from white, green, pink, grey, red or brown. Rhyolite lava is highly viscous and frequently forms lava domes.

Rift valley/rift zone regions where crustal plates diverge or spread apart—commonly associated with volcanic activity. Rift valleys include the Taupo Volcanic Zone (New Zealand), the Great Rift Valley in Africa and the Þingvellir Rift in Iceland (see volcanic rift).

Ring of fire the Pacific 'Ring of Fire' (RoF) is an interlinked system of volcanic arcs that overlies crustal plate boundaries. Movement of these plates either through spreading or by subduction is the cause of earthquakes and volcanic activity.

Risk Management (in volcanic environments) identification, assessment and implementation of suitable strategies to minimise the potential for accidents and injuries. Should include education of tourists about the potential hazards from volcanic activity and how to stay safe.

Scoria highly vesicular volcanic rock including pumice, whereby pumice is less dense than scoria. The vesicular texture develops when lava solidifies and the gas remains trapped inside the rock. Also known as spatter or cinder.

Scoria cone see cinder cone.

Seafloor Spreading caused by plates that are diverging or separating at their boundaries thereby creating new sections of oceanic crust (lithosphere) at mid-ocean ridges.

Shield volcanoes produced by multidirectional lava flows which form gentle sloping landforms similar to a broad shield—can occur on plate boundaries as well as over hot spots. Examples include Mauna Loa, Mauna Kea (Hawai'i), Cascades (Oregon), Erta Ale (Ethiopia), Hierro (Canary Islands), Skjaldbreiður (Iceland), Rangitoto (New Zealand).

Solfatara see fumarole.

Spatter cone see cinder cone.

Stratovolcano or composite volcano symmetrical and steep-sided volcanic cones built up of alternating layers of erupted lava flows, tephra and other pyroclastic material. Stratovolcanoes represent the majority of volcanoes worldwide and are known on one hand for their beautiful shaped cones but on the other hand also for their explosive eruptions due to built-up gas pressure.

Strombolian eruption characterised by mild to powerful explosive activity with frequent small eruptions of basaltic magma typically lasting a few seconds with one burst of exploding lava after another. This type of eruption is named after the Italian volcano Stromboli and produces large amounts of lava spatter or scoria including lapilli, tephra and bombs. Strombolian eruptions are the most visually attractive volcanic activity but should only be viewed from a safe distance.

Subduction [zone] interaction at convergent plate boundaries where continental or oceanic plates move underneath an adjoining plate (subduct) and disappear into the mantle thereby forming a trench (e.g. Marianas). Volcanic arcs such as Indonesia and Japan are located above subduction zones.

Sub-glacial volcano also called Moberg or Tuya—created by volcanic eruptions underneath a glacier and typically shaped like a table mountain. If the eruption causes the overlying ice layer to melt the release or outburst of this melt water (see Jökulhlaup) can cause a major catastrophe similar to the disaster in Iceland after the eruption of the volcano Grímsvötn in 1996. Other examples include Tuya Butte (Canada), Herðubreið (Iceland), Viedma (Argentina) and several in Antarctica.

Sub-marine or subaqueous volcano frequently occurring at mid-ocean ridges, e.g. Hawai'ian Emperor Seamount Chain and the Home Reef in Tonga—sometimes also forming small islands such as Surtsey (Iceland).

Super volcano or Mega volcano recently introduced popular terms for potentially dangerous volcanoes with underlying large magma chambers. Although rare, these volcanoes can produce [super] eruptions of a magnitude that would affect vast areas and cause unimaginable destruction and long term environmental damage. Potential examples include Mt Toba (Sumatra/Indonesia), Taupo (New Zealand), Yellowstone (Wyoming/USA) and Tambora (Indonesia).

Surtseyan eruption high energy hydrovolcanic eruption caused and sustained by [sea] water entering an active volcanic vent. Large eruption columns containing highly fragmented pyroclasts can rise tens of kilometres into the air. This type of eruption was named after an Icelandic volcano that in 1963 created a small island (Surtsey) south of Iceland which remains to this day.

Tephra collective term for air fall deposits ejected from a volcanic vent during an eruption. Tephra is commonly known as volcanic ash but in fact consists of pyroclastic fragments such as pulverised rock, blocks, bombs and lapilli. Tephra from an eruption column or plume can be deposited over great distances. Particle size can range from 'ash' (<2 mm) and lapilli (>2 mm) to blocks and bombs (>64 mm).

Trachyte fine-grained alkaline extrusive igneous rock with a generally rough surface texture.

Tuff (welded, unwelded) rock formed from compacted, consolidated or lithified pyroclastic material and mineral fragments. Volcanic material from collapsed eruption columns and pyroclastic surges such as fragmented tephra, glass shards, pumice and other debris deforms on impact and at high temperatures into 'welded tuff' (see ignimbrite). In the absence of sufficient heat the pyroclastic material is loosely consolidated and compacted without being fused together resulting in 'unwelded tuff'.

Tuff cone volcanic cone with steep sides formed by phreatomagmatic eruptions.

Tuff ring low rim of accumulated phreatomagmatic air fall deposits surrounding a crater.

Tumulus small mound or volcanic blister formed by upward pressure of slow-moving lava beneath a solidified crust. Tumuli are common on pahoehoe flows.

Volcanic Explosivity Index (VEI) devised by Chris Newhall and Stephen Self in 1982 the VEI is a relative measure to determine the size of a volcanic eruption based on its magnitude, intensity and destructive power. The VEI is divided into an open-ended scale starting at 0 with 8 currently representing the most powerful and destructive volcanic events in history. Different eruption types and their intensity are correlated to the VEI scale whereby Hawaiian eruptions are classed as VEI 0–1, Strombolian as 1–2, Vulcanian as 2–3, Peléan as 3–4, Plinian as 4–6, Ultra-Plinian as 6–7 and VEI 8 is classed as super-volcanic.

Volcanic arc chain of volcanic islands, commonly featuring strong seismic activity due to underlying tectonic processes. Examples include the Japanese island chains, the Aleutian Volcanic Arc, the Marianas, the Kuril Islands, the Lesser Antilles, the Kermadic Islands, the Tonga Islands, the Sunda Island Arc, the Solomon Islands and the Philippines.

Volcanic belt large geographical region of volcanic and geothermal activity including the Pacific Ring of Fire, Garibaldi VB (Canada), Andean VB, Trans-Mexican VB, Taupo VB (New Zealand), Okhotsk-Chukotka VB (Russia), Pontide VB (Turkey), Wrangell VB (Alaska), Central Iranian VB.

Volcanic bomb pyroclastic fragments of magma ejected during a volcanic eruption, which take on various shapes while airborne or when hitting the ground while still soft. Volcanic bombs are named according to their shape and texture including bread crust bombs, spindle bombs and ribbon bombs.

Volcanic field an area prone to volcanic activity which can contain cinder cones, maars, lava domes and geothermal features. Some examples include the Auckland VF (New Zealand), the Oku VF in Cameroon, the Tuya VF of British Columbia in Canada and the Clear Lake VF in California.

Volcanic fissure/fissure volcano (see fissure eruptions) linear or elongated fractures of the ground from which lava erupts in curtains of fire. They can be up to several kilometres long and do not have a central crater or vent. Examples include the Krafla, Eldgjá and Laki Fires in Iceland, lava fountaining in Hawai'i, the Great Rift Valley in Africa, Tobalchik (Kamchatka) and Komagatake (Japan).

Volcanic hazards the risk to life and property from volcanic activity which includes eruptions, ash fall, gas emissions and toxic fumes, lahars and mudflows, lava flows and pyroclastic flows, extreme hot springs and other geothermal features such as geysers and fumarolic vents. Volcanic hazards are assessed based on their historical eruption records and the mapping of prehistoric deposits in comparison with other volcanoes of similar composition.

Volcanic plug or neck remnant volcanic landform consisting of solidified lava that originally was the conduit of a volcanic vent. More resistant to erosion than the surrounding walls, a plug remains as a solitary peak or neck after the majority of the original mountain has eroded away.

Volcanic rift a zone or region that features a series of fractures in the earth's crust. Unlike volcanic fissures these extended linear cracks can cover much larger areas and are caused by volcanic action when magma reaches the surface. Examples include the Canary Islands, Hawai'i, Iceland, the East African Rift, the Tonga Arc and Antarctica (see rift valley).

Vulcanian eruption named after volcanic eruptions observed on the Italian island Vulcano. During vulcanian activity, which can reach from moderate to powerful explosions, tephra, gas and pyroclasts are ejected at high velocity due to built up pressure of magmatic gases trapped beneath viscous magma. 'Moderate' Vulcanian eruption eruptions are also known as 'throat clearing' of a blocked volcanic vent, sometimes indicating the onset of further, more violent eruptive activity. Examples of volcanoes where Vulcanian type eruptions have been observed include Vulcano and Stromboli (Italy), Sakurajima (Japan), Krakatau (Indonesia), Arenal (Costa Rica), Soufrière Hills (Montserrat), Galeras (Colombia), Tavurvur Volcano (Rabaul, PNG).

Welded tuff see tuff or ignimbrite

Glossary compiled from various sources.

References

Allaby, A., & Allaby, M. (2003). *Oxford Dictionary of Earth Science*. Oxford: Oxford University Press.

Bullard, F. M. (1977) *Volcanoes of the Earth*. Revised Edition. Brisbane: University of Queensland Press.

Decker, R. W., & Decker, B. B. (1991). *Mountains of Fire: The Nature of Volcanoes*. Cambridge: Cambridge University Press.

Gallagher, R. (2003). Mud volcanoes: Mysterious phenomena fascinate scientists and tourists. *Azerbaijan International, 11*(2), 44–49.

Hudak, G. J. (2001). *Glossary of Volcanic Terms*. Retrieved December 5, 2013, from www.d.umn.edu/~rmorton/ronshome/Volcanology/VolcanicGlossary.html

Keary, P. (1996). *The New Penguin Dictionary of Geology*. London: Penguin Books.

Newhall, C. G., & Self, S. (1982). The volcanic explosivity index (VEI): An estimate of explosive magnitude for historical volcanism. *Journal of Geophysical Research, 87*(C2), 1231–1238.

Rosi, M., Papale, P., Lupi, L., & Stoppato, M. (2003). *Volcanoes: A Firefly Guide*. Buffalo, New York: Firefly Books Ltd.

Schmincke, H. U. (2006). *Vulkanismus*. Berlin: Springer.

Sigurdsson, H., Houghton, B., McNutt, S., Rhymer, H. & Stix, J. (Eds.). (2014). *Encyclopedia of Volcanoes* (2nd ed.). San Diego: Academic Press, An Imprint of Elsevier.

USGS (2013) *Geologic Glossary*. Retrieved October 2, 2013, from http://geomaps.wr.usgs.gov/parks/misc/glossarye.html

Volcano World (2013) *Glossary of Volcanic and Geologic Terms*. Retrieved October 2, 2013, from http://volcano.oregonstate.edu/education/glossary.html

Geographical Locations Index

P. Erfurt-Cooper (ed.), *Volcanic Tourist Destinations*, Geoheritage, Geoparks and Geotourism,
DOI: 10.1007/978-3-642-16191-9, © Springer-Verlag Berlin Heidelberg 2014

Index

P. Erfurt-Cooper (ed.), *Volcanic Tourist Destinations*, Geoheritage, Geoparks and Geotourism,
DOI: 10.1007/978-3-642-16191-9, © Springer-Verlag Berlin Heidelberg 2014

Q

R

S

T

MIX
Papier aus verantwortungsvollen Quellen
Paper from responsible sources
FSC® C105338

If you have any concerns about our products,
you can contact us on
ProductSafety@springernature.com

In case Publisher is established outside the EU,
the EU authorized representative is:
Springer Nature Customer Service Center GmbH
Europaplatz 3, 69115 Heidelberg, Germany

Printed by Libri Plureos GmbH
in Hamburg, Germany